Inorganic Membrane Reactors

Inorganic Membrane Reactors

Fundamentals and Applications

Xiaoyao Tan
Department of Chemical Engineering
Tianjin Polytechnic University
China

Kang Li
Department of Chemical Engineering
and Chemical Technology
Imperial College London
UK

This edition first published 2015

Registered Office
John Wiley & Sons, Ltd, The Atrium, Southern Gate, Chichester, West Sussex, PO19 8SQ, United Kingdom

For details of our global editorial offices, for customer services and for information about how to apply for permission to reuse the copyright material in this book please see our website at www.wiley.com.

Library of Congress Cataloging-in-Publication Data

Tan, Xiaoyao.
Inorganic membrane reactors : fundamentals and applications / Xiaoyao Tan, Kang Li.
pages cm
Includes bibliographical references and index.
ISBN 978-1-118-67284-6 (cloth)
1. Membrane reactors. I. Li, Kang, 1960– II. Title.
TP248.25.M45T36 2015
660′.2832–dc23

2014021348

A catalogue record for this book is available from the British Library.

Set in 10.5/13pt Sabon by SPi Publisher Services, Pondicherry, India

Printed in Singapore by C.O.S. Printers Pte Ltd

1 2015

Contents

Preface

Membrane reactors combine membrane functions such as product separation, reactant distribution, and catalyst support with chemical reactions in a single unit, leading to enhanced conversion and/or improved selectivity/yield. This combination also provides advantages compared with conventional reaction/separation systems – such as lower energy requirements, the possibility of heat integration, and safer operation. Such advantages lead potentially to compact and cost-effective design of reactor systems with substantial savings in processing costs. As one of the most effective solutions in chemical process intensification, membrane reactor technology has attracted substantial worldwide research and process development efforts in the last 30 years, and the subject is still currently undergoing rapid development and innovation.

Nowadays, membrane reactors have found commercial applications in biochemical processes (thus called membrane bioreactors) – for example, the production of fine chemicals via use of enzymes, large-scale biogas production from wastes, and environmental clean-up (wastewater treatment). These processes operate at low temperatures, seldom exceeding 60 °C, and thus polymeric membranes can be used. However, most chemical reactions of interest operate at temperatures well in excess of the limitation of polymeric membranes (100–150 °C). The development of inorganic membrane materials (zeolites, ceramics, and metals) with inherent high-temperature structural and chemical stability has broadened the application potential of membrane reactors toward the (petro)chemical industry. Many of these materials can be applied at elevated temperatures (up to 1000 °C), allowing their application in catalytic processes. Nevertheless, inorganic membrane reactors have not found any large-scale commercial applications so far, which implies that there are still a lot hurdles to their practical application. The challenges involve not only fundamentals but also engineering aspects. On the contrary, many novel inorganic membranes and membrane reaction processes have been developed in recent years. Review articles and book chapters have also emerged on all types of inorganic membrane reactors in the past few years. The present book aims to describe the fundamentals of various types of inorganic membrane reactors, demonstrate the extensive applications in a variety of chemical reaction processes, and elucidate the limitations, challenges, and potential areas for future innovation. It is nevertheless not the intention to provide a complete overview of all the relevant literature, but rather to highlight the current state and advances in inorganic membrane reactors. It is hoped that this book will serve as a useful source of information for both novices learning about membrane reactors and scientists and engineers working in this field.

The book starts with a general description of inorganic membranes and membrane reactors, followed by a detailed description of each type – porous, zeolite, metallic, and ionic transport ceramic membrane reactors – in separate chapters. We are conscious of the importance of producing membranes with the desired properties of good separation and selectivity, and enough robustness to withstand the severe operating conditions often encountered in industrial practice. For each type of membrane reactor, all the important topics such as the membrane transport principle, membrane fabrication, configuration and operation, current and potential applications are described comprehensively. A summary of critical issues and hurdles for each membrane reaction process is also provided to help readers focus on the key problems in making the technology succeed commercially. As the modeling methodology contributes significantly to the knowledge-based development of membrane reactors and engineering, a separate chapter involving a general modeling process for membrane reactors and their applications to typical reaction systems is included at the end of the book.

This book may serve as introductory material for novices learning about membrane reactors, and also as a reference for professionals. Novices can grasp the elementary concepts, and professionals can familiarize themselves with the most recent developments in the area. The audience for this book will be industrial and institutional researchers, scientists and engineers with an interest in membrane reactors, and also senior undergraduate and postgraduate students pursuing advanced membrane separation/reaction courses.

We are deeply indebted to many colleagues and students for the completion of this book. A number of people contributed to this book, including Jian Song, Zhigang Wang, Zhaobao Pan, Nan Liu, Haiping

Pan, Yang Liu, Zhentao Wu, Franciso Garcia Garcia, Nur Othman, and Ana Gouveia Gil, and they are acknowledged for their assistance with various aspects.

Our special thanks go to several members of staff at John Wiley: Emma Strickland, Sarah Keegan, and Audrey Koh for their patience and advice, and Rebecca Stubbs for initiating the book. Also Jayavel Radhakrishnan of SPi Global.

Finally, we would like to acknowledge financial support provided by the Royal Academy of Engineering in the UK for writing the book and our ongoing research funding in ceramic membranes and catalysis provided by EPSRC in the UK and NSFC in China.

1

Fundamentals of Membrane Reactors

1.1 INTRODUCTION

A membrane reactor (MR) is a device integrating a membrane with a reactor in which the membrane serves as a product separator, a reactant distributor, or a catalyst support. The combination of chemical reactions with the membrane functions in a single step exhibits many advantages, such as preferentially removing an intermediate (or final) product, controlling the addition of a reactant, controlling the way for gases to contact catalysts, and combining different reactions in the same system. As a result, the conversion and yield can be improved (even beyond the equilibrium values), the reaction conditions can be alleviated, and the capital and operational costs can be reduced significantly. Different reactions usually require different types of membranes. Membrane reactors are also operated in different modes. The purpose of this chapter is to introduce readers to the main concepts of membranes and membrane reactors. The principles, structure, and operation of inorganic membrane reactors are presented below.

1.2 MEMBRANE AND MEMBRANE SEPARATION

A membrane is defined as a region of discontinuity interposed between two phases [1]. It restricts the transport of certain chemical species in a specific manner. In most cases, the membrane is a permeable or semi-permeable

Inorganic Membrane Reactors: Fundamentals and Applications, First Edition. Xiaoyao Tan and Kang Li.

medium and is characterized by permeation and perm-selectivity. In other words, the membrane may have the ability to transport one component more readily than others due to the differences in physical and/or chemical properties between the membrane and the permeating components.

1.2.1 Membrane Structure

Membranes can be classified according to different viewpoints – for example, membrane materials, morphology and structure of the membranes, preparation methods, separation principles, or application areas. In general, the most illustrative means of classifying membranes is by their morphology or structure, because the membrane structure determines the separation mechanism and the membrane application. Accordingly, two types of membranes may be distinguished: symmetric and asymmetric membranes. **Symmetric membranes** have a uniform structure in all directions, which may be either porous or non-porous (dense). A special form of symmetric membrane is the liquid immobilized membrane (LIM) that consists of a porous support filled with a semi-permeable liquid or a molten salt solution. **Asymmetric membranes** are characterized by a non-uniform structure comprising a selective top layer supported by a porous substrate of the same material. If the selective layer is made of a different material from the porous substrate, we have a **composite membrane**. Figure 1.1 shows schematically the principal types of membranes.

Sometimes, the porous support itself may also possess different pores and exhibit an asymmetric structure. Figure 1.2 depicts the cross-section of an asymmetric membrane where the structural asymmetry is clearly observed [2]. The asymmetric membranes usually have a thin selective

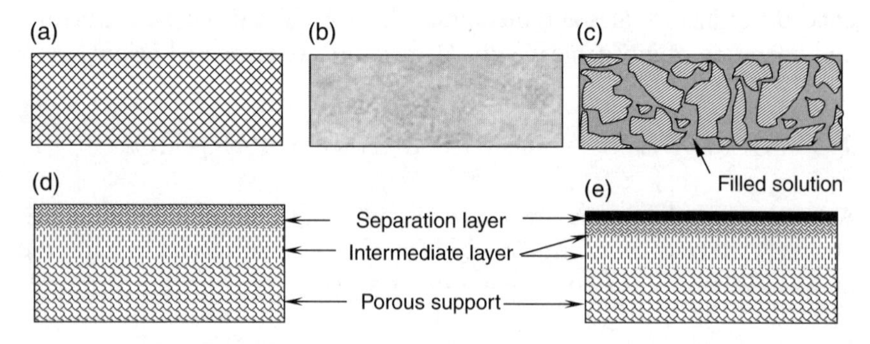

Figure 1.1 Schematic diagrams of the principal types of membranes: (a) porous symmetric membrane; (b) non-porous/dense symmetric membrane; (c) liquid immobilized membrane; (d) asymmetric membrane with porous separation layer; (e) asymmetric membrane with dense separation layer.

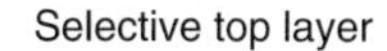

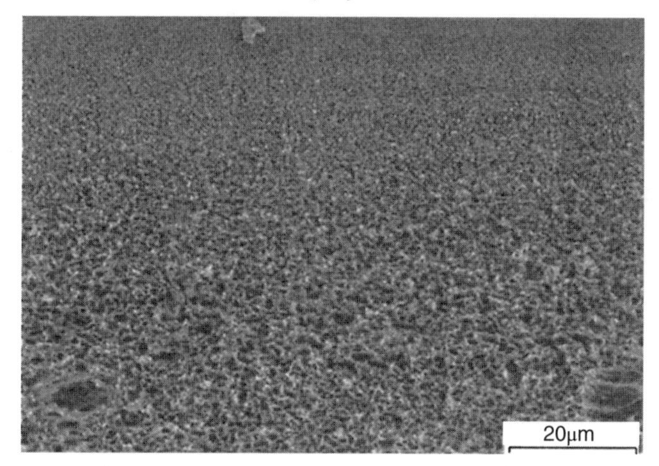

Figure 1.2 Cross-sectional SEM image of an asymmetric membrane. Reproduced from [2]. With permission from Elsevier.

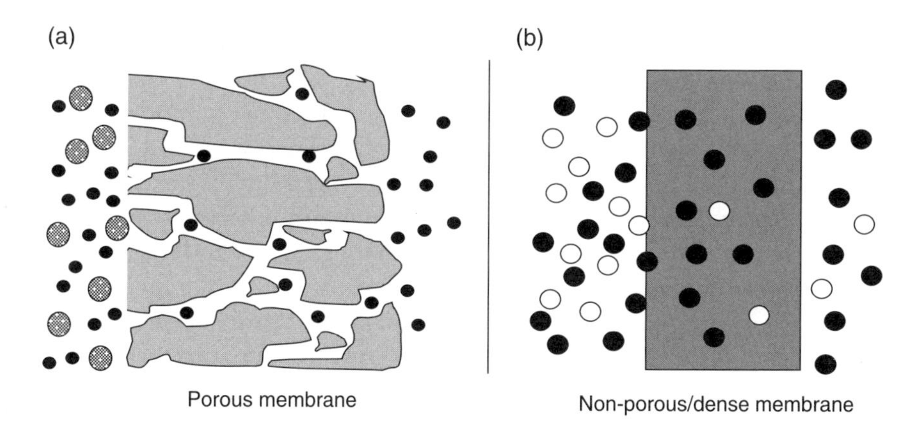

Figure 1.3 Schematic drawing of the permeation in porous and dense membranes.

top layer to obtain high permeation flux and a thick porous support to provide high mechanical strength. The resistance of the membrane to mass transfer is largely determined by the thin top layer.

Based on the membrane structure and separation principle, membranes can also be classified into porous and dense (non-porous) membranes, as depicted schematically in Figure 1.3. The porous membranes have a porous separation layer and induce separation by discriminating between particle (molecular) sizes (Figure 1.3(a)). The separation characteristics (i.e., flux and selectivity) are determined by the dimensions of the pores in the separation layer. The membrane material is of crucial importance for chemical, thermal, and mechanical stability but not for flux and rejection. The non-porous/dense membranes have a dense separation layer, and separation is achieved through differences in solubility or reactivity and the mobility of various species in the membrane. Therefore, the intrinsic properties of the membrane material determine the extent of selectivity and permeability. The LIMs can be considered as a special dense membrane since separation takes place via the filled liquid semi-permeable phase, although a porous structure is contained within the membrane.

1.2.2 Membrane Separation

The **membrane separation** process is characterized by the use of a membrane to accomplish a particular separation. Figure 1.4 shows the concept of a membrane separation process. By controlling the relative transport rates of various species, the membrane separates the feed into two streams: the **retentate** and the **permeate.** Either the retentate or the permeate can be the product of the separation process.

The performance of a membrane in separation can be described in terms of permeation rate or permeation flux (mol m^{-2} s^{-1}) and permselectivity. The permeation flux is usually normalized per unit of pressure (mol m^{-2} s^{-1} Pa^{-1}), called the **permeance,** or is further normalized per unit of thickness (mol m m^{-2} s^{-1} Pa^{-1}), called the **permeability,** if the thickness of the separation layer is known. In many cases only a part of the separation layer is active, and the use of permeability gives rise to larger values than the real intrinsic ones. Therefore, in case of doubt, the flux values should always be given together with the (partial) pressure of the relevant components at the high-pressure (feed) and low-pressure (permeate) sides of the membrane as well as the apparent membrane thickness.

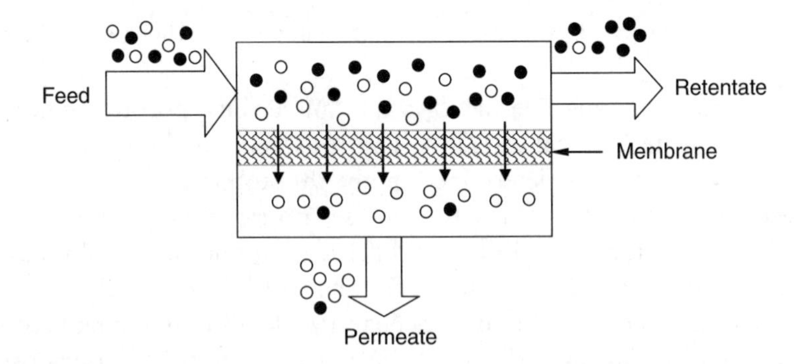

Figure 1.4 Schematic drawing of the membrane separation process.

The **permeation flux** is defined as the molar (or volumetric or mass) flow rate of the fluid permeating through the membrane per unit membrane area. It is determined by the driving force acting on an individual component and the mechanism by which the component is transported. In general cases, the permeation flux (J) through a membrane is proportional to the driving force; that is, the flux–force relationship can be described by a linear phenomenological equation:

$$J = -L \cdot \frac{dX}{dx} \tag{1.1}$$

where L is called the phenomenological coefficient and dX/dx is the driving force, expressed as the gradient of X (temperature, concentration, pressure, etc.) along the coordinate (x) perpendicular to the transport barrier. The mass transport through a membrane may be caused by convection or by diffusion of an individual molecule – induced by a concentration, pressure, or temperature gradient – or by an electric field. The driving force for membrane permeation may be the chemical potential gradient ($\Delta\mu$) or the electrical potential gradient ($\Delta\phi$) or both (the electrochemical potential is the sum of the chemical potential and the electrical potential). In case the concentration gradient serves as the driving force, the transport equation can be described by Fick's law:

$$J_A = -D_{Am} \cdot \frac{dc_A}{dx} \tag{1.2}$$

where D_{Am} ($m^2\ s^{-1}$) is the diffusion coefficient of component A within the membrane. It is a measure of the mobility of the individual molecules in the membrane and its value depends on the properties of the species, the chemical compatibility of the species, the membrane material, and the membrane structure as well. In practical diffusion-controlled separation processes, useful fluxes across the membrane are achieved by making the membranes very thin and creating large concentration gradients across the membrane.

For the pressure-driven convective flow, which is most commonly used to describe flow in a capillary or porous medium, the transport equation may be described by Darcy's law:

$$J_A = -Kc_A \cdot \frac{dp}{dx} \tag{1.3}$$

where *dp/dx* is the pressure gradient existing in the porous medium, c_A is the concentration of component A in the medium, and K is a coefficient reflecting the nature of the medium. In general, convective-pressure-driven membrane fluxes are high compared with those obtained by simple diffusion. More details of the transport mechanisms in membranes can be found elsewhere [3].

The **perm-selectivity** of a membrane toward a mixture is generally expressed by one of two parameters: the **separation factor** and **retention.** The separation factor is defined by

$$\alpha_{A/B} = \frac{y_A/y_B}{x_A/x_B} \tag{1.4}$$

where y_A and y_B, x_A and x_B are the mole fractions of components A and B in the permeate and the retentate streams, respectively.

The **retention** is defined as the fraction of solute in the feed retained by the membrane, which is expressed by

$$R = \left(1 - \frac{c_p}{c_f}\right) \times 100\% \tag{1.5}$$

where c_f and c_p are the solute concentrations in the feed and the permeate, respectively. For a selective membrane, the separation factors have values of 1 or greater whereas values of the retention are l or less.

Membrane separations are driven by pressure, concentration, or electric field across the membrane and can be differentiated according to type of driving force, molecular size, or type of operation. Common membrane processes include microfiltration, ultrafiltration, nanofiltration/reverse osmosis, gas separation, pervaporation, and dialysis/electrodialysis [3, 4]. Some processes have been applied extensively for separation and purification of gas and liquid mixtures in industry.

1.2.3 Membrane Performance

The membrane performance can be evaluated using permeability, selectivity, and stability. Ideally, a membrane with both high selectivity and permeability is required, but the attempt to maximize one factor will usually compromise the other. Comparatively, selectivity is a more important characteristic of a membrane because low permeability can be compensated to a certain extent by an increase in membrane surface area, whereas low selectivity leads to multi-stage processes which in most cases are not economical compared with established conventional processes.

The permeability and selectivity of a membrane are determined by the material and the structure of the membrane, which essentially determine the separation mechanism and application. Asymmetric membranes are mostly applied in practical applications because they have a thin selective layer to obtain high permeation fluxes and a thick porous support to provide high mechanical strength. For a certain mass separation, the type of membrane and the driving force required depend on the specific properties of the chemical species in the mixture.

In addition to permeability and selectivity, the following membrane stabilities are also required in various industrial applications:

- chemical resistance
- mechanical stability
- thermal stability
- stable operation.

Although the stability of a membrane depends on the membrane structure to some extent, it is mainly determined by the nature of the membrane material. For example, the upper temperature limit of polymeric membranes never exceeds 500°C, but inorganic materials can withstand very high temperatures and are inherently more stable at high temperatures and with various chemicals such as aggressive organic compounds and liquids with extreme pH values.

1.3 INORGANIC MEMBRANES

1.3.1 Types of Inorganic Membranes

Inorganic membranes are made of inorganic materials such as metals, ceramics, zeolites, glasses, carbon, and so on. Actually, inorganic membranes usually consist of several layers from one or more different inorganic materials. Details of inorganic membranes with respect to their syntheses, characterizations, transport theories, and scaling-up problems have been well reviewed and summarized by several authors [5, 6].

Inorganic membranes may be of either symmetric or asymmetric structure. Symmetric membranes often have considerable thickness to obtain sufficient mechanical strength. This is unfavorable for obtaining large fluxes, which usually require thin separation layers. In order to obtain high fluxes, most applicable inorganic membranes possess a multi-layered asymmetric structure as shown in Figure 1.5(a). A porous substrate with large pores (1–15 μm for low flow resistance) but sufficient mechanical

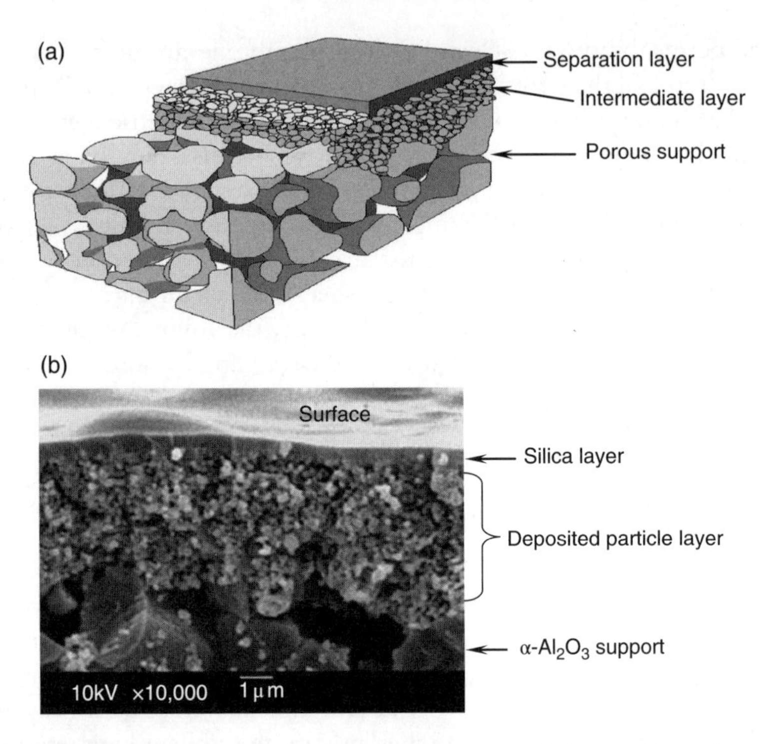

Figure 1.5 (a) Schematic representation of the asymmetric inorganic membrane and (b) cross-sectional SEM image of the silica/Al_2O_3 composite membrane. Reproduced from [7]. With permission from Elsevier.

strength is used to support a thin selective layer for separation. Commonly used materials for the macroporous support include Al_2O_3, ZrO_2, TiO_2, Si_3N_4, carbon, glass, stainless steel, and so on. Figure 1.5(b) shows the pore structure of a silica membrane supported on a cylindrical α-Al_2O_3 porous tube (outer diameter (OD) 10 mm; thickness 2 mm; average pore size 1 μm) [7].

In general, it is difficult to produce a thin separation layer directly on top of a support with large pores because the precursor system from which the separation layer is made will penetrate significantly into the pores of the support (e.g., the small particles from which small-pore membranes are made will penetrate much larger pores), leading to an increase in flow resistance. Furthermore, the thin layers covering large pores are mechanically unstable and would crack or peel off easily. A practical solution is to produce a graded structure by adding one or more intermediate layers with gradually decreasing layer thickness and pore size between the bulk support and the separation layer. An intermediate layer is also applied to

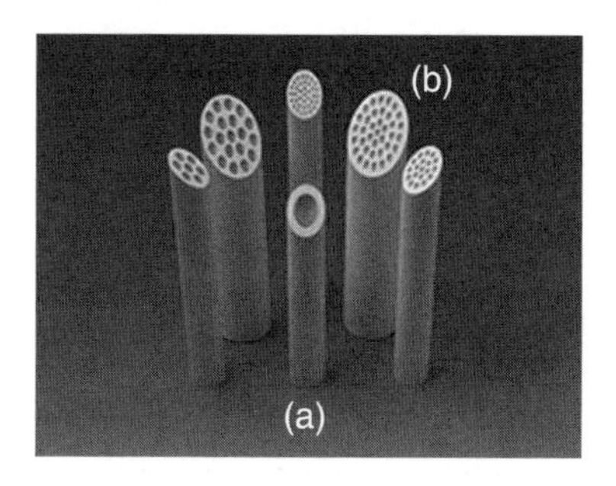

Figure 1.6 Tubular (a), monolithic (b), and hollow fiber (c) inorganic membranes.

match the thermal expansion difference of the membrane with the substrate, and as a buffer zone in case of chemical incompatibility during the membrane preparation process.

The separation layer may be dense (non-porous), such as Pd or Pd-alloy membranes for hydrogen separation and mixed (electronic, ionic) conducting oxide membranes for oxygen separation, or porous, such as metal oxides, silicalite, or zeolite membranes. Inorganic membranes are generally named for this separation layer, since it determines the properties and application of the membrane. The flux and selectivity of inorganic membranes are mainly determined by the quality of the separation layer, which is required to be defect-free and as thin as possible.

In addition to the planar geometry, inorganic membranes can also be produced in flat disk, tubular (dead-end or not), monolithic multi-channel, or hollow fiber configurations as shown in Figure 1.6. Disk membranes are often used in the laboratory because they can easily be fabricated by the conventional pressing method. In the case of tubes, they can be assembled in a module containing a number of tubes connected to a single manifold system.

The multi-channel monolithic form is developed to increase the mechanical robustness and the surface area-to-volume ratio, which gives more separation area per unit volume of membrane element. In the monolithic membranes, the monolith bulk is a porous support and the separation layer is produced on the inner surface of the channels. Therefore, feed is introduced in the channels and the permeate is obtained from the membrane wall, as shown in Figure 1.7. The surface area-to-volume ratio of the multi-channel monolithic membrane ranges from 130–400 $m^2\ m^{-3}$ compared with 30–250 $m^2\ m^{-3}$ for tubes. Honeycomb multi-channel monolithic membranes can even reach up to 800 $m^2\ m^{-3}$ of surface area-to-volume ratio.

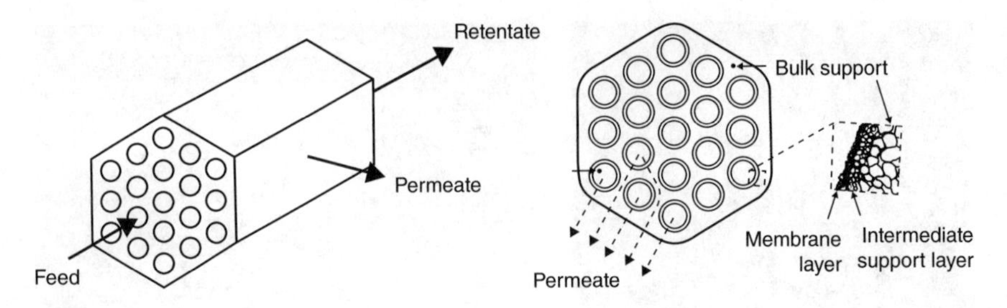

Figure 1.7 Schematic picture of a multi-channel monolithic membrane. Reproduced from [8]. With permission from Elsevier.

Table 1.1 Types of inorganic membranes

Material	Structure		Configuration
Metal or alloys (Pd, Ag, Ni)	Symmetric/composite	Dense	Tube; plate; hollow fiber
Stainless steel	Symmetric	Porous	Tube; hollow fiber
Metal oxides (Al_2O_3, ZrO_2, TiO_2, etc.)	Symmetric/asymmetric	Porous/mesoporous	Tube; hollow fiber; monolith
Glass	Symmetric	Mesoporous	Hollow fiber; tube
Silica (SiO_2)	Composite	Microporous	Plate; tube; hollow fiber; monolith
Zeolites (NaA, ZSM-5, etc.)	Composite	Microporous	Plate; tube; hollow fiber
Carbon	Symmetric/asymmetric	Microporous	Tube; hollow fiber
Mixed ionic–electronic ceramic conductors	Symmetric/asymmetric/composite	Dense	Tube; disk; plate; hollow fiber
ZrO_2- or CeO_2-based ionic conductors	Symmetric/asymmetric	Dense	Tube; disk; plate; hollow fiber
LIM (molten salt)	Symmetric	Dense	Tube; disk

It is possible to increase the surface area-to-volume ratio of tubular membranes by decreasing their diameter. If the diameter of the membrane tube is reduced to a certain level, it is then called a hollow fiber membrane. Such hollow fiber membranes usually have an internal diameter ranging from 40–300 µm and wall thicknesses of 10–100 µm, and can provide surface area-to-volume ratios of more than $3000\,m^2\,m^{-3}$ [4].

A variety of inorganic membranes are summarized in Table 1.1. The metal membranes mainly include palladium-based membranes for hydrogen permeation and silver-based membranes for oxygen permeation. Currently, the commercially available inorganic membranes are porous membranes made from alumina, silica and titania, glass, and stainless steel. These membranes are characterized by high permeability, but low selectivity. ZrO_2- or CeO_2-based membranes are solid oxide electrolytes and their permeability depends on their ionic conductivity.

1.3.2 Fabrication of Inorganic Membranes

As inorganic membranes have a multi-layered asymmetric structure consisting of porous support, intermediate layers and a selective separation layer, the fabrication of inorganic membranes is a multi-step process, as illustrated in Figure 1.8.

The fabrication starts with the preparation of porous substrates, with which the shape and configuration of the final membrane products can be determined. The porous substrate is critical for the quality of the membrane itself, because it not only provides sufficient mechanical strength but also takes effect on the permeability and selectivity of the membrane. Therefore, the commercial availability of high-quality substrates is a critical issue in the further development of membrane separation units.

Porous substrates are mostly made from ceramics like α-Al_2O_3, but also from other materials such as metal or glass. They are formed by shaping inorganic powders and consolidation of the green body by

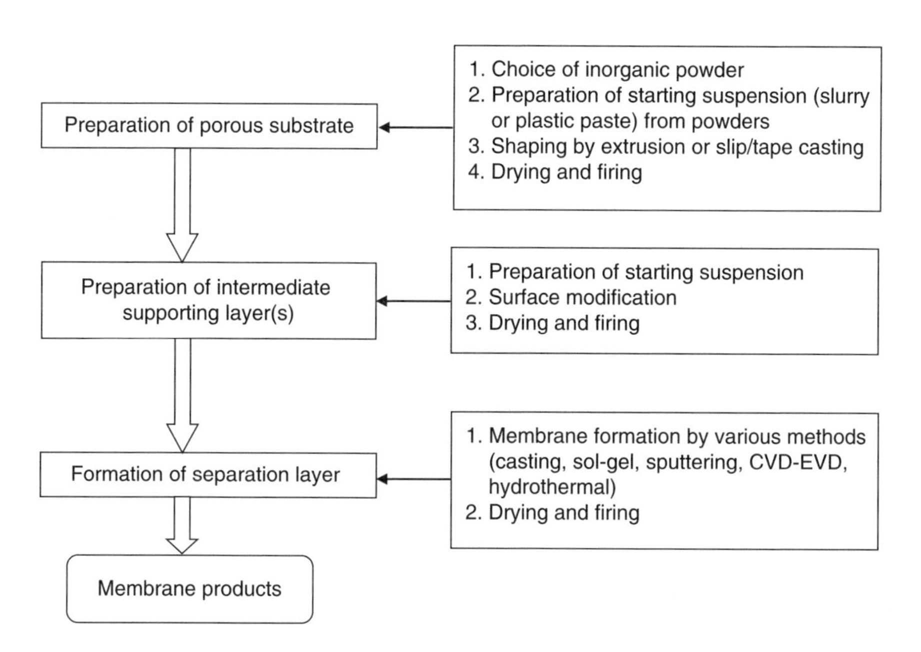

Figure 1.8 Fabrication process of inorganic membranes.

sintering. Four main stages are included in the fabrication process: choice of inorganic powder, paste/slurry preparation, shaping into green body, and firing into a porous substrate at high temperature. The particle size and morphology of the inorganic powders, the composition and homogeneity of the powder suspension, and the drying and firing conditions have considerable influence on the quality of the porous substrates [8]. Depending on the tubular or flat sheet (or disk) configuration, porous substrates can be prepared by the well-established techniques of slip casting, tape casting, pressing, or extrusion [4]. Figure 1.9 shows schematically the extrusion apparatus for shaping mono- or multi-channel tubular substrates. Different shapes are obtained by changing the geometry of the die (e.g., number of channels, diameter of channels, and external diameter of tubes).

Porous substrates should have a smooth surface with constant and homogeneous characteristics (wettability) and a narrow pore size distribution. Pores much larger than average and grains broken out of the surface, or irregularities in the porous substrate, may result in defects in the separation layer applied on it. Therefore, surface modification or formation of intermediate layer(s) is necessary to prepare a thin and defect-free membrane.

The preparation of intermediate layers is actually a process to produce a porous layer with smaller pores than those in the bulk support. Since

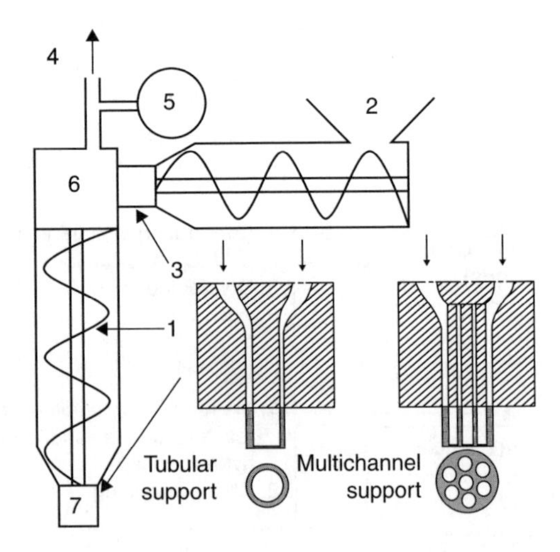

Figure 1.9 Schematic view of the extrusion apparatus for the fabrication of tubular membranes (porous substrates): (1) endless screw; (2) paste inlet; (3) compression; (4) vacuum; (5) pressure gauge; (6) vacuum chamber; (7) die. Reproduced from [8]. With permission from Elsevier.

it is difficult to have thermostable powder particles smaller than 5–6 nm, an intermediate layer with pore diameters below 2 nm cannot be produced by packing of spherical or plate-shaped particles. Sometimes, more than one intermediate layer has to be produced to form a graded structure with gradually decreasing layer thickness and pore size between the bulk support and the separation layer.

The separation layer, either porous or dense, can be formed using different methods such as sol-gel and template routes, hydrothermal synthesis, chemical vapor deposition (CVD), or physical sputtering, depending on the membrane material and its application. These membrane preparation methods will be described in the following chapters of this book for different membranes and membrane reactors. We note that the preparation of inorganic membranes involves a multi-step high-temperature treatment process. Therefore, inorganic membranes are much more expensive than polymeric ones.

1.3.3 Characterization of Inorganic Membranes

Inorganic membrane performances are determined by the membrane structure and the material properties. Information on pore size, shape, distribution, connectivity, and porosity for porous membranes and gas tightness, crystal structure, and surface properties for dense ceramic membranes is of importance to predict the separation performances of the membranes. The membranes developed have to undergo a series of characterization tests using techniques based on adsorption, X-ray diffraction (XRD), scanning electron microscopy (SEM), transmission electron microscopy (TEM), and so on. The characterization of inorganic membranes generally refers to three aspects: evaluation of porous features (porosity, pore size and distribution, tortuosity, etc.); microstructure and morphology (pore shape, surface, cross-section); and transport or reaction properties (permeability, selectivity, reactivity). Table 1.2 summarizes the common characterization methods for inorganic membranes.

1.3.4 Applications of Inorganic Membranes

Compared with their polymeric counterparts, inorganic membranes are characterized by high chemical and thermal resistances and high mechanical stability, and thus can be applied in a harsh environment. However, inorganic membranes also exhibit the shortcoming of high cost, because

Table 1.2 Characterization methods for inorganic membranes

Properties	Methods
Pore size	Bubble point
Porosity	Archimedes; Burnauer, Emmett, and Teller (BET)
Pore size distribution	Mercury porosimetry; thermoporometry
Tortuosity of pores	Permeation of pure water
Surface morphology	SEM, TEM, atomic force microscopy (AFM), scanning transmission microscopy (STM)
Cross-sectional morphology/thickness	SEM, STM
Crystalline phase	XRD
Grain structure/boundary	SEM, TEM
Mechanical strength	Three-point-bending method
Thermal expansion	Thermal dilatometer
Permeation rate	Pure gas or liquid permeation test
Selectivity	Separation factor or retention curves
Reactivity/catalytic activity	Reaction test

of their long and complicated production route in which multi-step high-temperature treatment is required. Therefore, inorganic membrane applications should preferably be found in areas where polymer membranes cannot or do not perform well. The application areas of inorganic membranes mainly include [8]:

- separation in food, beverage, and biotechnology fields;
- purification in environmental protection;
- energy conversion in solid oxide fuel cells;
- separation and reaction in petroleum and chemical industries.

1.4 INORGANIC MEMBRANE REACTORS

1.4.1 Basic Principles of Membrane Reactors

In a conventional design of chemical production process, the reaction and separation functions are carried out by two different processing units, as illustrated in Figure 1.10(a), where a membrane separator is used for product separation. If the membrane is placed inside the reactor to carry out the separation function as shown in Figure 1.10(b), it then becomes a membrane reactor. Since the reaction and separation proceed simultaneously, the separation of products can be accomplished in the reactor unit itself, or at least the downstream separation load can be

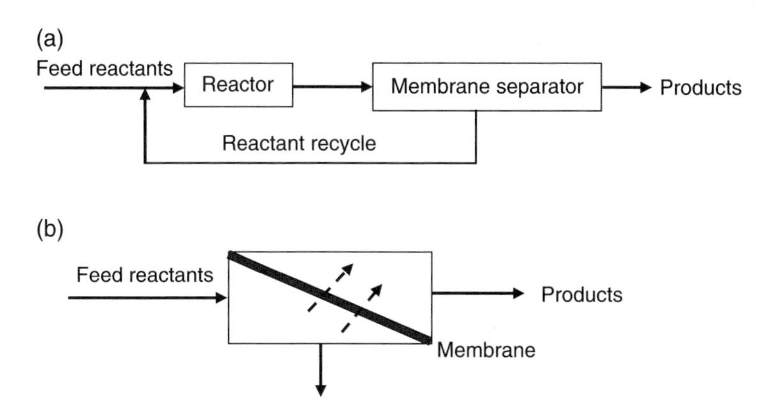

Figure 1.10 (a) Conventional reactor and membrane separator; (b) membrane reactor.

reduced. As a result, the chemical process will become much simpler and the operational costs will thereby be reduced drastically. Moreover, the reaction can be enhanced significantly due to the combination of membrane functions. Based on the above discussions, an MR is a device integrating a membrane with a reactor in which the membrane functions as a separator or a reaction interface to enhance the reaction process.

Both organic (polymeric) and inorganic membranes have been used in MRs. Since organic membranes cannot withstand high temperature, organic MRs are applied mainly in biochemical processes (thus called membrane bioreactors), such as for the production of fine chemicals via the use of enzymes and for large-scale biogas production from waste and environmental clean-up. Most chemical reactions of interest operate at temperatures well in excess of the limitation of polymeric membranes; inorganic membranes with inherent high-temperature structural and chemical stability are appropriate candidates for use in catalytic MRs. The main requirement is to produce membranes with the desired properties of good separation and selectivity and enough robustness to withstand the severe operating conditions often encountered in industrial practice.

Membrane reactors promote a reaction process based on the following three routes, with the membranes having different effects:

1. The membrane serves as a product extractor (Figure 1.11(a,b)). For thermodynamically limited equilibrium reactions, the removal of at least one of the products by the membrane from the reaction zone makes the reaction equilibrium shift to the product side with the single-pass conversion increased (Figure 1.11(a)). On the contrary, if the intermediate product is taken out through the membrane, the

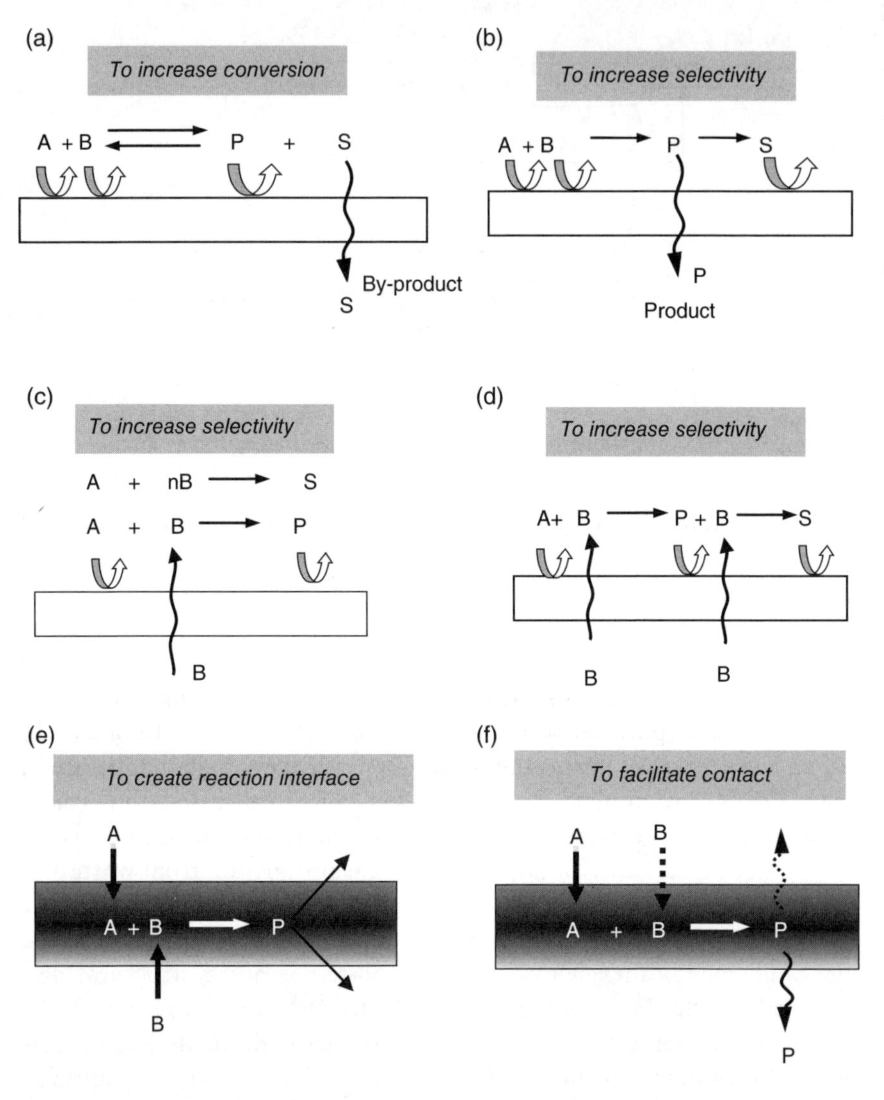

Figure 1.11 Principles of membrane reactors to enhance the reaction process: (a,b) membrane as a product extractor; (c,d) membrane as a reactant distributor; (e,f) membrane as an active contactor.

undesired side-reactions or the secondary reaction of products can be suppressed, leading to improved selectivity (Figure 1.11(b)).

2. The membrane serves as a reactant distributor (Figure 1.11(c,d)). A reactant is added to the reaction zone in a controlled manner through the membrane. As a consequence, the side reactions are limited, leading to increased selectivity and yield. In addition, it is

possible to use low-purity feed instead of pure reactant to obtain higher selectivity and reduce capital investment and operation costs.

3. The membrane serves as an active contactor (Figure 1.11(e,f)). Reactants are supplied to the catalyst by the controlled diffusion in the membrane, hence a well-defined reaction interface (or region) between two reactant streams is created. The reactants can be provided from one side or from opposite sides of the membrane. Furthermore, more reactive sites can be provided due to the easy access of reactants to the catalyst, and thus the catalyst's efficiency can be increased greatly.

A synergy may be created due to the use of MRs. Since the chemical reaction can be enhanced by membrane separation, it is possible to attain a given conversion at less severe conditions of temperature and pressure. This implies that hot spots may be avoided and the MR may be energy-efficient and relatively safe in operation. Furthermore, the reduction in temperature leads to a decrease in catalytic deactivation from coke deposition and sintering, and thus provides improved catalyst life and/or less frequent regeneration requirements. In addition, the MR may also allow hot separation of products and eliminate the need for quenching a reaction to prevent back-reactions.

1.4.2 Incorporation of Catalyst in Membrane Reactors

Most chemical reactions take place in the presence of a catalyst. There are four ways to incorporate catalysts in the membranes of MRs, as illustrated in Figure 1.12 [9].

(a) **Catalyst physically separated from an inert membrane**
In this case, the membrane compartmentalizes the reactor and functions for separation but is not involved directly in the catalytic reaction (called an "inert membrane"). The catalyst pellets are usually packed or fluidized on the inert membrane (Figure 1.12(a)), which acts as an extractor for fractionation of products and/or as a distributor for controlled addition of reactants. This incorporation of catalyst is most popular in practical use and can easily be operated. Since the catalyst is physically separated from the membrane, the separation function of the membrane and the activity of the catalyst can be modulated independently. Catalysts are generally placed on the separation layer

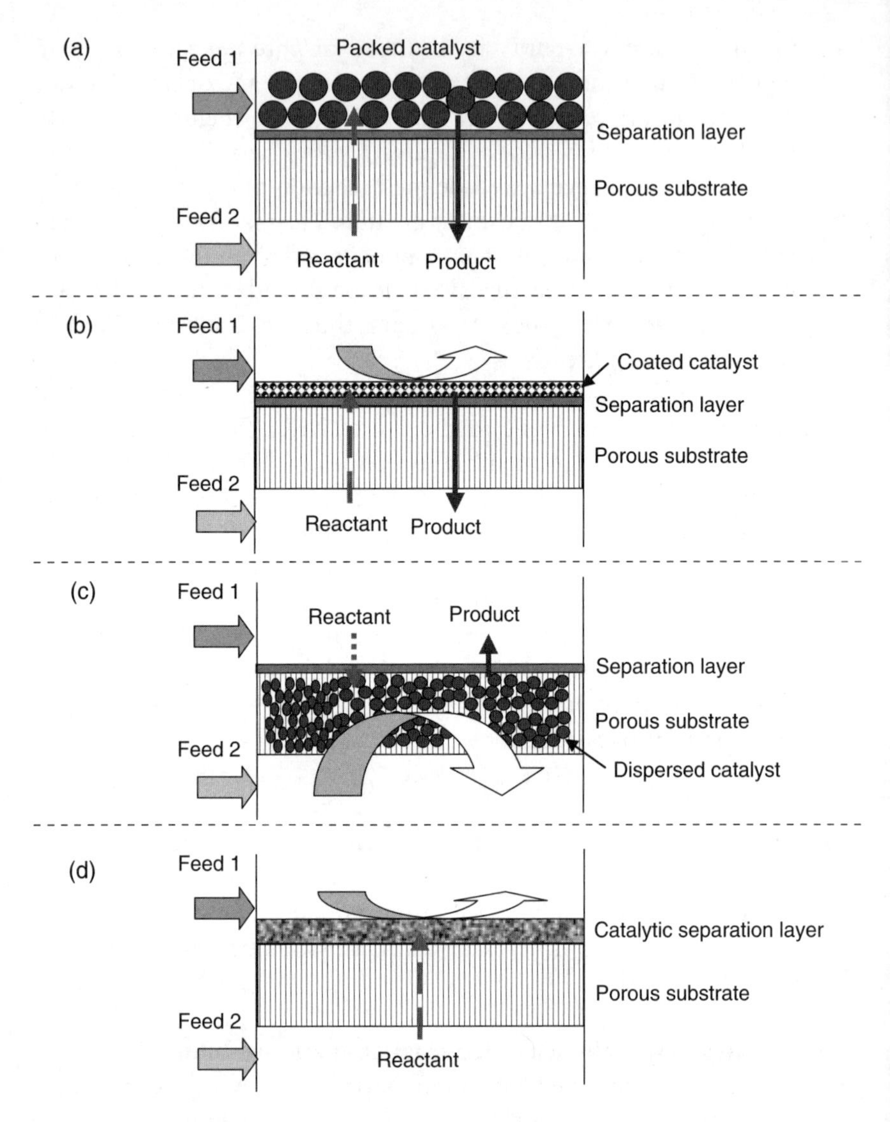

Figure 1.12 Membrane/catalyst combination modes in MRs: (a) catalyst separated from the membrane; (b) catalyst coated on the membrane; (c) catalyst dispersed inside the membrane; (d) membrane as catalyst.

side but can also be placed on the opposite side of the membrane, especially if the catalysts may do harm to the separation layer.

(b) Catalyst coated on the membrane surface

In this case, the catalyst is coated on the membrane surface using a catalyst paste (Figure 1.12(b)). The catalyst layer is generally porous and is integrated with the membrane into a single body.

If the catalyst is reactive with the separation layer of the membrane, direct contact between them must be avoided.

(c) **Catalyst dispersed in the membrane porous structure**
In Figure 1.12(c), the catalyst is dispersed in the porous substrate of the membrane to form a membrane catalyst. One of the reactants or products traverses through the membrane into or out of the reaction zone. In classical reactors, the reaction conversion is often limited by the diffusion of reactants into the pores of the catalyst or catalyst carrier pellets. If the catalyst is inside the pores of the membrane, the combination of the open pore path and transmembrane pressure provides easier access of the reactants to the catalyst as shown in Figure 1.13. As a result, the access of reactants to the catalyst is improved and the catalytic efficiency can be increased greatly. If the membrane thickness and porous texture, as well as the quantity and location of the catalyst in the membrane, are adapted to the kinetics of the reaction, the membrane catalyst can be 10 times more active than that in the form of pellets [10, 11].

(d) **Inherently catalytic membranes**
In some cases, the membrane material is inherently catalytic and the membrane serves as both catalyst and separator, controlling the two important functions of the reactor simultaneously as shown in Figure 1.12(d). A number of meso- and microporous inorganic membrane materials have catalytic properties, such as titania and zeolites with acid sites. As an example, mesoporous TiO_2

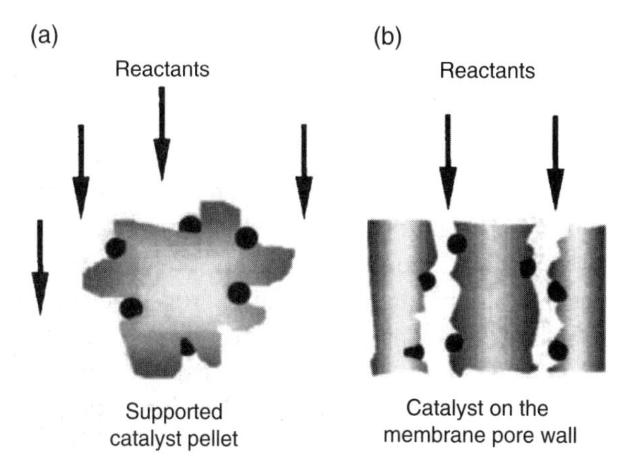

Figure 1.13 Comparison of the contact of the reactant with the catalyst in (a) a catalyst-supported pellet and (b) a catalyst-loaded membrane. Reproduced from [10]. With permission from Elsevier.

photocatalytic membranes prepared by the sol-gel process are used for the continuous degradation of volatile organic compounds (VOCs) in water [12]. Most perovskite membranes for oxygen permeation also exhibit good catalytic activity toward the oxidation of hydrocarbons [13].

In general, the membranes modified by coating a catalyst on the membrane surface or loading a catalyst inside the membrane's porous structure are called "catalytic membranes." In most cases, catalytic membranes do not need to be perm-selective but are required to be highly active for the reactions considered and to have a sufficiently low overall permeability so that they are operated in the diffusion-controlling regime. For catalytic membranes, the catalytic composition and activity as well as the porous texture have to be optimized for the reactions considered and kept stable in use. However, it is difficult to modulate these properties of catalytic membranes. Therefore, very few studies have been performed on catalytic MRs in application. It is noteworthy that in some cases extra catalysts will be applied in the catalytic MRs to promote the reactions considered.

1.4.3 Configuration of Membrane Reactors

Unlike conventional reactors, MRs have two compartments separated by the membrane and thus are characteristic of at least three inlets/outlets for the feed and product streams. They are designed and fabricated based on the membrane configuration and the application conditions.

Tubular MRs are applied mostly in laboratory studies and industrial applications. Figure 1.14 shows schematically the structure of a tubular MR. The catalyst may be placed on either the tube side or the shell side. Two feed streams are introduced concurrently or countercurrently on opposite sides of the membrane. Usually, a sweep gas (feed 2) is employed on the permeate side to reduce the build-up of products and therefore reduce the potential rise in film mass transfer resistance on the permeate side of the membrane. The ends of the membranes are sealed by glazing or by very fine powders. The membrane tube is sealed by carbon or graphite strings pressed against Swagelok-type compression fittings. In industrial reactors, a shell and tube configuration with an assembly of single tubes or multi-channel monoliths may be incorporated into a large shell.

Disk/flat sheet MRs are applied mostly in research work because they can be fabricated easily in the laboratory with a small amount of

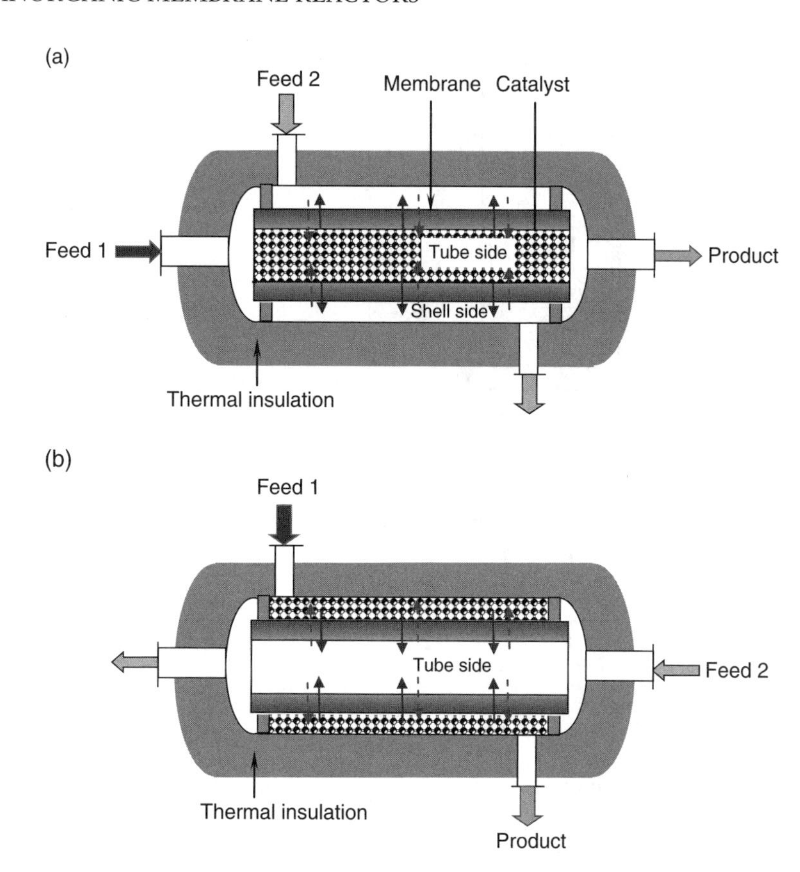

Figure 1.14 Schematic diagram showing tubular membrane reactors: (a) catalyst on tube side with concurrent flow; (b) catalyst on shell side with countercurrent flow.

membrane material. Figure 1.15 shows the structure of a disk/flat sheet MR. The membrane disk is mounted between two vertical ceramic or quartz tubes. Pyrex or gold gaskets are used to obtain effective seals between the disk and the walls of the tubes at high temperatures by placing the assembly in compression with the use of spring clamps. The catalyst is usually packed on the membrane or coated on the membrane surface.

In general, hollow fiber membranes can be assembled into reactors following the same procedures as for tubular MRs. Hollow fibers offer a much greater packing density, but suffer from poor mechanical strength. Figure 1.16 illustrates schematically the configuration of a hollow fiber MR [13, 14]. A quartz shell is used to house the hollow fibers. Two pairs of gas inlet/outlet fittings and alumina thermocouple sleeve assemblies are housed in quartz end-caps that fit closely to the inner wall of the shell tubing. The hollow fiber membranes are placed in a pair of quartz tubes

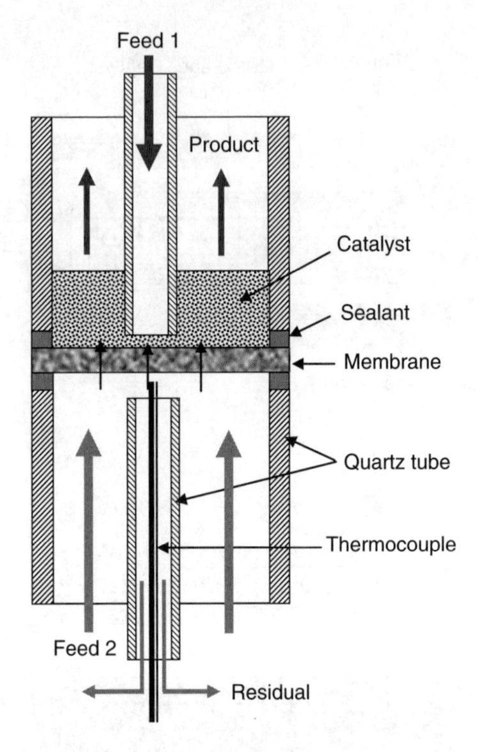

Figure 1.15 Typical structure of the disk/flat sheet membrane reactor.

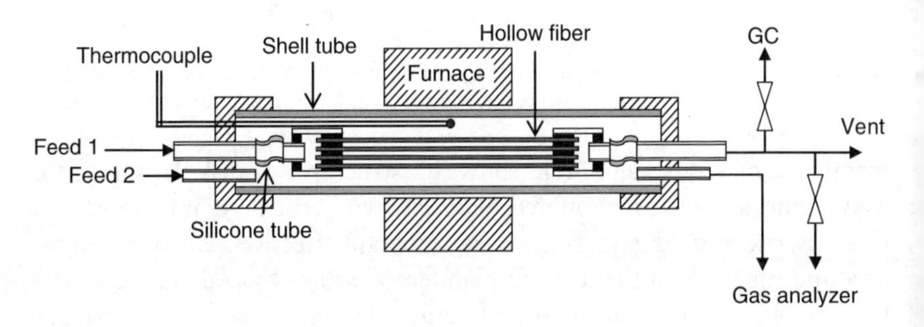

Figure 1.16 Schematic diagram of the hollow fiber membrane reactor. Reproduced with permission from [16]. Copyright © (2013), Woodhead Publishing (Elsevier).

with flexible silicone tubes to connect to the lumen-side gas inlet/outlet of the reactor and to offset the thermal expansion of the hollow fibers occurring during operation. A seal is achieved by using high-temperature water-based glass/ceramic sealant. A thermocouple that can be moved along the length of the module is inserted inside an alumina sleeve, the tip of which is positioned close to the center of the hollow fibers, allowing

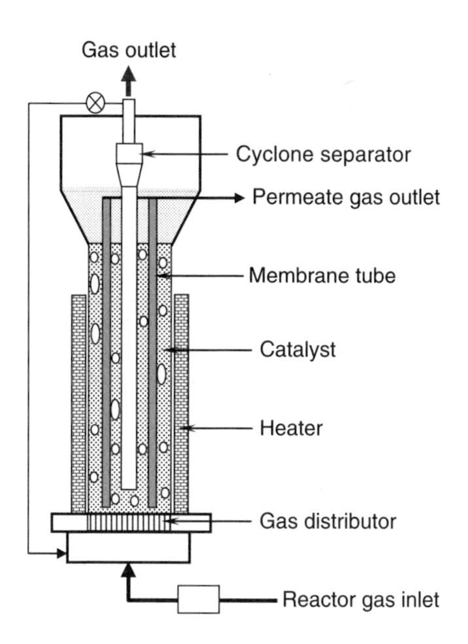

Figure 1.17 Schematic representation of the FBMR. Reproduced from [15]. With permission from Elsevier.

the temperature profile to be recorded during operation. A custom-made furnace with a short can be used to heat the reactor so that the sealing points are kept away from the high-temperature zone. The catalyst, if used, is usually packed in the fiber lumen for convenient operation.

In addition to the above-mentioned MR configurations, other MRs including catalyst fluidized bed membrane reactors (CFBMRs), electrolyte membrane reactors (EMRs), and membrane microreactors (MMRs) have also been developed and investigated. Figure 1.17 shows a typical representation of the fluidized bed MR for hydrogen production [15]. Hydrogen-permeable membranes are placed vertically in a catalyst bed. A mixture of steam and hydrocarbon gas is fed into the bottom of the reactor to fluidize the particulate catalyst. Hydrogen is separated through perm-selective membranes. Details of these special MRs will be given in the following chapters of this book.

1.4.4 Classification of Membrane Reactors

So far, MRs have been investigated extensively in the literature. These MRs can be categorized based on the properties of the membrane or the reactor itself. A summary of different types of MRs is given in Table 1.3.

Table 1.3 Types of membrane reactors

Membrane reactor type	Description	Acronym
Packed-bed membrane reactor	Additional catalysts are packed in the membrane reactor	PBMR
Fluidized-bed membrane reactor	Catalysts in the reactor are present in a fluidized mode	FBMR
Inert membrane reactor	The membrane does not participate directly in the reaction	IMR
Catalytic membrane reactor	The membrane functions as both catalyst and separator	CMR
Catalytic non-selective membrane reactor	The membrane is not selective but serves as a catalytic site for reactions	CNMR
Flow-through catalytic membrane reactor	Catalytic reactions take place while the reactants flow through the membrane	FTCMR
Membrane microreactor	Membrane is integrated with the microreactor having a characteristic length of <1 mm	MMR
Electrolyte membrane reactor	An external electrical circuit is applied to complete reactions	EMR

By far the most commonly referred to reactor is the packed bed membrane reactor (PBMR), in which the reaction function is provided by a packed bed of catalysts in contact with the membrane. The membrane itself is not catalytic or at least not intentionally so, but is used to add or remove certain species from the reactor. If the membrane is highly perm-selective, this configuration appears ideal for situations where two complementary reactions take place on either side of the membrane – the product of the reaction on one side acting as a reactant on the other side, while the endothermicity of one reaction is compensated by the exothermicity of the other. When the catalysts at work are present in a fluidized mode, the reactor is then called a fluidized bed membrane reactor (FBMR).

In catalytic membrane reactors (CMRs), the reactions take place directly on the membrane and the membrane functions as both a catalyst and a separator/distributor. This requires that the membrane material has intrinsic catalytic activity or that it is modified by the addition of active components. Some of the commonly utilized inorganic (such as metal oxide and zeolite) and metal membranes are intrinsically catalytically active. In other cases, the catalysts can be integrated with the membrane into a single body by being coated on the membrane surface or deposited inside the membrane porous structure. In case the membrane does not participate in the reaction directly, but is used to add or remove certain species from the reactor, this is called an inert membrane reactor (IMR).

In some applications the membrane is not required to be perm-selective but only to provide reactive sites. Such devices are called catalytic non-perm-selective membrane reactors (CNMRs). The reactants can flow into the membrane from opposite sides, and the membrane's role is to provide a controlled reactive interface. If the reactants flow through the membrane from one side to the other while the reaction takes place instantly, such a reactor is also called a flow-through catalytic membrane reactor (FTCMR).

When the membrane is integrated in a microreactor having a characteristic length of <1 mm, it is called an MMR. For EMRs, an external electrical circuit is necessary for reactions to proceed. Electrical power can be co-generated with the production of chemicals in the EMRs.

REFERENCES

[1] Hwang, S.-T. and Kammermeyer, K. (1975) *Membranes in Separations*, John Wiley & Sons, Inc., New York.
[2] Nunes, S.P. and Peinemann, K.-V. (2006) *Membrane Technology in the Chemical Industry*, Wiley-VCH, Weinheim.
[3] Baker, R.W. (2004) *Membrane Technology and Applications*, 2nd edn, John Wiley & Sons, Ltd, Chichester.
[4] Li, K. (2007) *Ceramic Membranes for Separation and Reaction*, 1st edn, John Wiley & Sons, Ltd, Chichester.
[5] Bhave, R.R. (1991) *Inorganic Membranes: Synthesis, Characteristics and Applications*, Springer-Verlag, New York.
[6] Zaman, J. and Chakma, A. (1994) Inorganic membrane reactors. *Journal of Membrane Science*, 92, 1–28.
[7] Asaeda, M. and Yamasaki, S. (2001) Separation of inorganic/organic gas mixtures by porous silica membranes. *Separation and Purification Technology*, 25, 151–159.
[8] Burggraaf, A.J. and Cot, L. (1996) *Fundamentals of Inorganic Membrane Science and Technology*, Elsevier Science B.V., Amsterdam.
[9] Coronas, J. and Santamaria, J. (1999) Catalytic reactors based on porous ceramic membranes. *Catalysis Today*, 51, 377–389.
[10] Julbe, A., Farrusseng, D. and Guizard, C. (2001) Porous ceramic membranes for catalytic reactors – overview and new ideas. *Journal of Membrane Science*, 181, 3–20.
[11] Zaspalis, V.T., Van Praag, W., Keizer, K., Van Ommen, J.G., Ross, J.R.H. and Burggraaf, A.J. (1991) Reaction of methanol over catalytically active alumina membranes. *Applied Catalysis*, 74, 205–222.
[12] Tsuru, T., Kan-no, T., Yoshioka, T. and Asaeda, M. (2006) A photocatalytic membrane reactor for VOC decomposition using Pt-modified titanium oxide porous membranes. *Journal of Membrane Science*, 280, 156–162.
[13] Tan, X., Li, K., Thursfield, A. and Metcalfe, I.S. (2008) Oxyfuel combustion using a catalytic ceramic membrane reactor. *Catalysis Today*, 131, 292–304.

[14] Tan, X., Pang, Z., Gu, Z. and Liu, S. (2007) Catalytic perovskite hollow fibre membrane reactors for methane oxidative coupling. *Journal of Membrane Science*, 302, 109–114.

[15] Deshmukh, S.A.R.K., Heinrich, S., Mörl, L., van Sint Annaland, M. and Kuipers, J.A.M. (2007) Membrane assisted fluidized bed reactors: Potentials and hurdles. *Chemical Engineering Science*, 62, 416–436.

[16] Tan, X. and Li, K. (2013) Dense ceramic membranes for membrane reactors, in *Handbook of Membrane Reactors, Volume I – Fundamental Materials Science, Design and Optimisation* (ed A. Basile), Woodhead Publishing Limited, Cambridge, pp. 271–297.

2

Porous Membrane Reactors

2.1 INTRODUCTION

Inorganic membranes can be categorized into two types: dense (non-porous) and porous. The porous membranes are made mainly from ceramics (such as Al_2O_3, ZrO_2, TiO_2, Si_3N_4, etc.), glass, carbon, silica, and porous stainless steel (PSS), or a combination of these materials. They are characteristic of an interconnected, tortuous, and randomly oriented pore network with constrictions and dead-ends which are formed by packing of particles. Since the large-scale production of porous membranes with consistent quality is relatively mature, with low production costs, studies in porous membrane reactors have been conducted extensively. Furthermore, a variety of porous membranes with both innovative materials and microstructure makes them popular in different chemical reaction processes, not only as a separator or distributor, but also as a substrate for other microporous or dense membranes.

Porous membranes are characterized by high permeability, but low selectivity. The transport mechanisms in porous membranes can be viscous flow, Knudsen diffusion, surface diffusion, capillary condensation, and/or molecular sieving, depending on the membrane pore size and its surface characteristics. The performance of porous MRs is very much dependent on the membrane structures. Close adherence to a rigid protocol is necessary to obtain membranes of consistent quality.

This chapter starts with a general description of porous membranes and separation mechanisms related to the pore structure. A variety of porous membranes – including glass, ceramic, carbon, and silica – and their MRs will be illustrated individually. Considerable attention will be focused on

---

Inorganic Membrane Reactors: Fundamentals and Applications, First Edition. Xiaoyao Tan and Kang Li.

the preparation, operation, and application of various porous MRs. Finally, the challenges and prospects of the porous MRs will be discussed.

2.2 GAS PERMEATION IN POROUS MEMBRANES

2.2.1 Types of Porous Membranes

Porous membranes usually have a multi-layered structure consisting of a separation layer to fulfill the actual membrane function and a porous support comprising at least one porous structure. The porous support provides general mechanical stability and has larger pores compared with the separation layer, with a permeability 10 times higher than that of the separation layer. Each layer of the membrane must have a thickness 100–1000 times the pore size in it. The higher the membrane selectivity required, the more substrate layers are needed. Figure 2.1 shows a layered structure of the zirconia membrane formed from nanocrystallites of less than 10 nm.

According to the IUPAC definition, porous membranes can be classified into macro-, meso-, and microporous membranes based on the pore size of the separation layer. The separation layer and the substrates can be made from different materials. Common inorganic porous membranes are listed in Table 2.1.

Different types of pores may co-exist in the porous membranes: isolated pore, dead-end pore, cylindrical pore, constricted pore, and conical

Figure 2.1 Scanning electron microscopy image of the microporous zirconia membrane. Reproduced from [6]. With permission from Elsevier.

Table 2.1 Types of inorganic porous membranes

Membrane type	Pore size (average diameter)	No. of separation layers, thickness	Materials	Application
Support	$d = \sim 5\,\mu m$	~1.5 mm	Ceramics; glass; stainless steel	
Macroporous membrane	$50\,nm < d < 2\,\mu m$	1–3 layers, ~25 μm	Al_2O_3; ZrO_2; Si_3N_4; and so on	Micro/ ultrafiltration
Mesoporous membrane	$2\,nm < d < 50\,nm$	4 layers 2–3 μm	Al_2O_3; ZrO_2; TiO_2, SiO_2; and so on	Ultrafiltration
Microporous membrane	$d < 2\,nm$	5 layers 100 nm	Carbon; SiO_2; and so on	Nanofiltration/ gas separation/ pervaporation

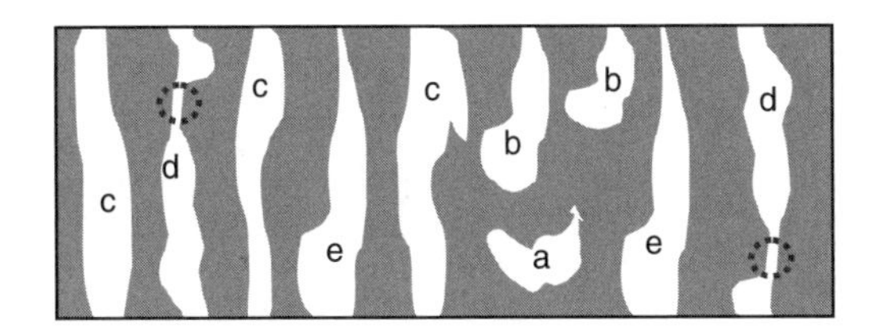

Figure 2.2 Schematic representation of the main types of membrane pores (a) isolated; (b) dead-end; (c) straight cylindrical; (d) constricted; (e) conical.

pore [1], as shown schematically in Figure 2.2. The formation of these pores of various types is related to the fabrication process. The isolated and dead-end pores do not contribute to permeation under steady conditions. The dead-end pores contribute to the porosity as measured by the adsorption technique or mercury porosimetry, but do not contribute to the effective porosity in permeation. The permeation pores may be channel-like or slit-shaped. The permeation pores refer to those running from one side of the membrane to the other. Pore constrictions are important for flow resistance, especially when capillary condensation and surface diffusion phenomena occur in systems with a relatively large internal surface area. A membrane with conical pores is asymmetric and combines a low flow resistance (large pores across a considerable fraction of the membrane thickness) with a relatively large selectivity (small pores on the top side of the membrane).

The transport properties of porous membranes are highly related to the parameters of porosity, pore size distribution, pore shape, interconnectivity, orientation, and roughness of the surface. The separation activity of multi-layer asymmetric membranes is determined mainly by the pore structure of the top layer.

2.2.2 Transport Mechanisms

The typical gas transport mechanisms in porous membranes are: molecular diffusion and viscous flow, capillary condensation, Knudsen diffusion, surface diffusion, and configurational or micropore-activated diffusion. The contributions of these different mechanisms depend on the properties of both the membrane and the gas under the operating temperature and pressure. Figure 2.3 illustrates schematically the gas transport mechanisms in a single membrane pore.

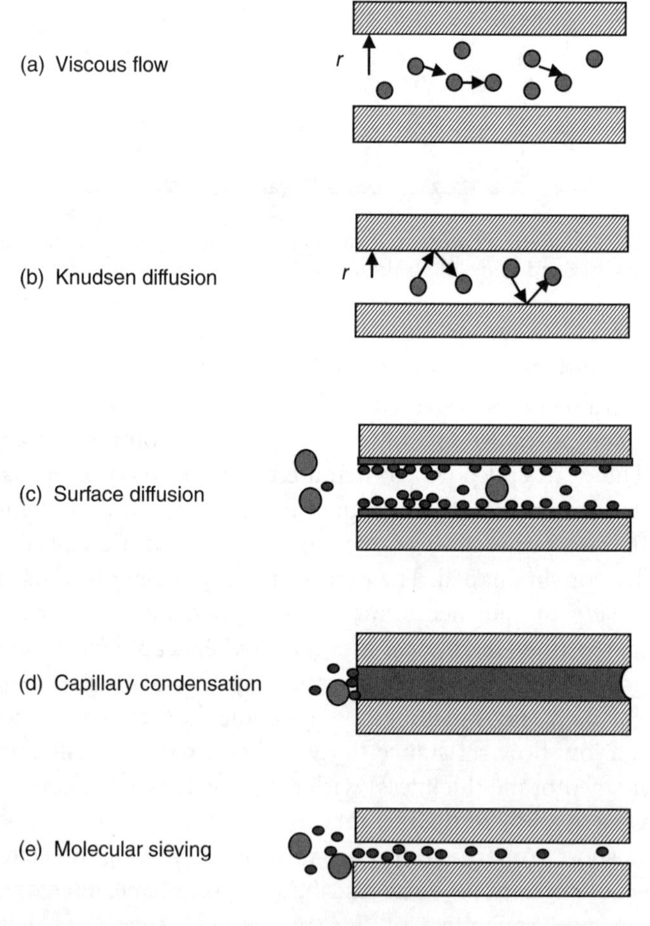

Figure 2.3 Gas transport mechanisms in membrane pores.

When the pore radius is much greater than the mean free path of permeating gas molecules (i.e., $r/\lambda > 3$), **viscous flow** occurs if a pressure gradient is applied to such a pore regime. The mean free path is defined as

$$\lambda = \frac{RT}{\sqrt{2} \cdot \pi d^2 N_A p} \tag{2.1}$$

where R is the gas constant ($8.314\,\mathrm{J\,mol^{-1}\,K^{-1}}$), T the temperature (K), p the gas pressure (Pa), d the diameter of a gas molecule (m), and N_A is the Avogadro number ($6.02 \times 10^{23}\,\mathrm{mol^{-1}}$). The flux through the membrane under viscous flow can be described by the Hagen–Poiseuille equation (hence, viscous flow is also called **Poiseuille flow**):

$$J_V = -\frac{r^2}{8\mu}\frac{p}{RT}\frac{dp}{dx} \tag{2.2}$$

where r is the pore radius (m), μ the dynamic viscosity (Pa s), and dp/dx represents the pressure gradient. The minus sign shows that the direction of diffusion is down the pressure gradient.

When $r/\lambda \leq 0.05$, the molecules collide on average more frequently with the pore wall than with one another and **Knudsen diffusion** occurs under a pressure gradient. This mechanism is often predominant in macroporous and mesoporous membranes. The flowing flux in a capillary pore can be described by the Knudsen equation:

$$J_K = -\frac{2}{3}\left(\frac{8RT}{\pi M}\right)^{0.5}\frac{r}{RT}\frac{dp}{dx} \tag{2.3}$$

where M is the molecular weight of gas ($\mathrm{kg\,mol^{-1}}$).

In the intermediate range between viscous flow and Knudsen flow, that is, $0.05 < r/\lambda \leq 3$, a gas molecule may travel a certain distance and collide with another one after the last collision with the pore wall. As a result, the velocity of gas molecules at the wall surface is not zero. This mechanism – combining both viscous flow and Knudsen diffusion – is thus called **slip flow**. The slip effect is negligibly small when $r >> \lambda$ but becomes significant when r is close to λ. A correction has to be applied to the viscous flow with a wall velocity to describe the permeation flux as

$$J_s = -\frac{r}{2}\left(\frac{\pi}{8MRT}\right)^{0.5}\frac{dp}{dx} \tag{2.4}$$

Surface flow occurs when the gas molecules are strongly adsorbed on the pore walls and migrate along the pore surface. Usually this is the case with gases which condense rather easily at moderate temperature–pressure conditions. The flux is generally described by a Fick's law type of expression:

$$J_{surf} = -\rho_{mem} D_s \frac{dq}{dx} \tag{2.5}$$

where ρ_{mem} is the density of the porous membrane (kg m^{-3}), D_s is the surface diffusion coefficient (m^2 s^{-1}), and q is the adsorption capacity (mol kg^{-1}). At room temperature, typical surface diffusion coefficients are in the range 1×10^{-3} to 1×10^{-4} cm^2 s^{-1}, intermediate between the diffusion coefficients of molecules in gases and liquids [2]. With the assumption of local adsorption equilibrium (adsorption processes are fast), Eq. (2.5) can be rewritten in terms of pressure as

$$J_{surf} = -\rho_s D_s \frac{dq}{dp} \cdot \frac{dp}{dx} \tag{2.6}$$

The relationship between adsorption capacity q and pressure can be expressed by an adsorption isotherm such as a Langmuir equation.

As the temperature increases, most species will desorb from the surface and surface diffusion will become less important. At low temperature and high pressure, the surface coverage (occupancy) can become larger than unity and **multi-layer diffusion** may occur.

Capillary condensation occurs when the pores are small enough and the gas is a condensable vapor so that the whole pore is filled with liquid. Accordingly, condensation can occur in small pores even if the partial pressure of the gas is below the vapor pressure:

$$\frac{RT}{V_m} \ln \frac{p}{p_s} = -\frac{2\sigma_s \cos\theta}{r} \tag{2.7}$$

where p_s is the normal gas vapor pressure, σ_s the surface tension of condensed fluid in the pore (N m^{-1}), θ the contact angle between condensed fluid and the pore wall, and V_m is the molar volume of the condensed liquid (m^3 mol^{-1}). It can be predicted that the smaller the pore radius, the lower the pressure at which capillary condensation starts. Owing to the capillary condensation, gas-phase diffusion through the pores can be blocked, leading to reduced permeation flux but improved selectivity.

Molecular sieving occurs when the pore diameter is small enough to permit only the smaller molecules to permeate, while the larger ones are excluded from entering the pore. This type of process is frequently

Table 2.2 Transport mechanisms in porous membranes

Transport type	Pore diameter	Characteristics	Membrane
Viscous flow	>20 nm	No selectivity	Substrate
Molecular diffusion	>10 nm	No selectivity	Mesoporous membrane
Knudsen flow	2–10 nm	Selectivity $\propto 1/\sqrt{M}$	Meso- and microporous membrane
Surface diffusion		Highly dependent on temperature	Meso- and microporous membrane
Capillary condensation		Dependent on gas condensability and temperature	
Molecular sieving	<1.5 nm	Very high selectivity	Microporous membrane

referred to as shape-selective diffusion. This mechanism is characterized by a strong temperature dependence and a sharp decline in permeability for larger gas molecules.

Among these mechanisms, viscous flow is non-selective while Knudsen diffusion is selective to smaller molecules. At high temperature, gas adsorption becomes weak and thus the surface diffusion and capillary condensation may be negligible. In fact, the perm-selectivity in microporous membranes is a complex function of the temperature, pressure, and gas composition. Therefore, it is necessary to evaluate the perm-selectivity of the porous membranes using a gas mixture under similar operating conditions [3]. Table 2.2 gives an overview of the transport mechanisms in porous membranes. Note that the perm-selectivity is not always a key factor in MRs.

2.2.3 Gas Permeation Flux through Porous Membranes

Gas permeation through the porous membranes may be driven by pressure or concentration gradient. Under a pressure or concentration gradient, gas will permeate through the membrane in a convective or a diffusive flow, respectively. In general, the pressure-driven convective fluxes are much higher than the concentration-driven diffusion fluxes.

2.2.3.1 Pressure-Driven Convection Flux

It is well known that the pore network of real porous membranes is very complicated, and geometrical effects play an important role in gas permeation. Different transport mechanisms may take place in the individual pores with different radius.

In most cases, only the Knudsen flow, slip flow, and viscous flow are taken into account – especially at high temperature. The permeation flux expressions, Eqs (2.2)–(2.4), can be modified to account for the number of capillaries per unit volume (porosity ε) and the complexities of the structure (tortuosity τ). For example, the viscous flux may be expressed as

$$J_V = \frac{\varepsilon}{\tau}\frac{r^2}{8\mu}\frac{\bar{p}}{RT}\cdot\frac{\Delta p}{L} \tag{2.8}$$

with mean pressure $\bar{p} = (p_u + p_d)/2$ and pressure difference $\Delta p = p_u - p_d$, in which p_u and p_d are the upstream and downstream pressure, respectively. L is the length of the capillaries (m). The tortuosity values generally fall in the region of $2<\tau<5$, but larger values (6–7) were also reported for membranes consisting of a packing of plate-shaped (boehmite, γ-Al_2O_3) particles [4].

For the Knudsen diffusion, the pore walls have a certain roughness and this causes diffuse reflections. Accordingly, Eq. (2.2) is then modified as

$$J_K = \frac{2r}{3}\frac{\varepsilon}{\tau\theta_K}\left(\frac{8}{\pi RTM}\right)^{0.5}\cdot\frac{\Delta p}{L} \tag{2.9}$$

where θ_K is the reflection factor of the pore surface. For smooth surface, $\theta_K = 1$, otherwise it is larger than 1.

It is noted that Eqs (2.8) and (2.9) are based on the average pore radius of the membrane. In fact, the pores in membranes usually have different sizes. For simplification, the membrane pores are considered to be a group of capillaries with different radius in a single length, $L = \tau\delta$, in which δ is the membrane thickness, as shown in Figure 2.4.

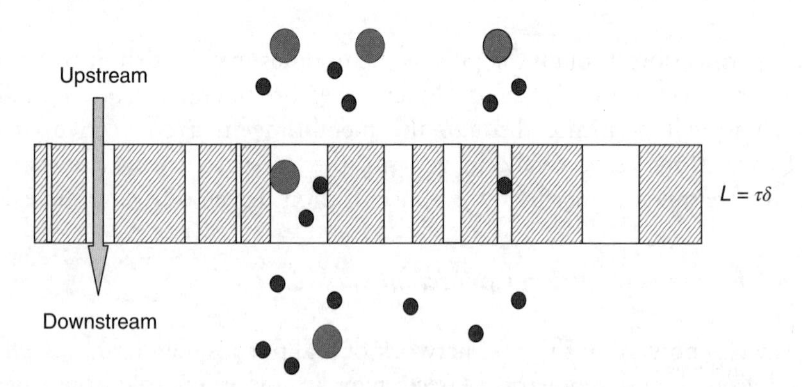

Figure 2.4 Schematic diagram of the ideal pore structure of porous membranes.

The pore size distribution can be described by a distribution density function of pore number

$$g(r) = \frac{1}{N} \cdot \frac{dn}{dr} \tag{2.10}$$

where N is the total number of all pores, and n is the number of pores ranging within $(r, r+dr)$.

For example, the log-normal distribution function is usually used to describe the pore distribution of asymmetric membranes [5], that is

$$g(r) = \frac{1}{\sqrt{2\pi}r}\left[\ln\left(1+\sigma^2\right)\right]^{-0.5} \times \exp\left[-\frac{\left(\ln(r/r_m)\left(1+\sigma^2\right)^{0.5}\right)^2}{2\ln\left(1+\sigma^2\right)}\right] \tag{2.11}$$

where r_m is the mean pore radius (m) and σ is the dimensionless standard deviation of the membrane pore size.

The cross-sectional area of all the pores may be obtained by

$$A_p = N\int_0^{\infty} \pi r^2 g(r)dr \tag{2.12}$$

Therefore, the porosity can be represented by the ratio of the pore's cross-sectional area to the membrane surface area A_m:

$$\varepsilon = \frac{A_p}{A_m} = \frac{N}{A_m}\int_0^{\infty} \pi r^2 g(r)dr \tag{2.13}$$

As a result, the permeation flux through a porous membrane can be given by

$$J_t = \frac{N}{A_m}\left[\int_0^{0.05\lambda} J_K \pi r^2 g(r)dr + \int_{0.05\lambda}^{3\lambda} J_s \cdot \pi r^2 g(r)dr + \int_{3\lambda}^{r_{\max}} J_V \cdot \pi r^2 g(r)dr\right] \tag{2.14}$$

Substitution of Eqs (2.2)–(2.4) and (2.13) into Eq. (2.14) gives

$$J_t = \frac{\varepsilon}{\tau\delta} \cdot \frac{C_1\int_0^{0.05\lambda} r^3 g(r)dr + C_2\int_{0.05\lambda}^{3\lambda} r^3 g(r)dr + C_3\bar{p}\int_{3\lambda}^{r_{\max}} r^4 g(r)dr}{\int_0^{r_{\max}} r^2 g(r)dr} \cdot \Delta p \tag{2.15}$$

where

$$C_1 = \frac{2}{3}\left(\frac{8}{\pi MRT}\right)^{0.5}, C_2 = \frac{1}{2}\left(\frac{\pi}{8MRT}\right)^{0.5}, C_3 = \frac{1}{8\mu RT} \quad (2.15a)$$

Eq. (2.15) shows that the Knudsen flow will play a more important role at low pressure and high temperature than at high pressure and low temperature.

2.2.3.2 *Concentration-Driven Diffusion Flux*

Under a concentration gradient, the permeation flux through a porous membrane can be expressed by Fick's law:

$$J_i = -D_{ei}\frac{dc_i}{dx} \quad (2.16)$$

where c_i (mol m^{-3}) is the concentration of component i and D_{ei} (m^2 s^{-1}) is the effective diffusion coefficient in the membrane. On considering the pore number per unit volume (porosity ε) and the complexity of the pore structure (tortuosity τ):

$$D_{ei} = \frac{\varepsilon D_i}{\tau} \quad (2.16a)$$

in which D_i is the diffusion coefficient of component i.

In the transition region the transport resistances are assumed to be in series, as expressed by the Bosanquet equation:

$$\frac{1}{D_i} = \frac{1}{D_{mi}} + \frac{1}{D_{Ki}} \quad (2.17)$$

where D_{mi} and D_{Ki} are the diffusion coefficients for continuum and Knudsen diffusion, respectively. The Knudsen diffusion coefficient is given by the gas kinetic expression:

$$D_{Ki} = \frac{2r}{3}\left(\frac{8RT}{\pi M_i}\right)^{1/2} \quad (2.18)$$

The molecular diffusion coefficient of gas in the pair *i–j* is given by

$$D_{m,ij} = \frac{4.36 \times 10^{-5} T^{3/2} \left(1/M_i + 1/M_j\right)^{0.5}}{p\left(V_{m,i}^{1/3} + V_{m,j}^{1/3}\right)^2} \tag{2.19}$$

For the membrane with a pore size distribution, the overall permeation flux of component *i* through the membrane may be given by

$$J_i = \frac{N}{A_m} \int_0^{\infty} \left(D_{ei} \frac{\Delta c_i}{\tau \delta} \right) \cdot g(r) \cdot \pi r^2 dr \tag{2.20}$$

Substituting Eqs (2.13) and (2.16) into the above equation gives

$$J_i = \frac{\varepsilon D_{mi}}{\tau \delta} \cdot \frac{\int_0^{\infty} \frac{r^3}{r+\alpha} \cdot g(r) dr}{\int_0^{\infty} r^2 g(r) dr} \cdot \Delta c_i \tag{2.21}$$

where

$$\alpha = \frac{D_{mi}}{\frac{2}{3}\sqrt{8RT/\pi M_i}} \tag{2.21a}$$

For the composite membrane consisting of a separation layer with thickness δ_1 and a substrate with thickness δ_2, as shown in Figure 2.5, the permeation flux can be described by the serial resistance model.

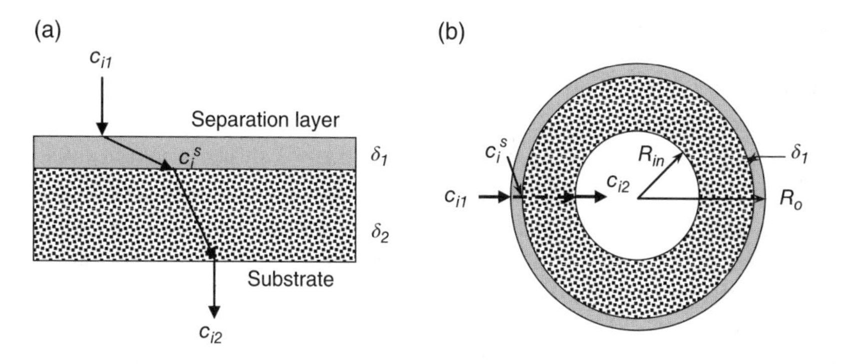

Figure 2.5 Gas permeation in the composite porous membranes: (a) planar membrane; (b) tubular membrane.

For a planar porous membrane, the permeation flux for component *i* can be given by

$$J_i = \frac{D_{ei}^{mem}}{\delta_1}\left(c_{i,1} - c_i^I\right) = \frac{D_{ei}^{sub}}{\delta_2}\left(c_i^I - c_{i,2}\right) \tag{2.22}$$

or

$$J_i = \frac{c_{i,1} - c_{i,2}}{\dfrac{\delta_1}{D_{ei}^{mem}} + \dfrac{\delta_2}{D_{ei}^{sub}}} \tag{2.22a}$$

where D_{ei}^{mem} and D_{ei}^{sub} are the effective diffusion coefficients in the separation layer and the substrate, respectively; c_i^I is the concentration on the interface between the two layers.

Similarly, the permeation flux through the tubular porous membrane can be given by

$$J_i = \frac{c_{i,1} - c_{i,2}}{\dfrac{R_m \ln\left(R_o/(R_o - \delta_1)\right)}{D_{ei}^{mem}} + \dfrac{\ln\left((R_o - \delta_1)/R_{in}\right)}{D_{ei}^{sub}}} \tag{2.23}$$

where R_m is the logarithm mean radius of the membrane tube, $R_m = \dfrac{R_o - R_{in}}{\ln(R_o/R_{in})}$, in which R_o and R_{in} are the outer radius and inner radius of the membrane tube, respectively.

2.3 PREPARATION OF POROUS MEMBRANES

Membrane preparation generally aims at the development of thin films on porous supports (Figure 2.6(a)). Such membranes may be of high flux but usually have limited mechanical and thermal-shock resistance. Close adherence to a strict synthesis protocol is necessary to obtain such membranes with consistent quality. Large-scale production usually induces high-cost membranes and limits their range of industrial applications. Besides, the porous membranes can also be prepared by depositing the membrane material inside the pores of porous supports to reduce the substrate pore size (Figure 2.6(b)). Such infiltrated composite membranes have low permeability but good thermochemical resistance and low sensitivity to the presence of defects, and are easily reproducible.

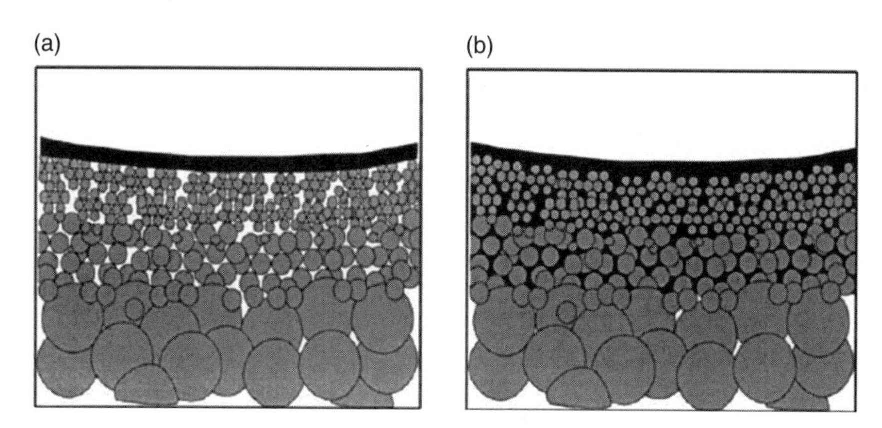

Figure 2.6 Schematic representation of (a) a thin supported layer on an asymmetric support and (b) a composite membrane infiltrated in an asymmetric support. Reproduced from [17]. With permission from Elsevier.

A variety of techniques have been developed to produce various porous membranes – such as particle dispersion and dip-coating, sol-gel processing, chemical vapor deposition, phase separation and leaching, pyrolysis, anodic oxidation, and so on [6, 7]. A good method should be of simple operation, good control of membrane quality (no cracks and pinholes, homogeneous thickness, narrow pore size distribution, and sufficient mechanical strength), as well as low cost.

2.3.1 Dip-Coating Method

The dip-coating technique is widely used for the preparation of porous ceramic membranes [1]. Figure 2.7 illustrates the flow sheet of a dip-coating process. The prepared coatings may be adjustable between 100 nm and 100 μm in thickness and the pore size covers from micropore to mesopore and part of the macropore range.

To begin with, a suspension/sol is prepared by dispersing submicrometer ceramic powders in a liquid or by synthesizing the solid particles in situ in a liquid. More than one polymeric compound serving as binder, surfactant, lubricant, and plasticizer is used to impart suspension/sol stability, adjusting the rheology of suspension and improving the properties of the green coating with respect to drying behavior and mechanical properties.

Prior to dip-coating, the porous substrate has to be polished or coated with a layer of smaller particles so that the surface roughness and the average pore size of the substrate are reduced. The pretreated substrate

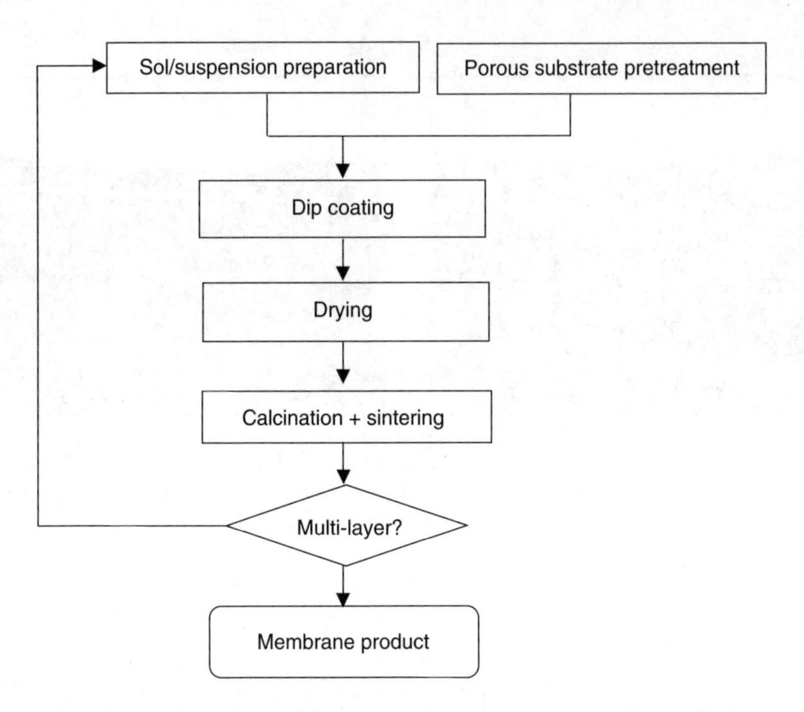

Figure 2.7 Flow sheet of the preparation of porous ceramic membranes.

is then dipped into a ceramic sol/suspension and subsequently withdrawn from that sol/suspension, followed by drying and sintering for consolidation of the wet coating. The commercially available ceramic membranes consist of several macroporous support layers with successively decreasing pore diameters. Such a structure favors very much avoiding crack formation and eliminating significant infiltration of the deposited layer into the underlying support layer.

The properties and quality of the membrane products depend critically on the support quality, the concentration and structure of the precursor solution, and also the details of the drying and calcination process. The following are necessary to obtain defect-free membranes:

- a homogeneous support;
- substrate cleaning and filtration of coating liquids;
- sufficient de-aeration of the suspension/sol;
- a decrease in the thermal expansion coefficient difference between coating and substrate;
- not too thick a coating, which may lead to drying shrinkage tensile stresses;
- avoiding dust and other foreign particulates in air or coating fluids.

2.3.2 Sol-Gel Method

The sol-gel process is one of the most appropriate methods for the preparation of functional oxide layers. There are two sol-gel routes: one is based on colloid chemistry in aqueous media, and the other on the chemistry of metal organic precursors in organic solvents. Figure 2.8 shows the sol-gel process for membrane formation.

The first stage involves the preparation of a sol having an appropriate rheological behavior adapted to the porous substrate using molecular precursors, either metal salts or metal organics. The sol is coated by dip-coating or spin-coating on porous supports. Condensation reactions

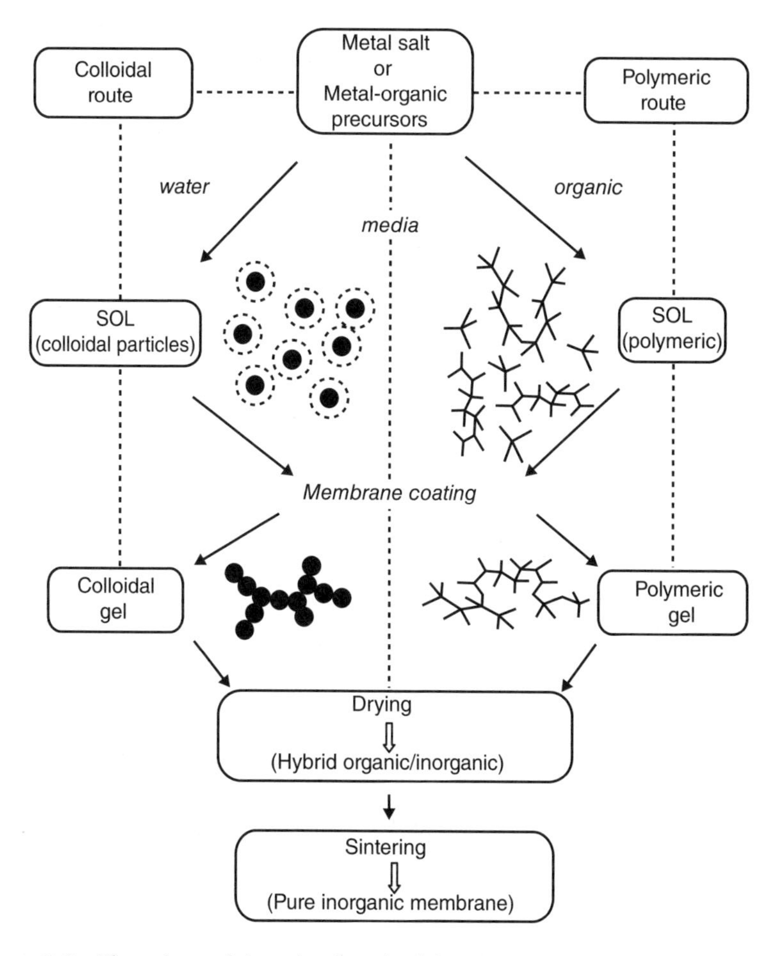

Figure 2.8 Flow sheet of the sol-gel method for inorganic membrane preparation. Reproduced from [1]. With permission from Elsevier.

take place, with the formation of colloids or clusters that will form gels at the final stage. The presence of dust particles, as well as a partial gelation in the sol, must be avoided in order to prevent the formation of defects and pinholes in the membrane. The drying and sintering steps will determine the nature of the membrane. A drying treatment is performed in an intermediate temperature range (80–350°C) to remove residual organics. Finally, the consolidation of the membrane will be performed through viscous or conventional sintering depending on the amorphous or crystalline structure of the membrane material.

During sol-gel processing of inorganic membranes, sols and gels evolve in a different way depending on the category of precursors used in the process. This evolution has a great influence on the porous structure of the final membrane materials. Several common membranes prepared by the sol-gel method are reported in Table 2.3.

The main problem with sol-gel-derived membranes is the thermal and hydrothermal instability (pore growth, grain coarsening, phase transition, defect development) above 400–500°C. Mixed oxide systems or stabilizing agents can improve the membrane stability in some cases, where precise control of the process parameters (synthesis of designed precursors, control of the hydrolysis–condensation reactions, etc.) is required [11, 12]. The challenges for sol-gel-derived porous membranes involve the preparation of thin perm-selective films and high-quality but low-cost underlying macroporous supports.

2.3.3 Chemical Vapor Deposition Method

The CVD method is a technique to produce microporous membranes by deposition of a solid inside the pores of mesoporous membranes, thus reducing the pore size through chemical reactions in a gaseous medium at an elevated temperature. The reactions to deposit a solid inside the pores of mesoporous membranes can be thermal decomposition, oxidation, or hydrolysis of precursors, as summarized in Table 2.4.

Figure 2.9 shows a CVD system consisting of a gas metering system, a heated reaction chamber, and a system for the treatment and disposal of exhaust gases [13]. The aluminum or silicon vapor precursor (e.g., $AlCl_3$) in a carrier gas is introduced into the reaction chamber that is heated to the desired temperature in advance. This vapor precursor will be chemsorbed on the pore surface to form intermediate species (e.g., $–O–AlCl_2$). Subsequently, the pores are evacuated by vacuum to remove all the precursors in the gas phase, and then exposed to water

Table 2.3 Typical porous membranes prepared by the sol-gel route

Membrane	Porous substrate	Raw material	Peptizing agent	Additive	Membrane thickness (μm)	Membrane pore size (nm)	Ref.
γ-Al_2O_3	α-Al_2O_3	Aluminum butoxide ($Al(OC_4H_9)_3$)	Inorganic acids (HNO_3)	Polyvinyl alcohol (PVA)	4	4–10	[8]
ZrO_2	α-Al_2O_3	Zirconyl chloride ($ZrOCl_2 \cdot 8H_2O$)	Oxalic acid/ HCl	PVA	1.2	4.3	[9]
TiO_2	α-Al_2O_3	Ti-tetra-isopropoxide (TTI) ($Ti(OC_3H_7)_4$)	HNO_3	PVA or hydroxypropyl cellulose (HPC)	1	3.5–5	[8]
SiO_2	γ-/α-Al_2O_3	Tetraethoxysilane (TEOS)	HCl	Alkyltriethoxysilane	0.03	0.3–0.4	[10]

Table 2.4 Porous membranes prepared by the CVD method

Membrane	Operation route	Chemical reactions
Al_2O_3	Thermal decomposition	$2Al(OC_3H_7)_2 \rightarrow Al_2O_3 + 6C_3H_6 + 3H_2O$
	Hydrolysis	$2AlCl_3 + 3H_2O \rightarrow Al_2O_3 + 6HCl$
SiO_2	Oxidation	$SiH_4 + O_2 \rightarrow SiO_2 + H_2O$
TiO_2	Co-reduction	$TiCl_4 + 2BCl_3 + 5H_2 \rightarrow TiB_2 + 10HCl$

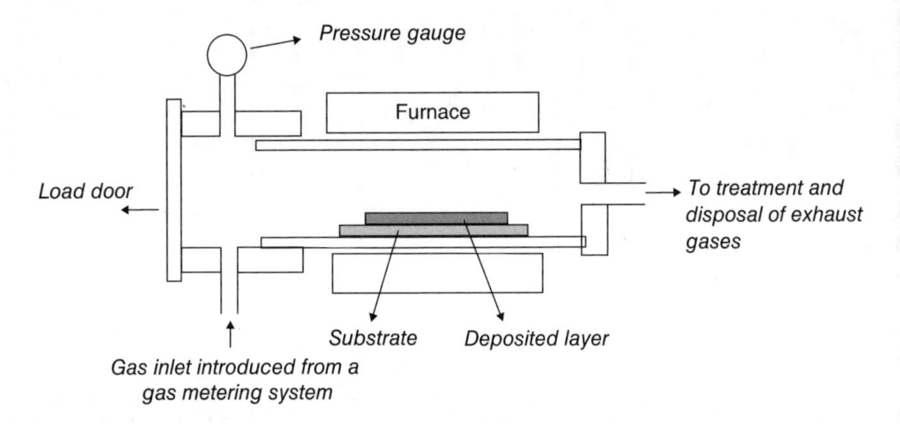

Figure 2.9 Schematic illustration of CVD. Reproduced with permission from [13]. Copyright © 2007, John Wiley & Sons, Ltd.

vapor which reacts with the surface intermediate species to form an alumina or silica layer.

The films deposited by CVD can grow in different ways depending on the conditions. One way is by the homogeneous deposition mechanism – that is, film growth starts on the pore walls and then gradually increases in thickness leading to pore narrowing as shown in Figure 2.10(a). For this mechanism, the substrate is required to have a smooth surface with constant and homogeneous characteristics (wettability) and preferably a relatively narrow pore size distribution. The ideal approach is to deposit a solid of fractal structure, as shown in Figure 2.10(b). In order to minimize the mass-transfer resistance, the deposit along the permeation flow direction should be as narrow as possible, as shown in Figure 2.10(c).

2.3.4 Phase Inversion Method

The phase inversion method is developed to prepare hollow fiber inorganic membranes. Since the porous substrate and the separation layer are formed in a single step for this method, the preparation process can be

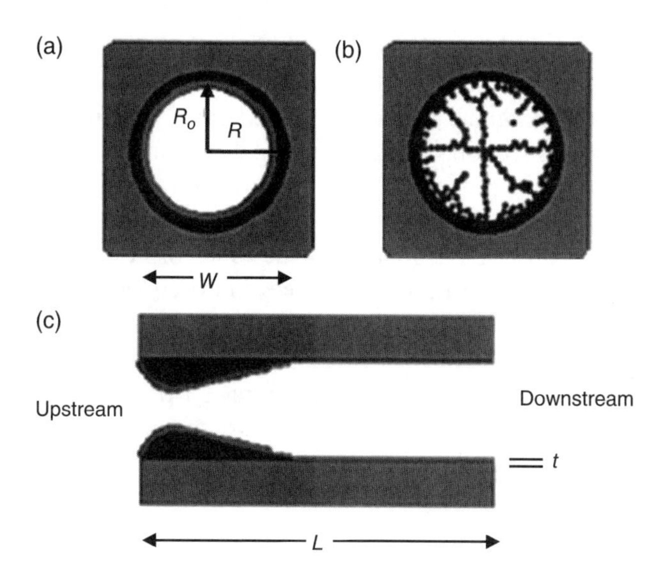

Figure 2.10 Schematic illustration of modifying the pore size of mesoporous membranes to obtain microporous membranes. Reproduced from [7]. With permission from Elsevier.

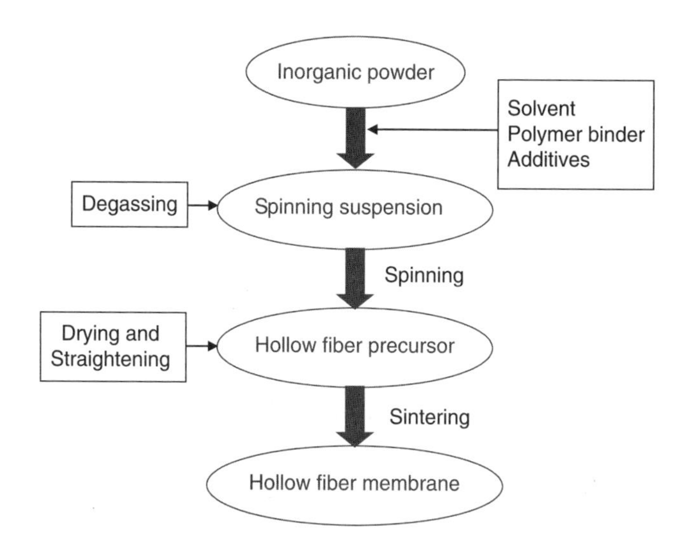

Figure 2.11 Phase inversion/sintering process to fabricate inorganic hollow fibres. Reproduced from [14]. With permission from Elsevier.

much simplified, with the membrane costs being reduced remarkably. Figure 2.11 illustrates the phase inversion/sintering process to fabricate inorganic hollow fiber membranes [14]. It is a three-step process, including: (i) preparation of a spinning suspension; (ii) spinning of inorganic hollow fiber precursors; and (iii) high-temperature sintering.

The spinning suspension is composed of a polymer binder, inorganic powders, solvent, and additives (dispersant, plasticizer, etc.). All the components (except the inorganic powders) are used to pre-design the membrane morphology (cross-sectional structures) and to facilitate the spinning and phase inversion of the polymer binder. The suspension is pushed through a tube-in-orifice spinneret into a coagulation bath while the internal coagulant is simultaneously passed through the inner orifice to form hollow fiber precursors. Owing to the phase inversion occurring in the polymer binder, an asymmetric structure consisting of a porous structure and a relatively dense layer can be formed. Sintering is conducted to remove the organic components and to bond the inorganic particles into hollow fiber membranes. The general morphology of the hollow fibers can be retained, but the microstructure may be changed during the sintering process. The densification and sintering can be modified by including some additives in the suspension.

The microstructure of the hollow fiber membranes can be well tailored by modulation of the suspension composition and the preparation parameters. However, it is still a great challenge to design precisely and control the macro- and microstructure of the inorganic hollow fiber membranes because too many factors – such as the particle size and its distribution, the shape and surface property of inorganic powders, the composition and viscosity of the spinning suspension, the spinning conditions (spinning rate, air gap, internal coagulant, etc.), and the sintering parameters (sintering temperature, dwelling time, heating rate) – can solely or jointly affect the formation of membrane structures. Therefore, a well-designed phase inversion/sintering process coupled with an optimal spinning suspension is the key to obtain asymmetric hollow fibers with the desired permeation characteristics and excellent mechanical strength as well.

2.3.5 Other Preparation Methods

Other methods to prepare porous membranes include pyrolysis for carbon membranes, heat treatment and leaching for mesoporous glass membranes, and anodization for alumina membranes. The microporous carbon membranes are prepared by coating a polymeric precursor such as polyfurfuril alcohol and polycarbosilane on porous substrates, followed by controlled pyrolysis under N_2 atmosphere [15]. The carbon membrane structure is determined by the fabrication variables, including the polymeric solution concentration, solvent extraction, heating rate, and pyrolysis temperature [16].

2.4 POROUS MEMBRANES FOR CHEMICAL REACTIONS

2.4.1 Membrane Materials

The variety of porous membranes, in terms of both materials and microstructures, makes them popular in different chemical reaction processes. They are applied mostly in a tubular configuration. However, the hollow fiber represents a trend for future development due to its remarkably high area/volume ratio. Table 2.5 summarizes the conventional inorganic porous membranes used in membrane reactors.

Porous vycor glass is a mesoporous membrane with average pore diameter around 4 nm. It is usually used as a porous substrate for the deposition of other inorganic films. Compared with other ceramic membranes, the glass membranes suffer from fragility and unsuitability beyond 800°C. The self-supporting, mechanically resistant vycor glass membranes usually have a thickness around 1 mm. Hollow fiber glass membranes are commercially available but are plagued by mechanical stability problems.

Alumina occupies the dominant position in the research and development of ceramic membranes due to its high chemical and mechanical stability as well as low cost. Anodic alumina membranes are ideally suited

Table 2.5 Inorganic porous membranes used in MRs

Membrane	Structure	Characteristics	Applications
Vycor glass	Mesoporous	• Low mechanical strength • Tubular and capillary	• Porous substrate • PBMR
Al_2O_3	Macroporous Mesoporous	• High chemical stability • Low cost • Tubular and plate	• Porous substrate • Catalyst host • PBMR and CMR
ZrO_2	Mesoporous	• High mechanical stability • Composite	• PBMR
TiO_2	Mesoporous	• Photocatalytic activity • Composite	• CMR
Stainless steel	Macroporous	• Good mechanical strength • Tubular and plate	• Porous substrate
Silica (SiO_2)	Mesoporous Microporous	• High selectivity • Low hydrothermal stability	• Membrane separator • Membrane reactor
Carbon	Mesoporous Microporous	• Low stability in oxidative environment • Tubular and capillary	• Membrane separator • CMR

for fundamental studies of transport and reaction because they have straight non-intersecting pores, but confront the problem of insufficient mechanical strength. The commercially available porous alumina membranes are mostly α-Al_2O_3 tubes with various dimensions. In order to improve the membrane perm-selectivity, the macroporous alumina membranes are usually deposited with γ-Al_2O_3 films within the macropores or on the top surface by use of the conventional sol-gel method. ZrO_2 and TiO_2 membranes are generally also developed on the tubular α-Al_2O_3 porous substrate following the same pattern as the alumina membrane.

Macroporous ceramic membranes are utilized as a means of bringing together reactants flowing on opposite sides and to create a well-controlled reactive interface. For selective catalytic reductions, a sharp reaction front can be created within the membrane and thus the slip of reactants may be restricted. Earlier efforts also involve the use of macroporous alumina membranes as efficient contactors for various gas/liquid reactions, where the membrane functions to increase the gas–liquid–solid interface through the high surface of the porous membrane acting as a catalyst support. The primary applications of mesoporous ceramic membranes are for selectivity-limited reactions such as partial oxidation of hydrocarbons, in which the desired products may react further with oxygen to produce undesirable total oxidation products. In such MRs, the total oxidation can be weakened by feeding the two reactants separately on either side of the membrane rather than co-feeding them on either side. In addition, the porous ceramic membranes can be made into catalytic MRs by loading catalysts in the porous structure.

PSS membranes are characteristic of high mechanical stability but low perm-selectivity. Commercially available PSS membranes are mostly of tubular configuration, and are often used as the porous substrate for other membranes.

Silica (SiO_2) membranes are microporous, generally in thin films supported on ceramic or PSS porous substrates. Dense SiO_2 membranes are permeable to hydrogen molecules but not to other gases, hence they are usually applied to dehydrogenation reactions. The main problems involve poisoning and coking, as well as densification (stability), especially in the presence of steam.

Carbon membranes span the full range from molecular sieves to mesoporous membranes. They can provide a high surface area for loading catalysts for reactions, and thus show great potential for use in catalytic MRs. However, the carbon membrane cannot be utilized in oxidative atmospheres. Even under reductive atmospheres, caution should be employed in the presence of various metal impurities.

2.4.2 Membrane Functions

In porous MRs, the membrane may function as an extractor, a distributor, or an active contactor, as listed in Table 2.6. The extractor mode corresponds to the earlier applications of MRs and has been applied to increase the conversion of a number of equilibrium-limited reactions, such as alkane dehydrogenation, by selectively extracting the hydrogen produced. Other H_2-producing reactions – such as water gas shift (WGS), steam reforming of methane, and the decomposition of H_2S and HI – have also been investigated successfully with the MR extractor mode. The H_2 perm-selectivity of the membrane and its permeability are two important factors controlling the efficiency of the processes [17].

The distributor mode is typically adapted to consecutive reaction systems such as partial oxidation and oxidative dehydrogenation of hydrocarbons, or oxidative coupling of methane, where there is a favorable kinetic effect regarding the partial pressure of the distributed reactant. For these reactions, a low partial pressure of oxygen favors the selective oxidation reaction versus the deep oxidation to CO and CO_2. As a reactant, sufficient oxygen feed is essential for high conversions. By using a porous membrane to distribute oxygen addition, the local oxygen partial pressures can be lowered while maintaining the total amount of additive oxygen, leading to greater selectivities with respect to the conventional co-feed arrangements. In addition, the use of a membrane for the distribution of oxygen in oxidation processes can also produce a safer operation with reduced formation of hot spots and a lower probability of runaway. The avoidance of hot spots will further give additional increments of selectivity by suppressing undesired reactions that take place at high

Table 2.6 Membrane functions in the porous membrane reactors

Functions	Description	Application
Extractor	The removal of product(s) increases the reaction conversion by shifting the reaction equilibrium	Dehydrogenation; decomposition (H_2S, HI); H_2 production
Distributor	The controlled addition of reactant(s) limits side reactions: to increase selectivity by optimizing the reactant concentration profile; to mitigate the temperature rise in exothermic reactions	Partial oxidation; oxidative dehydrogenation of hydrocarbons; oxidative coupling of methane
Contactor	Controlled diffusion of reactants to the catalyst leads to an engineered catalytic reaction zone: to increase conversion and selectivity	Gas–liquid–solid reactions; VOC oxidation; alkene hydrogenation

temperatures, and in any case helps to extend catalyst life. In addition, the distribution of oxygen also allows a wider range of operating conditions: by distributing the oxygen feed it is possible to operate at overall hydrocarbon-to-oxygen ratios that would be within the explosive region if the same composition was fed at the entrance of a fixed bed reactor. The O_2 perm-selectivity of the membrane is an important economic factor, because air can be used instead of pure oxygen as the oxidant.

In the active contactor mode, the membrane acts as a diffusion barrier and does not need to be perm-selective but usually to be catalytically active. This concept can be operated with a forced flow or opposing reactant feed. Forced flow contactors have been applied to the removal of VOCs by catalytic combustion. The oxidation reaction rate can be increased greatly due to the improved access of reactants to catalysts. For the alkene hydrogenation triphasic reactions, the selectivity can be improved greatly when both the alkene and H_2 are forced to pass through a microporous catalytic membrane, because the back-mixing of initial products is prevented. The opposing feed contactors are applied to both equilibrium and irreversible reactions, if the reaction is sufficiently fast compared with mass transport (diffusion rate of reactants in the membrane). In such a case the reaction can be confined in a finite reaction zone inside the porous structure, avoiding the slip of reactants to the opposite side. Further, the reactants arrive at the reaction zone in a stoichiometric ratio, which helps to reduce unwanted side reactions. This concept has been demonstrated experimentally for reactions requiring strict stoichiometric feeds, such as the Claus reaction, or for kinetically fast, strongly exothermic heterogeneous reactions such as partial oxidations.

2.5 CATALYSIS IN POROUS MEMBRANE REACTORS

2.5.1 Catalyst in Membrane Reactors

For most chemical reactions, a catalyst is required to obtain the desired product. The catalyst performance in MRs can be influenced positively or negatively by the use of a membrane. It is well known that the withdrawal of hydrogen during dehydrogenation in an MR will likely favor coking processes, promoting deactivation of the catalyst. This implies that MRs require catalysts with improved stability. However, in the Fischer–Tropsch synthesis, the use of membranes to extract water from the reaction zone may protect the catalyst from poisoning and increase selectivity by suppressing undesirable side reactions, since water deactivates Fe-based catalysts and moreover tends to oxidize the metallic Co- and Fe-based catalysts. For selective hydrogenation reactions, lowering the H_2 concentration at the catalyst bed inlet with a membrane distributor may avoid local deactivation. Therefore, the catalysts for MRs require a specific design so as to have adapted characteristics and properties for high performance [18]. In order to draw the full benefit of the separation–reaction synergism, highly active catalysts that are able to comply with the high extraction ability of the membrane need to be developed. In extractors, the membrane should present high permeance and selectivity, and the catalyst should present a level of efficiency in keeping with that of the membrane. Also, the catalyst should be able to withstand the specific reactive mixture generated by product removal, which may originate deactivation phenomena. In distributors, the presence of a reactant composition gradient all over the MR may change the nature of the active phase. In flow-through MRs, the reactor performance may even be limited mainly by the catalyst rather than the membrane separation property.

As mentioned in Chapter 1, the catalyst in porous MRs may just be placed on the membrane (illustrated in Figure 1.12(a)). The reaction takes place in the catalyst phase and the membrane only serves either as a product extractor or as a reactant distributor but does not participate directly in chemical reactions. It is not always easy to obtain a true inert membrane since the porous membrane materials such as alumina, silica, titania, zirconia, zeolite or the components used to modify membrane permeation properties (e.g., pore-filling materials) can make a contribution to reactions. In order to reduce non-selective catalytic activity, the membrane used in selective oxidation reactions often has to be modified significantly by using controlled sintering to reduce surface area, or by doping with alkaline compounds to decrease surface acidity [19].

When the catalyst is coated on the membrane surface or dispersed inside the membrane pores, the membrane body will exhibit catalytic activity and participate directly in chemical reactions. To obtain high performance, the porous membrane should provide high surface area and strong adhesion of the catalyst. The reactants permeate from one side or opposite sides of the membrane into the catalyst layer where reactions take place. The catalyst in porous membranes may benefit from the better transfer and membrane active role in promoting the contact of reactants. An appropriate thickness of the catalytic layer is necessary to enhance the reaction selectivity [20].

2.5.2 Catalyst Deposition in Porous Membranes

Catalytic membranes can be prepared by deposition of a catalytically active phase (e.g., a metal) in the inert porous membranes through various techniques commonly used for conventional catalyst preparation – such as wet impregnation, monolayer metal complexation, and ion exchange – followed by heat treatment (activation) [17]. The conventional activation steps can also be avoided by direct deposition of solvated metal microclusters [21]. The structural characteristics of the porous substrate, and in particular the relative laminar and Knudsen contributions to permeation, have a strong influence on the membrane performance [22]. They make use of the membrane structure to optimize the access of disfavored reactants, or to control and rule the residence time and contact of species in the active zone. Furthermore, the porous membrane is required to present a rather homogeneous structure to avoid heterogeneities in the reactant-to-catalyst contact, and also to facilitate CMR operation control.

Figure 2.12 shows the γ-Al_2O_3/α-Al_2O_3 bimodal supported catalytic membrane prepared by the sol-impregnation method [23]. First, the α-Al_2O_3 porous support is soaked in a boehmite sol solution with a concentration of 10 wt% for 1 h, followed by drying at room temperature and calcination at 550°C in air to convert impregnated boehmite sols into γ-Al_2O_3, resulting in a bimodal support. The catalysts are then impregnated in the bimodal support by soaking the support in a 1.5 wt%

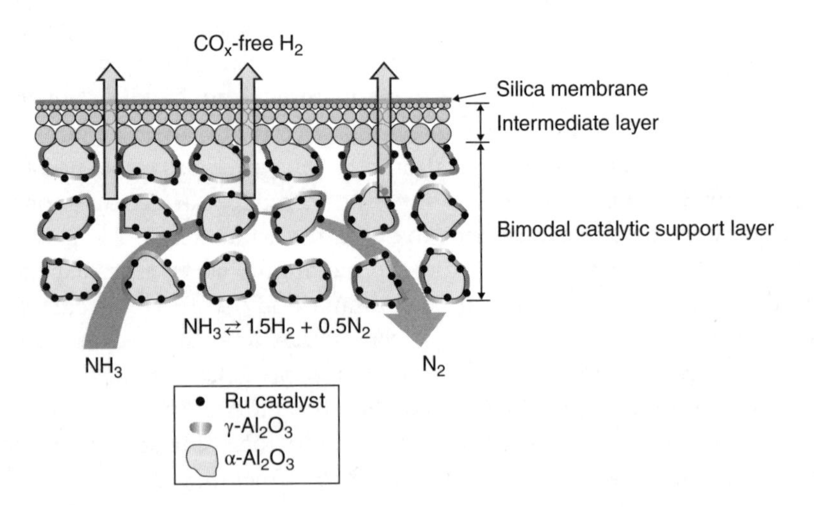

Figure 2.12 Schematic diagram of the γ-Al_2O_3/α-Al_2O_3 bimodal supported catalytic membrane for ammonia decomposition. Reproduced from [23]. With permission from Elsevier.

$Ru(NO)(NO_3)_3$ solution. After drying, the support is calcined at 550°C in air to obtain a bimodal catalytic support ($Ru/\gamma\text{-}Al_2O_3/\alpha\text{-}Al_2O_3$). Finally, a homogeneous silica membrane is prepared on the outside surface of the bimodal catalytic support using polyhedral oligomeric silsesquioxane (POSS) as precursor, by the sol-gel method.

The integration of the selective and catalytic functions into one single layer usually demands contradicting material properties. For example, to achieve high selectivity the diffusion of products inside the material should be low, whereas efficient use of the catalytic properties requires the diffusion of products to be high. This conflicting demand on material properties can be avoided by accommodating the selective and catalytic features in two different distinct layers that are in close physical contact. This approach also allows independent optimization of the selective and catalytic properties.

A non-uniform and precisely controlled position of the catalyst within the membrane pores can impact the reactor performance positively [17]. A modeling study on the first-order reaction in a catalytic membrane indicates that a Dirac-delta function of the concentration of catalyst in the membrane, placed at the feed side, allows the highest conversions. In other words, it is better to promote the reaction as close to the membrane as possible (on its surface), letting the rest of the membrane work as a mere separator of some of the reaction products [24]. The membrane structure is critical to the preparation of non-uniform catalytic membranes. If a sufficiently homogeneous membrane structure is present, simple impregnation may be sufficient to obtain a controlled, non-uniform distribution of active materials.

2.6 OPERATION OF POROUS MEMBRANE REACTORS

2.6.1 Packed Bed Membrane Reactors

The PBMRs are most popular in practical applications because of their easy operation. Figure 2.13 shows a typical tubular PBMR for dehydrogenation reactions, where the catalysts are packed in the membrane tube. In experimental operations, an inert sweep gas is often used flowing on the shell side to generate a hydrogen partial pressure gradient across the membrane (Figure 2.13(a)). In order to obtain pure hydrogen as a product (for hydrogen production), the MR may be operated at high pressure instead of using a sweep gas. In this case, the catalyst is usually packed on the shell side of the MR and the hydrogen is "pressurized" into the membrane tube (Figure 2.13(b)).

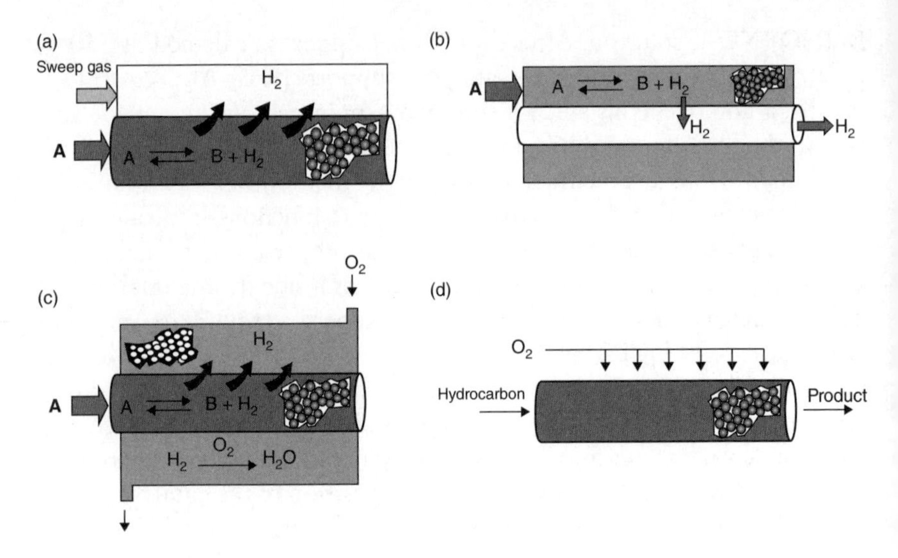

Figure 2.13 PBMRs for dehydrogenation reactions: (a) sweeping mode; (b) pressurized mode; (c) coupling operation mode; (d) distribution mode. Reproduced from [25]. With permission from Elsevier.

Equilibrium displacement can be enhanced through reaction coupling in bifunctional MRs. Two complementary reactions take place on both sides of the membrane. The product of the reaction on one side is permeated and used as a reactant on the other side (chemical coupling, e.g., dehydrogenation/hydrogenation or dehydrogenation/combustion reactions), or the endothermicity of one reaction is compensated by the exothermicity of the other (thermal coupling, exothermic/endothermic processes). The reactions often use different catalysts, which are packed on opposite sides of the membrane tube. Figure 2.13(c) shows the coupling of a dehydrogenation (endothermic) reaction and a combustion (exothermic) reaction. This operation mode may enhance the per-pass conversion of both the reactions. Furthermore, the exothermic combustion reaction could in principle supply the heat required for the endothermic dehydrogenation reaction. Considerable energy savings compared with indirect heating of the reactor have been proved [24]. However, on the contrary, the number of degrees of freedom for control purposes is decreased compared with the single reaction mode [25]. Figure 2.13(d) shows PBMRs with the membrane to distribute a reactant to a fixed bed of catalyst packed on the opposite side. This operation mode is often applied in selective oxidation processes where a low partial pressure of oxygen favors the selective oxidation reaction versus deep oxidation to CO and CO_2.

2.6.2 Catalytic Membrane Reactors

Figure 2.14 shows the operation modes of catalytic MRs. In Figure 2.14(a), the porous catalytic layer is deposited on the membrane surface. Reactant A permeates through the porous substrate to the catalytic layer where the reaction between A and B takes place. The partial pressure of reaction A is lowered due to the resistant diffusion layer. The advantages of this configuration lie in the uniform distribution of reactants throughout the reactor and the negligible pressure drop. Figure 2.14(b) shows the operation of the MR with the catalyst dispersed inside the membrane pores. The key to the success of such a reactor is the ability to obtain a sufficiently well-defined sharp distribution of the active component across the membrane thickness. The reactants diffuse from opposite sides of the membrane to a confined reaction zone in a stoichiometric ratio. The location of the reaction interface within the membrane depends on

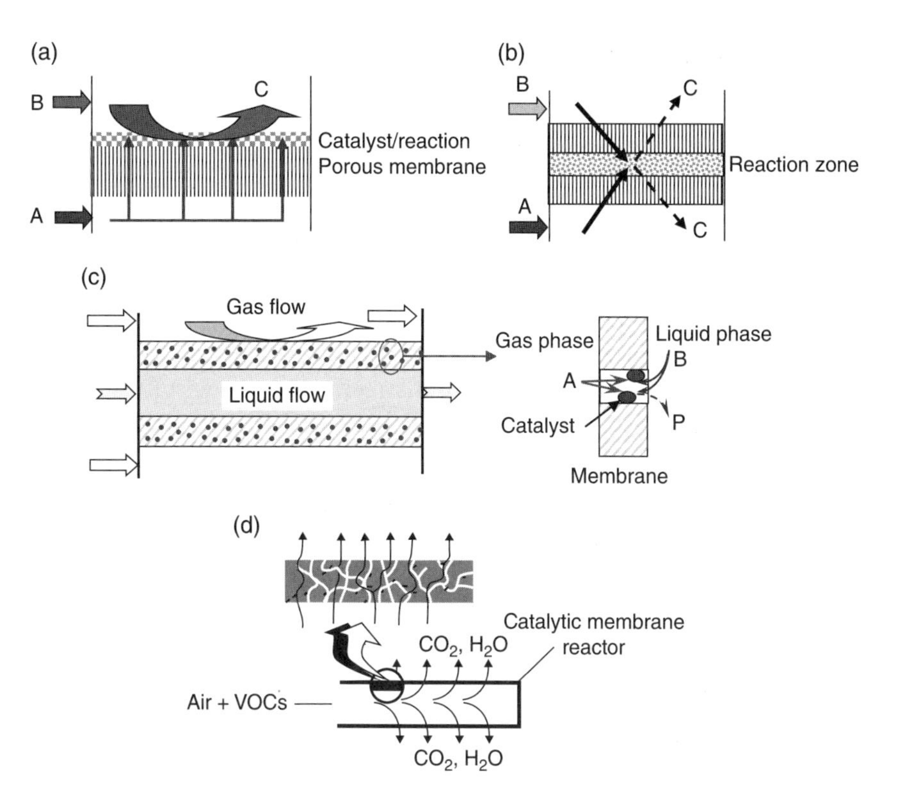

Figure 2.14 Configurations of the CMRs: (a) catalyst-coated CMR; (b) catalyst-dispersed CMR; (c) gas–liquid–solid three-phase CMR; (d) FTCMR. Reproduced from [25]. With permission from Elsevier.

the relative bulk concentrations and the intra-particle diffusion coefficients of the two reactants. This reaction mode is applied for oxidation processes, providing safer operation by avoiding any premixing of reactants. In the case where a pressure difference is applied over the membrane, the products can be shifted preferentially toward the low-pressure side, which gives a lower residence time in the reaction zone and reduces further the reaction of valuable partial oxidation products.

Figure 2.14(c) shows the tubular catalytic membrane reactor for gas–liquid–solid reactions. The volatile reactant (gas phase) passes in the shell side while the non-volatile reactant (liquid phase) passes in the tube side. The reactant dissolved in the liquid diffuses through the liquid-filled porous structure of the membrane to the catalytically active sites and reacts with the reactant from the gas phase. This configuration is very useful for systems where there is a large external mass transfer resistance on the gas side, as in a conventional gas–liquid system in a catalytic packed bed. Since the liquid-filled pores of the catalytic zone of the membrane are in close proximity to the gas interphase, the volatile reactant can react with the liquid in direct contact with the catalyst without the need for diffusion through a liquid film. Furthermore, it does not require a perm-selective membrane and avoids the problem of catalyst recovery that appears in slurry reactors. This configuration has typically been applied for hydrogenation or oxidation processes. To eliminate pore diffusion and achieve a very short contact time, the dissolved reactants can be pumped through an asymmetric ceramic membrane or just a ceramic support coated with catalytically active metals.

Figure 2.14(d) shows the operation of the FTCMRs typical for VOC removal by oxidation, where an air stream containing VOCs flows through the catalytic membrane in a single pass with the aim of reaching complete conversion. The porous ceramic membrane (usually rendered catalytic by impregnation with catalysts) is applied in a dead-end configuration, thereby forcing all the reactants to flow through it. As the reactant flows convectively through the membrane, it will be in intensive contact with the catalytic sites while the mass transport resistance may be negligible. The number of collisions between the reactant and catalytic sites inside the pores can be amplified noticeably by decreasing the pore diameters to small values, so that Knudsen diffusion prevails. Such a combination of Knudsen diffusion and the FTCMR configuration results in a unique and fundamental feature: all the reactant molecules are guaranteed to have multiple contacts with the catalyst surface. It was found that the catalytic activity of the membrane catalyst is 10 times higher than the activity of the same catalyst when packed [25].

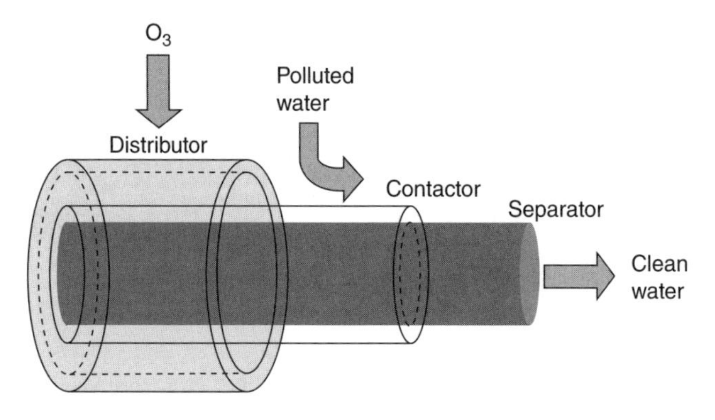

Figure 2.15 Schematic drawing of the coupled membrane reactor. Reproduced from [26]. With permission from Elsevier.

2.6.3 Coupling of Membrane Functions

Coupling of different membrane functions may be an effective way to improve the performance of MRs. Figure 2.15 shows a coupled ozone MR that combines synergistically ozone oxidation, adsorption, and membrane separation to treat pollutants in water [26]. It is composed of a PSS membrane as ozone distributor, a porous ceramic membrane clad with a thin layer of high-surface-area coating as reaction contactor, and a thin zeolite membrane deposited on the inner surface of the ceramic tube as water separator. The membrane distributor is designed to generate minute ozone bubbles in water, thus enhancing the mass transfer rate and oxidation of organic pollutants. The contactor serves the purpose of increasing the interfacial contact area between the gas-phase ozone and dissolved organics. With the help of adsorbent material, the contactor will also adsorb and trap the organics within the reaction zone, allowing for more complete remediation. The membrane separator produces clean water and concentrates the organics within the reaction zone, resulting in better reaction kinetics. Such an MR represents an order of magnitude improvement over traditional semi-batch reactor design and is capable of complete conversion of recalcitrant endocrine disrupting compounds (EDCs) in water at residence times of less than 3 minutes.

2.6.4 Non-uniform Distribution of Membrane Permeability

The scenario occurring in a porous MR with a homogeneous porous membrane as the O_2 distributor for the partial oxidation of alkane is

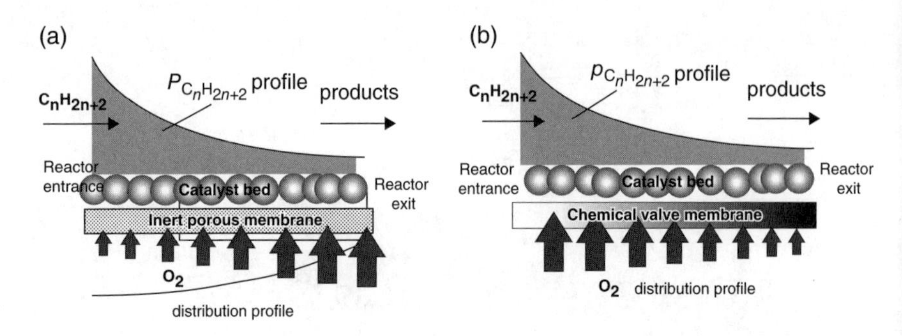

Figure 2.16 Schematic representation of the O_2 profile generated in a tubular membrane reactor for the partial oxidation of alkanes: (a) with a classical inert uniform porous membrane and (b) with the chemical valve membrane. Reproduced from [17]. With permission from Elsevier.

schematized in Figure 2.16(a) [17]. The progressive alkane consumption from the entrance to the exit of the reactor decreases the alkane/O_2 ratio continuously. This phenomenon is enhanced by the influence of the pressure drop generated by the catalyst bed, which tends to increase the O_2 permeation rate near the end of the reactor due to the higher pressure gradient in that zone. The evolution of the alkane/O_2 ratio along the reactor alters the catalyst efficiency and decreases the reaction yield.

Two solutions have been proposed to overcome this problem: (i) co-feeding a part of the oxygen with alkane in the reaction zone or (ii) using a membrane with a thickness increasing from the entrance to the exit of the reactor. The non-uniform membrane will create a pattern of moderately decreasing permeability in the direction of flow along the membrane length, as shown in Figure 2.16(b). As a result, the alkane/O_2 ratio along the reactor gets more uniform and the selectivity of the desired product is increased. Such enhancement of the packed bed reactor performance by modifying the oxygen feed distribution has been demonstrated experimentally for methane coupling and for oxidative dehydrogenation of butane [27]. The simulation performed on the catalytic methanol oxidative dehydrogenation to formaldehyde in PBMR indicates that, compared with a uniform membrane, optimizing the feed distribution may result in a two- to threefold increase in the overall reactor productivity [28]. In fact, any MR can potentially benefit from the adjustment of the membrane permeation pattern to the kinetic parameters of the reaction system.

A chemical valve membrane with permeability controlled by the redox properties of the gas phase has been proposed to better regulate the O_2 flux in the reactor [17]. It can be prepared by depositing V_2O_5 and $AlPO_4$ crystallites inside the pores of a tubular α-Al_2O_3 substrate. The reversible redox properties of V_2O_5 crystallites, which will transform to V_2O_3 under reducing atmosphere, may influence highly membrane properties such as grain morphology, pore size distribution, pore volume, and specific surface area. Consequently, the permeation behavior of the membrane for oxygen – as well as the reducing gases such as alkane – can be changed depending on the reaction atmosphere. Such a chemical valve membrane is very attractive for the autoregulation of O_2 concentration in MRs for the partial oxidation of alkanes (e.g., oxidation of propane to acrolein or acrylic acid). Since the membrane can adapt locally its permeance, rapidly and reversibly, as a function of the reaction atmosphere, it can help in optimizing the alkane/O_2 ratio all along the reactor (locally and upon use), limiting the effects of hot spots and improving the reactor efficiency.

2.7 APPLICATIONS OF POROUS MEMBRANE REACTORS

Porous membranes can be utilized for both equilibrium and selectivity-limited reactions. The former mainly involve various dehydrogenation reactions, while the latter are oxidative dehydrogenation and partial oxidation reactions. Besides, other applications related to porous MRs are illustrated and summarized in the following sections.

2.7.1 Dehydrogenation Reactions

This is the most widely studied class of reactions in MRs of all kinds. It is well known that the dehydrogenation of organic compounds is generally an endothermic reaction and controlled by the thermodynamic equilibrium

$$A \Leftrightarrow B + H_2, \Delta H_r > 0$$

In order to achieve a reasonably economic conversion per pass, the dehydrogenation is usually performed at high temperatures. This may lead to catalyst deactivation by coking, formation of undesired by-products, and loss of energy efficiency as well. When using MRs to perform dehydrogenation, the reaction can be promoted by removal of the hydrogen produced through a membrane, thus driving the equilibrium reaction shift to the product side. As a result, higher conversions can be achieved (or the operation temperature for a given conversion can be lowered). The milder operating conditions are perceived to be beneficial, not only as an energy-saving technology but also possibly as a remedy for coking and catalyst deactivation problems inherent in many commercial processes.

Table 2.7 Dehydrogenation reactions in porous membrane reactors

Reaction	Membrane	Catalyst/ configuration	Operation conditions	Ref.
$C_6H_{12} \Leftrightarrow C_6H_6 + 3H_2$	SiO_2/ Al_2O_3-capillary	Pt(2 wt%)-Al_2O_3; CMR	$P = 0.3$ MPa; $T = 583$ K; no sweep gas	[29]
$C_3H_8 \Leftrightarrow C_3H_6 + H_2$	SiO_2/α-Al_2O_3 tube	Cr_2O_3@ γ-Al_2O_3; PBMR	$P = 0.1$ MPa; $T = 500°C$; N_2 as sweep gas	[30]
$C_4H_{10} \Leftrightarrow C_4H_8 + H_2$	Carbon capillary	Cr_2O_3@ γ-Al_2O_3; PBMR	$P = 0.1$ MPa; $T = 450$ or 500°C; N_2 as sweep gas or vacuum	[31]
$2CH_3OH \Leftrightarrow HCOOCH_3 + H_2$	Cu–SiO_2/ α-Al_2O_3 tube	CMR	$P = 101$ kPa; $T = 240°C$	[32]
H_2S decomposition $H_2S \Leftrightarrow S + H_2$	ZrO_2–SiO_2/ α-Al_2O_3 tube	IMR	$P = 0.11$–0.25 MPa; $T = 923$–1023 K	[33]
	SiO_2/Al_2O_3 capillary	HD-cat; PBMR	$P = 101$ kPa; $T = 873$ K; N_2 as sweep gas	[34]
NH_3 decomposition $2NH_3 \Leftrightarrow N_2 + 3H_2$	SiO_2/α-Al_2O_3 tube	Ru@γ-Al_2O_3; CMR	$P = 101$ kPa; $T = 450°C$; N_2 as sweep gas	[35, 36]

Table 2.7 summarizes some of the dehydrogenation reactions studied in the literature using porous MRs. As can be seen, the membranes for dehydrogenation reactions are primarily microporous silica due to it high hydrogen perm-selectivity. Meso- or macroporous membranes are not suitable for dehydrogenation reactions because their low perm-selectivity may lead to much loss of the reactants. The catalysts for dehydrogenation may be packed in the reactor or loaded inside the membrane wall. All the studies demonstrate enhanced conversions or even beyond-equilibrium values.

2.7.2 Reforming Reactions for Hydrogen Production

Hydrogen has been utilized in a number of industries – including fertilizers, oil refining, and petrochemicals – and has also gained significant interest as a future clean energy carrier, with major advantages possible when used with new fuel cell technology. The majority of hydrogen is produced industrially from natural gas via methane steam reforming (MSR) or from coal via gasification and water gas shift (WGS) reactions.

$$\text{MSR: } CH_4 + H_2O \Leftrightarrow CO + 3H_2, \Delta H^0_{r298} = 206.2\,\text{kJ mol}^{-1}$$

$$\text{WGS: } CO + H_2O \Leftrightarrow CO_2 + H_2, \Delta H^0_{r298} = -41\,\text{kJ mol}^{-1}$$

Owing to its highly endothermic nature, the MSR process normally requires a high operating temperature, in the range of 850–900°C, so as to achieve near-equilibrium conversions. Using a H_2-selective membrane to withdraw the H_2 produced from the reaction zone, the equilibrium of the reforming reaction can be shifted significantly toward the product side, leading to enhanced conversions and a significantly reduced amount of catalyst and reaction temperature.

Methane dry reforming (MDR) becomes increasingly attractive because it allows us to reduce greenhouse gases (both CH_4 and CO_2) in the atmosphere.

$$\text{MDR: } CH_4 + CO_2 \Leftrightarrow 2CO + 2H_2, \Delta H^0_{r298} = 247\,\text{kJ mol}^{-1}$$

It is a highly endothermic process and requires severe operational temperatures (800–1000°C) to reach high conversion levels in conventional fixed bed reactors. These severe operating conditions will result in catalyst deactivation by coke deposition due to deep cracking of methane, which is thermodynamically favored at high temperatures. In the MRs, it is possible to achieve either a higher conversion than the traditional process at a fixed temperature, or the same conversion but at lower temperatures. Catalyst deactivation can be suppressed by lowering the temperature, as coke deposition via methane decomposition is thermodynamically limited.

Methanol is an attractive source of hydrogen due to its high H:C ratio, its production from sustainable sources (biomass, sugar beets), and its ease of storage as a liquid at atmospheric pressure. The methanol steam reforming (methanol-SR) process is considered to be a more suitable system for fuel cells because higher methanol conversions can be achieved at a low reaction temperature in the range of 250–350°C, and the products contain lower concentrations of CO.

$$\text{Methanol-SR: } CH_3OH + H_2O \Leftrightarrow CO_2 + 3H_2, \ \Delta H^0_{r298} = 49.57\,\text{kJmol}^{-1}$$

Ethanol is particularly appealing, since it is less toxic, easy to handle and distribute, and readily available. Recently, much attention has been paid to ethanol as a CO_2-neutral and renewable energy source because it can be produced from biomass.

$$\text{Ethanol-SR: } C_2H_5OH + 3H_2O \Leftrightarrow 2CO_2 + 6H_2, \Delta H^0_{r298} = 173.5\,\text{kJ mol}^{-1}$$

Dimethyl ether (DME) is a potential clean fuel and energy source of the next generation since it burns without producing NOx, smoke, or particulates. Steam reforming of DME may be another promising process as the hydrogen source for polymeric membrane fuel cells.

$$(CH_3)_2O + 3H_2O \Leftrightarrow 2CO_2 + 6H_2, \Delta H^0_{r298} = 135\,\mathrm{kJ\,mol^{-1}}$$

A number of studies have been reported on the MRs for hydrogen production by various reforming reactions, which are summarized in Table 2.8. Most research uses microporous silica membranes and a PBMR configuration. PSS substrate is usually used for high mechanical strength. A key point is to use optimum reaction conditions for a comparable hydrogen removal rate to production rate. The membrane characteristics for H_2 permeance and selectivity over time are a vital consideration in the commercial feasibility of this technology.

2.7.3 Partial Oxidation Reactions

Partial oxidation means that the organic reactants are oxidized into intermediate products instead of CO_2. One typical class of partial oxidation reactions is the oxidative dehydrogenation of alkanes to produce unsaturated hydrocarbons:

$$C_nH_{2n+2} + \frac{1}{2}O_2 \Leftrightarrow C_nH_{2n} + H_2O$$

This reaction is unlimited by thermodynamic equilibrium, energy saving, and tolerates catalyst deactivation. However, it also encounters some drawbacks of low selectivity due to undesired reactions forming carbon oxides, formation of hot spots due to exothermicity, and the problem of flammability limits. One approach to improve the performance of the reaction system is to use MRs with the membrane serving as an oxygen distributor, as shown in Figure 2.13(d). The catalyst is packed in the membrane tube while the oxidant is fed across the membrane into the reaction zone. As a result, the local oxygen partial pressures are kept low, which favors suppressing the deep oxidation reactions. Furthermore, the controlled addition of O_2 also avoids the explosion mixture and thus a wider range of operating conditions can be performed. A number of oxidative dehydrogenation reactions have been investigated in MRs, as summarized in Table 2.9. Other partial oxidation reactions include

Table 2.8 Reforming reactions for hydrogen production in porous membrane reactors

Reaction	Membrane	Catalyst/configuration	Operation conditions	Ref.
Methane steam reforming $CH_4 + H_2O \Leftrightarrow CO + 3H_2$ $\Delta H^0_{r298K} = 206.2\,kJ\,mol^{-1}$	$SiO_2/\alpha\text{-}Al_2O_3$ tube	$Ni@MgAl_2O_4$; PBMR	$P = 1–20$ atm; $T = 773–923$ K; Ar as sweep gas	[37]
	$Ni\text{-}SiO_2/\alpha\text{-}Al_2O_3$ tube	$Ni@\gamma\text{-}Al_2O_3$; CMR	$P = 500$ kPa; $T = 500°C$; Ar as sweep gas	[38, 39]
Methane dry reforming $CH_4 + CO_2 \Leftrightarrow 2CO + 2H_2$ $\Delta H^0_{r298K} = 247\,kJ\,mol^{-1}$	Al_2O_3 tube	5 wt% Ru-cat; CMR	$P = 1.2$ bar; $T = 350–500°C$; no sweep gas	[40]
	SiO_2/Vycor glass tube	1 wt% $Rh–Al_2O_3$; PBMR	$T = 848–973$ K; Ar as sweep gas	[41]
Water gas shift reaction $CO + H_2O \Leftrightarrow CO_2 + H_2$ $\Delta H^0_{r298K} = -41.2\,kJ\,mol^{-1}$	$Co–SiO_2/\alpha\text{-}Al_2O_3$ plate	$Cu–Zn–Al_2O_3$; PBMR	$T = 250°C$; N_2 as sweep gas	[42, 43]
	SiO_2/PSS flat	$CuO–CeO_2$; PBMR	$P = 6$ bar; $T = 220–290°C$; no sweep gas	[44]
Methanol steam reforming $CH_3OH + H_2O \Leftrightarrow CO_2 + 3H_2$ $\Delta H^0_{r298K} = 49.57\,kJ\,mol^{-1}$	$Pt–SiO_2$/PSS disk	Cu–Zn; PBMR	$P = 1$ bar; $T = 170–230°C$; He as sweep gas	[45]
	Carbon hollow fiber	$CuO–ZnO–Al_2O_3$; PBMR	$P = 1$ bar; $T = 200°C$; H_2O or Ar as sweep gas	[46]
$C_2H_5OH + 3H_2O \Leftrightarrow 2CO_2 + 6H_2$ $\Delta H^0_{r298K} = 173.5\,kJ\,mol^{-1}$	$SiO_2/\gamma\text{-}Al_2O_3$ disk	Na–Co–ZnO; PBMR	$P = 1$ bar; $T = 523–623$ K; Ar as sweep gas	[47]
$(CH_3)_2O + 3H_2O \Leftrightarrow 6H_2 + 2CO_2$ $\Delta H^0_{r298K} = 135\,kJ\,mol^{-1}$	$\gamma\text{-}Al_2O_3$/PPS disk	$Cu–Al_2O_3$; PBMR	$T = 250–500°C$; coupled with WGS reaction	[48]

Table 2.9 Partial oxidation reactions performed in porous membrane reactors

Reaction	Membrane	Catalyst/configuration	Reaction conditions	Ref.
$C_2H_6 + 1/2O_2 \rightarrow C_2H_4 + H_2O$	Al_2O_3 tube	VO_x–γ-Al_2O_3 (1.4%); PBMR	$T = 450–650°C$; $F_{shell} : F_{tube} = 9:1$	[50]
$C_3H_8 + 1/2O_2 \rightarrow C_3H_6 + H_2O$	α-Al_2O_3 tube	V/γ-Al_2O_3; coated CMR	$T = 380–500°C$; $F_{C_3} : F_{O_2} = 5.8$ or 2	[20]
$C_4H_{10} + 1/2O_2 \rightarrow C_4H_8 + H_2O$	α-Al_2O_3 tube	V–Mg–O; PBMR	$T = 873\,K$; $F_{C_4} : F_{O_2} = 1:2$	[27]
	α-Al_2O_3 tube	V-MgO; impregnated CMR	$T = 550°C$; $F_{C_4} : F_{O_2} = 1:1$	[22]
$CH_3OH + 1/2O_2 \rightarrow HCHO + H_2O$	PSS tube	Fe–Mo oxide; PBMR	$T = 200–250°C$; $F_{HC} / F_{O_2} = 1:2$	[51]
OCM: $2CH_4 + O_2 \rightarrow C_2H_4 + 2H_2O$	SiO_2/α-Al_2O_3 tube	Sm_2O_3–MgO; PBMR	$T = 973\,K$; $F_{CH_4} : F_{O_2} = 0.5 - 5.5:1$	[52]
	γ-Al_2O_3/α-Al_2O_3 tube	Mn–W–Na–SiO_2; PBMR	$T = 805°C$; $F_{CH_4} : F_{O_2} = 6.5:3.7$	[53]
$CH_4 + 1/2O_2 \rightarrow HCHO + H_2O$	ZrO_2/α-Al_2O_3 tube	Mo–Co–B–O–SiO_2; PBMR	$T = 883–963\,K$; $F_{CH_4} : F_{O_2} = 7.5:1$	[54]
Propane→ acrolein $C_3H_8 + O_2 \rightarrow CH_2{=}CHCHO$	α-Al_2O_3 tube	V–MgO and BiMo oxide; PBMR	$T = 375–450°C$; $F_{C_3} : F_{O_2} = 1:1$	[55]
	SiO_2/γ-Al_2O_3/α-Al_2O_3 tube	Ag–Bi–V–Mo–O; PBMR	$T = 450–550°C$; $F_{C_3} : F_{O_2} = 1.6:1$	[56]
Ethylene epoxidation	PSS	Ag–Cs/α-Al_2O_3; PBMR	$T = 210–270°C$; $F_E : F_{O_2} = 1.5:1$	[49]
diesel fuel → syngas $C_xH_y + O_2 \rightarrow CO + H_2$	Y_2O_3 stabilized ZrO_2 (YSZ tube)	$La_{0.5}Sr_{0.5}Co(Fe)O_{3-\delta}$; PBMR	$T = 950°C$	[57]

oxidative coupling of methane (OCM), partial oxidation of methanol to formaldehyde, epoxidation of ethylene to ethylene oxide [49], and so on. In most cases, an improvement in selectivity of the desired products has been obtained.

For partial oxidation reactions, the structural characteristics of the membrane (in particular, the relative laminar and Knudsen contributions to permeation) have a strong influence on the reactor performance since oxygen supply is the main factor influencing product distribution [58]. By controlling the oxygen concentration and contact time profile, the porous MR can be optimized for various objectives. For example, three operation modes for the Al_2O_3 porous MR could be identified for the partial oxidation of ethane into ethylene: selective oxidation mode with low oxygen to hydrocarbon ratios; CO favoring mode indicated by high oxygen excess, short contact times, and moderate temperatures; and deep oxidation mode characterized by high oxygen supply, high temperatures, and long contact times [50]. To optimize the operation of MRs, a fundamental knowledge of reaction kinetics and the interaction of concentration and residence time effects is necessary.

2.7.4 Gas–Liquid–Solid Multiphase Reactions

Porous MRs can be used for gas–liquid or gas–liquid–solid multiphase reactions, where the membrane serves as a contactor or provides a reaction zone. A typical application is the wet oxidation of organic pollutants by ozone in water treatment [59]. Chemically inert ceramic membranes are used to resist the high oxidation potential of ozone. Compared with conventional bubble-based contacting processes, the membrane process demonstrates many advantages – such as a constant ozone dosage independent of the liquid flow rate, the potential to easily recycle the oxygen carrier gas, and the high transfer performance within a small installation volume. By application of a hydrophobic coating to the membrane surface, the diffusion resistance of membrane to ozone transfer can be reduced significantly.

Many gas–liquid–solid multiphase reactions, such as the hydration of propene, catalytic hydrogenation of nitrate, chloroform dehalogenation, and H_2O_2 synthesis by $H_2 + O_2$ reaction in solution, are generally limited by the diffusion of the volatile reactant. Retention of homogeneous catalysts and efficient gas–liquid mass transfer are the key properties of such reaction systems. Both can be well achieved in the contactor-type MRs.

Table 2.10 Gas–liquid–solid multiphase reactions in porous membrane reactors

Reaction	Membrane/ configuration	Catalyst	Operating conditions	Ref.
$C_3H_6 + H_2O \Leftrightarrow C_3H_7OH$	Carbon/flat disk	Pt/Al_2O_3	$T = 130°C$; $P = 2.1$ MPa	[60]
Hydrogenation of nitrate $2NO_2^- + 3H_2 \rightarrow N_2 + 2OH^- + 2H_2O$	TiO_2- or $ZrO_2/\alpha\text{-}Al_2O_3$ tube	$Pd\text{–}Cu/TiO_2$	$F_{CO_2}/F_{H_2} = 3$; $P = 2$ bar; $T = 22°C$	[61]
Dehalogenation of chloroform $CHCl_3 + 3H_2 \rightarrow CH_4 + 3HCl$	$\alpha\text{-}Al_2O_3$ tube	Pd/Al_2O_3	$T = 20°C$;	[61]
$H_2 + O_2(\text{dissolved}) \Leftrightarrow H_2O_2$	Carbon/$\alpha\text{-}Al_2O_3$ tube	Pd/C	$F_{H_2}/F_{O_2} = 1$; $P = 5$ bar	[61]

Table 2.10 summarizes the application of porous MRs in gas–liquid–solid multiphase reactions. Operation of the high-pressure membrane contactor requires precision measurement and control. The activity and selectivity of the process may be controlled through controlling the dosage of gas reactant.

2.7.5 Other Reactions

The condensation reactions between alcohol and carboxylic acid over acidic catalysts are thermodynamically reversible with water formed as by-product:

$$R_1 - OH + HOOC - R_2 \Leftrightarrow R_1COR_2 + H_2O$$

In MRs, liquid reactants are fed to one side of the membrane while vacuum is applied to the other side (pervaporation operation). Water permeates preferentially through the membrane and evaporates at the vacuum side, driving the esterification reaction toward the product side with the result of enhanced product yields. Beneficial aspects of pervaporation also include low energy consumption and the possibility of selecting the reaction temperature.

Some typical dehydration reactions performed in porous MRs with improved product yield are summarized in Table 2.11.

The decomposition of VOCs in contaminated air stream or flue gas by catalytic or photocatalytic oxidation represents a typical application of

Table 2.11 Porous membrane reactors for dehydration and VOC decomposition reactions

Reaction	Membrane	Catalyst/ configuration	Operating conditions	Ref.
$CH_3COOH + C_4H_9OH \Leftrightarrow CH_3COOC_4H_9 + H_2O$	$SiO_2/\gamma\text{-}Al_2O_3/\alpha\text{-}Al_2O_3$ hollow fiber	Zeolite H-USY; coated CMR	$T = 75°C$; pervaporation	[62]
$C_3H_8 + CO_2 \Leftrightarrow C_3H_6 + H_2O + CO$	SiO_2–TiO_2–ZrO_2 hybrid membrane tube	Pd–Cu/ MoO_3–SiO_2; PBMR	$T = 673\,K$; CO_2 as sweep gas	[63]
VOC decomposition $2CH_3OH + 3O_2 \rightarrow 2CO_2 + 4H_2O$	Pt–$TiO_2/\alpha\text{-}Al_2O_3$ tube	Pt–TiO_2; FTCMR	$T = 110$–$120°C$; $c_{MeOH} = 100\,ppm$	[64, 65]

porous MRs. Complete conversion can be achieved at low temperature but at the expense of a significant pressure drop. Industrial application requires optimization of the membrane structure to reduce the pressure drop. Otherwise, its use will be restricted to applications involving reactions simultaneous with gas filtration, where the pressure drop is already present [19].

2.8 PROSPECTS AND CHALLENGES

Porous membranes can be used as extractor, distributor, or contactor in MRs for a variety of applications. They are characteristic of high permeability, high chemical/thermal stability, and relatively low cost, making them more easily amenable to industrial application – especially in situations where the membrane selectivity is not a limiting factor. However, porous membranes exhibit low perm-selectivity to allow other species to penetrate through, in addition to the preferentially permeable component, leading to some loss of attainable yield. Furthermore, reactant permeation through the porous membrane impacts negatively on the economics of the MR processes because of the need for additional downstream separation. Considerable efforts in the past have been placed on the improvement of membrane perm-selectivity with high permeability still retained.

One important factor hindering further commercial development of MRs concerns the membrane themselves, which still need optimization and new developments. Multi-layer membranes with successively smaller pore diameter are attractive to couple different membrane functions in a compact manner. However, the preparation of such membranes needs a strict and standardized operation, and quality control is based on more fundamental studies and insights. It is noteworthy that the final cost of membranes may be determined largely by the quality and cost of the underlying macroporous support. Novel concept membranes – such as the chemical valve membrane for O_2 distribution or active contactor applications – show the constant evolution of membrane materials and performance. Owing to the high conversions obtained with the forced-flow contactor concept, attractive development can be expected for a number of small-scale applications such as the removal of gaseous atmospheric pollutants (e.g., VOCs or NOx) with compact installations [24].

Ceramic hollow fiber membranes represent a relatively recent development due to their high specific area per volume. However, they are mechanically too weak to withstand the kind of environments one expects to encounter in high-temperature catalytic reaction applications. Another major problem still to be solved is the creation of the connection between the ceramic material and the steel tubing of the rest of the reactor [14].

In addition to the availability of affordable, efficient, and long-term reliable membranes, necessary elements for the practical application of MR systems include optimal reactor design, up-scale process analysis, catalyst optimization, as well as reliable, chemically inert and cost-effective sealants. Searching for better membrane materials, developing effective membrane synthesis methods for better quality control – especially in large-scale production – and improving chemical and structural stability of the current membrane materials will continue to be the focus of active research in these areas.

NOTATION

Symbol	Description
A_p, A_m	cross-sectional area of membrane pores or membrane surface (m^2)
c_i	concentration of component *i* (mol m^{-3})
C_1, C_2, C_3	constants defined in Eq. (2.15)

D_{ei}	effective diffusion coefficient component i ($m^2\ s^{-1}$)
D	diffusion coefficient ($m^2\ s^{-1}$)
d	pore diameter (m)
J	molar flux ($mol\ m^{-2} s^{-1}$)
L	length of the capillary (m)
M	molecular weight of gas ($kg\ mol^{-1}$)
N	total number of membrane pores
N_A	Avogadro number, $6.02 \times 10^{23}\ mol^{-1}$
p	pressure (Pa)
p_s	normal gas vapor pressure (Pa)
q	adsorption capacity ($mol\ kg^{-1}$)
R	ideal gas constant, $8.314\ J mol^{-1} K^{-1}$
R_{in}, R_o	inner and outer radius of membrane tube (m)
R_m	logarithm mean radius, $R_m = (R_o - R_{in})/\ln(R_o/R_{in})$
r	radius of membrane pore (m)
r_m	mean radius of the pores (m)
T	temperature (K)
V_m	molar volume of condensed liquid ($m^3\ mol^{-1}$)

Greeks

α	constant defined in Eq. (2.21)
ε	porosity
λ	mean free path of molecule (m)
μ	gas viscosity (Pa s)
δ	membrane thickness (m)
θ	contact angle
θ_K	reflection factor of the pore surface
ρ	density ($kg\ m^{-3}$)
σ_s	surface tension ($N\ m^{-1}$)
τ	tortuosity

Superscripts and subscripts

I	interface
i	i, component
K	Knudsen
m	molecule
mem	membrane
p	pore
s	surface
sub	substrate
u	upstream

REFERENCES

[1] Burggraaf, A.J. and Cot, L. (1996) *Fundamentals of Inorganic Membrane Science and Technology*, Elsevier Science B.V, Amsterdam.

[2] Baker, R.W. (2004) *Membrane Technology and Applications*, 2nd edn, John Wiley & Sons, Ltd, Chichester.

[3] Tennison, S. (2000) Current hurdles in the commercial development of inorganic membrane reactors. *Membrane Technology*, 2000, 4–9.

[4] Leenaars, A.F.M. and Burggraaf, A.J. (1985) The preparation and characterization of alumina membranes with ultra-fine pores. Part 3. The permeability for pure liquids. *Journal of Membrane Science*, 24, 245–260.

[5] Li, K., Kong, J. and Tan, X. (2000) Design of hollow fibre membrane modules for soluble gas removal. *Chemical Engineering Science*, 55, 5579–5588.

[6] Cot, L., Ayral, A., Durand, J. *et al.* (2000) Inorganic membranes and solid state sciences. *Solid State Sciences*, 2, 313–334.

[7] Lin, Y.S. (2001) Microporous and dense inorganic membranes: Current status and prospective. *Separation and Purification Technology*, 25, 39–55.

[8] Van Gestel, T., Vandecasteele, C., Buekenhoudt, A. *et al.* (2002) Alumina and titania multilayer membranes for nanofiltration: Preparation, characterization and chemical stability. *Journal of Membrane Science*, 207, 73–89.

[9] Hao, Y., Li, J., Yang, X., Wang, X. and Lu, L. (2004) Preparation of ZrO_2–Al_2O_3 composite membranes by sol-gel process and their characterization. *Materials Science and Engineering: A*, 367, 243–247.

[10] Kusakabe, K., Sakamoto, S., Saie, T. and Morooka, S. (1999) Pore structure of silica membranes formed by a sol-gel technique using tetraethoxysilane and alkyltriethoxysilanes. *Separation and Purification Technology*, 16, 139–146.

[11] Araki, S., Kiyohara, Y., Imasaka, S., Tanaka, S. and Miyake, Y. (2011) Preparation and pervaporation properties of silica–zirconia membranes. *Desalination*, 266, 46–50.

[12] Aust, U., Benfer, S., Dietze, M., Rost, A. and Tomandl, G. (2006) Development of microporous ceramic membranes in the system TiO_2/ZrO_2. *Journal of Membrane Science*, 281, 463–471.

[13] Li, K. (2007) *Ceramic Membranes for Separation and Reaction*, 1st edn, John Wiley & Sons, Ltd, Chichester.

[14] Tan, X. and Li, K. (2011) Inorganic hollow fibre membranes in catalytic processing. *Current Opinion in Chemical Engineering*, 1, 69–76.

[15] Briceño, K., Iulianelli, A., Montané, D., Garcia-Valls, R. and Basile, A. (2012) Carbon molecular sieve membranes supported on non-modified ceramic tubes for hydrogen separation in membrane reactors. *International Journal of Hydrogen Energy*, 37, 13536–13544.

[16] Briceño, K., Montané, D., Garcia-Valls, R., Iulianelli, A. and Basile, A. (2012) Fabrication variables affecting the structure and properties of supported carbon molecular sieve membranes for hydrogen separation. *Journal of Membrane Science*, 415/416, 288–297.

[17] Julbe, A., Farrusseng, D. and Guizard, C. (2001) Porous ceramic membranes for catalytic reactors – overview and new ideas. *Journal of Membrane Science*, 181, 3–20.

[18] Miachon, S. and Dalmon, J.A. (2004) Catalysis in membrane reactors: What about the catalyst? *Topics in Catalysis*, 29, 59–65.

[19] Coronas, J. and Santamaria, J. (1999) Catalytic reactors based on porous ceramic membranes. *Catalysis Today*, 51, 377–389.

[20] Bottino, A., Capannelli, G. and Comite, A. (2002) Catalytic membrane reactors for the oxidehydrogenation of propane: Experimental and modelling study. *Journal of Membrane Science*, 197, 75–88.

[21] Vitulli, G., Pitzalis, E., Salvadori, P. *et al.* (1995) Porous Pt/SiO_2 catalytic membranes prepared using mesitylene solvated Pt atoms as a source of Pt particles. *Catalysis Today*, 25, 249–253.

[22] Alfonso, M.J., Menendez, M. and Santamaria, J. (2002) Oxidative dehydrogenation of butane on V/MgO catalytic membranes. *Chemical Engineering Journal*, 90, 131–138.

[23] Li, G., Kanezashi, M., Yoshioka, T. and Tsuru, T. (2013) Ammonia decomposition in catalytic membrane reactors: Simulation and experimental studies. *AIChE Journal*, 59, 168–179.

[24] Saracco, G., Neomagus, H.W.J.P., Versteeg, G.F. and van Swaaij, W.P.M. (1999) High-temperature membrane reactors: Potential and problems. *Chemical Engineering Science*, 54, 1997–2017.

[25] Zaman, J. and Chakma, A. (1994) Inorganic membrane reactors. *Journal of Membrane Science*, 92, 1–28.

[26] Kit Chan, W., Jouët, J., Heng, S., Lun Yeung, K. and Schrotter, J.-C. (2012) Membrane contactor/separator for an advanced ozone membrane reactor for treatment of recalcitrant organic pollutants in water. *Journal of Solid State Chemistry*, 189, 96–100.

[27] Ge, S.H., Liu, C.H. and Wang, L.J. (2001) Oxidative dehydrogenation of butane using inert membrane reactor with a non-uniform permeation pattern. *Chemical Engineering Journal*, 84, 497–502.

[28] Diakov, V. and Varma, A. (2004) Optimal feed distribution in a packed-bed membrane reactor: The case of methanol oxidative dehydrogenation. *Industrial & Engineering Chemistry Research*, 43, 309–314.

[29] Akamatsu, K., Ohta, Y., Sugawara, T. *et al.* (2009) Stable high-purity hydrogen production by dehydrogenation of cyclohexane using a membrane reactor with neither carrier gas nor sweep gas. *Journal of Membrane Science*, 330, 1–4.

[30] Weyten H., Keizer K., Kinoo A., Luyten J., Leysen R. (1997) Dehydrogenation of propane using a packed-bed catalytic membrane reactor, *AIChE Journal*, 43, 1819–1827.

[31] Sznejer, G. and Sheintuch, M. (2004) Application of a carbon membrane reactor for dehydrogenation reactions. *Chemical Engineering Science*, 59, 2013–2021.

[32] Guo, Y.L., Lu, G.Z., Mo, X.H. and Wang, Y.S. (2005) Vapor phase dehydrogenation of methanol to methyl formate in the catalytic membrane reactor with Cu/SiO_2/ceramic composite membrane. *Catalysis Letters*, 99, 105–108.

[33] Ohashi, H., Ohya, H., Aihara, M., Negishi, Y. and Semenova, S.I. (1998) Hydrogen production from hydrogen sulfide using membrane reactor integrated with porous membrane having thermal and corrosion resistance. *Journal of Membrane Science*, 146, 39–52.

[34] Akamatsu, K., Nakane, M., Sugawara, T., Hattori, T. and Nakao, S. (2008) Development of a membrane reactor for decomposing hydrogen sulfide into hydrogen using a high-performance amorphous silica membrane. *Journal of Membrane Science*, 325, 16–19.

[35] Li, G., Kanezashi, M. and Tsuru, T. (2011) Highly enhanced ammonia decomposition in a bimodal catalytic membrane reactor for CO_x-free hydrogen production. *Catalysis Communications*, 15, 60–63.

[36] Li, G., Kanezashi, M., Lee, H.R., Maeda, M., Yoshioka, T. and Tsuru, T. (2012) Preparation of a novel bimodal catalytic membrane reactor and its application to ammonia decomposition for COx-free hydrogen production. *International Journal of Hydrogen Energy*, 37, 12105–12113.

[37] Hacarlioglu, P., Gu, Y. and Oyama, S.T. (2006) Studies of the methane steam reforming reaction at high pressure in a ceramic membrane reactor. *Journal of Natural Gas Chemistry*, 15, 73–81.

[38] Tsuru, T., Morita, T., Shintani, H., Yoshioka, T. and Asaeda, M. (2008) Membrane reactor performance of steam reforming of methane using hydrogen-permselective catalytic SiO_2 membranes. *Journal of Membrane Science*, 316, 53–62.

[39] Tsuru, T., Shintani, H., Yoshioka, T. and Asaeda, M. (2006) A bimodal catalytic membrane having a hydrogen-permselective silica layer on a bimodal catalytic support: Preparation and application to the steam reforming of methane. *Applied Catalysis A: General*, 302, 78–85.

[40] Paturzo, L., Gallucci, F., Basile, A., Vitulli, G. and Pertici, P. (2003) An Ru-based catalytic membrane reactor for dry reforming of methane – its catalytic performance compared with tubular packed bed reactors. *Catalysis Today*, 82, 57–65.

[41] Prabhu, A.K. and Oyama, S.T. (2000) Highly hydrogen selective ceramic membranes: Application to the transformation of greenhouse gases. *Journal of Membrane Science*, 176, 233–248.

[42] Battersby, S., Duke, M.C., Liu, S., Rudolph, V. and da Costa, J.C.D. (2008) Metal doped silica membrane reactor: Operational effects of reaction and permeation for the water gas shift reaction. *Journal of Membrane Science*, 316, 46–52.

[43] Battersby, S., Smart, S., Ladewig, B. *et al.* (2009) Hydrothermal stability of cobalt silica membranes in a water gas shift membrane reactor. *Separation and Purification Technology*, 66, 299–305.

[44] Brunetti, A., Barbieri, G., Drioli, E., Granato, T. and Lee, K.H. (2007) A porous stainless steel supported silica membrane for WGS reaction in a catalytic membrane reactor. *Chemical Engineering Science*, 62, 5621–5626.

[45] Lee, D.-W., Nam, S.-E., Sea, B., Ihm, S.-K. and Lee, K.-H. (2006) Preparation of Pt-loaded hydrogen selective membranes for methanol reforming. *Catalysis Today*, 118, 198–204.

[46] Sá, S., Sousa, J.M. and Mendes, A. (2011) Steam reforming of methanol over a CuO/ZnO/Al_2O_3 catalyst part II: A carbon membrane reactor. *Chemical Engineering Science*, 66, 5523–5530.

[47] Lim, H., Gu, Y. and Oyama, S.T. (2010) Reaction of primary and secondary products in a membrane reactor: Studies of ethanol steam reforming with a silica–alumina composite membrane. *Journal of Membrane Science*, 351, 149–159.

[48] Park, S.J., Lee, D.W., Yu, C.Y., Lee, K.Y. and Lee, K.H. (2008) Dimethyl ether reforming in a mesoporous gamma-alumina membrane reactor combined with a water gas shift reaction. *Industrial & Engineering Chemistry Research*, 47, 1416–1420.

[49] Al-Juaied, M.A., Lafarga, D. and Varma, A. (2001) Ethylene epoxidation in a catalytic packed-bed membrane reactor: Experiments and model. *Chemical Engineering Science*, 56, 395–402.

[50] Klose, F. (2004) Operation modes of packed-bed membrane reactors in the catalytic oxidation of hydrocarbons. *Applied Catalysis A: General*, 257, 193–199.

[51] Diakov, V., Blackwell, B. and Varma, A. (2002) Methanol oxidative dehydrogenation in a catalytic packed-bed membrane reactor: Experiments and model. *Chemical Engineering Science*, 57, 1563–1569.

[52] Tonkovich, A.L.Y., Jimenez, D.M., Zilka, J.L. and Roberts, G.L. (1996) Inorganic membrane reactors for the oxidative coupling of methane. *Chemical Engineering Science*, 51, 3051–3056.

[53] Lu, Y.P., Dixon, A.G., Moser, W.R. and Ma, Y.H. (2000) Oxidative coupling of methane in a modified γ-alumina membrane reactor. *Chemical Engineering Science*, 55, 4901–4912.

[54] Yang, C., Xu, N.P. and Shi, J. (1998) Experimental and modeling study on a packed-bed membrane reactor for partial oxidation of methane to formaldehyde. *Industrial & Engineering Chemistry Research*, 37, 2601–2610.

[55] O'Neill, C. and Wolf, E.E. (2010) Partial oxidation of propane to acrolein in a dual bed inert membrane reactor. *Catalysis Today*, 156, 124–131.

[56] Kolsch, P., Noack, M., Schafer, R., Georgi, G., Omorjan, R. and Caro, J. (2002) Development of a membrane reactor for the partial oxidation of hydrocarbons: Direct oxidation of propane to acrolein. *Journal of Membrane Science*, 198, 119–128.

[57] Mundschau, M., Burk, C. and Gribblejr, D. (2008) Diesel fuel reforming using catalytic membrane reactors. *Catalysis Today*, 136, 190–205.

[58] Pantazidis, A., Dalmon, J.A. and Mirodatos, C. (1995) Oxidative dehydrogenation of propane on catalytic membrane reactors. *Catalysis Today*, 25, 403–408.

[59] Janknecht, P., Wilderer, P.A., Picard, C. and Larbot, A. (2001) Ozone-water contacting by ceramic membranes. *Separation and Purification Technology*, 25, 341–346.

[60] Lapkin, A.A., Tennison, S.R. and Thomas, W.J. (2002) A porous carbon membrane reactor for the homogeneous catalytic hydration of propene. *Chemical Engineering Science*, 57, 2357–2369.

[61] Centi, G., Dittmeyer, R., Perathoner, S. and Reif, M. (2003) Tubular inorganic catalytic membrane reactors: Advantages and performance in multiphase hydrogenation reactions. *Catalysis Today*, 79/80, 139–149.

[62] Peters, T.A., Benes, N.E. and Keurentjes, J.T.F. (2005) Zeolite-coated ceramic pervaporation membranes; Pervaporation–esterification coupling and reactor evaluation. *Industrial & Engineering Chemistry Research*, 44, 9490–9496.

[63] Zhong, S.H., Sun, H.W., Wang, X.T., Shao, H.Q. and Guo, J.B. (2006) Preparation of hybrid membrane reactors and their application for partial oxidation of propane with CO_2 to propylene. *Journal of Membrane Science*, 278, 212–218.

[64] Tsuru, T., Kan-no, T., Yoshioka, T. and Asaeda, M. (2006) A photocatalytic membrane reactor for VOC decomposition using Pt-modified titanium oxide porous membranes. *Journal of Membrane Science*, 280, 156–162.

[65] Wang, W.Y., Irawan, A. and Ku, Y. (2008) Photocatalytic degradation of Acid Red 4 using a titanium dioxide membrane supported on a porous ceramic tube. *Water Research*, 42, 4725–4732.

3

Zeolite Membrane Reactors

3.1 INTRODUCTION

Zeolite membranes are a type of porous membrane in which the pores are the inherent feature of crystalline structures. Unlike other porous ceramic membranes (e.g., SiO_2, Al_2O_3, and TiO_2) that have tortuous and randomly oriented pore networks formed by packing of oxide particles, zeolite membranes have a well-defined, uniform pore system of molecular dimensions. Zeolite membranes are stable at high temperatures, acidic or basic in nature, and exhibit hydrophilic or organophilic properties. Based on the unique properties of high thermal and chemical stabilities, selective adsorption, and molecular sieving (size exclusion), zeolite membranes have found extensive potential applications in gas, vapor, and liquid separations.

In addition to their separation properties, zeolite membranes can be tailor-made for catalytic reactions through ion exchange, dealumination/realumination, isomorphous substitution, and insertion of catalytically active phases such as transition-metal ions, complexes, basic alkali metals, or metal oxide clusters. Moreover, membranes composed of different types of zeolites provide different framework structures and pore sizes, making them ideally suitable for applications in membrane reactors [1, 2]. In recent years there has been significant interest in zeolite membranes for high-temperature CMR applications. This chapter gives a detailed description of zeolite membrane reactors, with special focus on the preparation of high-quality zeolite membranes and their applications in a variety of chemical reactions. A brief comment on the challenges faced with zeolite MRs will be provided at the end of the chapter.

Inorganic Membrane Reactors: Fundamentals and Applications, First Edition. Xiaoyao Tan and Kang Li.

3.2 PERMEATION IN ZEOLITE MEMBRANES

3.2.1 Types of Zeolite Membranes

Zeolites are crystalline microporous aluminosilicate solids with a regular three-dimensional pore structure that is stable at high temperatures. Their atomic structures are based on the three-dimensional frameworks of silica and alumina tetrahedral, where the silicon or aluminum ions are surrounded by four oxygen ions in a tetrahedral configuration. Clusters of tetrahedra form different box-like polyhedral units, further linked to build up various frameworks with different channel sizes.

When zeolites are grown as films, zeolite membranes are formed. Efforts to prepare polycrystalline zeolite membranes started in the late 1980s, but not until the early 1990s were MFI-type zeolite membranes (ZSM-5 and silicalite-1) successfully prepared with very good permeation and separation properties [3]. Since then, zeolite membranes have constantly attracted considerable attention because of their unique properties in terms of size uniformity, shape selective separation behavior, and good thermal/chemical stabilities. So far, more than 20 different types of zeolite membranes have been prepared – such as LTA, FAU, MOR, FER, MEL, CHA, DDR, and AFI – with significant separation interest [4, 5]. Table 3.1 lists a few typical zeolite membranes and their potential applications for separation of fluid mixtures.

3.2.2 Transport Mechanisms

Transport in zeolite membranes is a complex process that is governed by adsorption and diffusion. The mechanism depends strongly on pore size, pore network structure, size and shape of diffusing molecules, interac-

Table 3.1 Typical zeolite membranes and their potential applications

Type of zeolite membrane	Aperture-free diameter (nm)	Separation
MFI (ZSM-5 or silicate-1)	0.53 × 0.56	H_2, CO_2, H_2O, p-xylene
LTA (NaA)	0.42	H_2O, alcohol
FAU (NaX, NaY)	0.74	H_2O, alcohol, CO_2
MOR (mordenite)	0.65 × 0.70	H_2O
DDR (deca-dodecasil 3R)	0.36 × 0.44	H_2
CHA (SAPO-34)	0.38	CO_2

tions between transported species and membrane materials, as well as adsorbate–adsorbate interactions in the mixtures.

Zeolites discriminate between the components of gaseous or liquid mixtures depending on their molecular size. This molecular sieving behavior occurs when the size of the pore apertures of zeolite is similar to the dimensions of gas molecules. The permeability and selectivity are highly dependent on temperature. However, the perm-selectivity of zeolite membranes also depends on the chemical nature of the permeating molecules and the membrane material as well as the adsorption properties of the membrane, that is, a stronger adsorbing compound will permeate preferentially. Very high selectivity is also possible with gas mixtures due to selective adsorption, even if the zeolite pore size is significantly larger than the molecules. For example, although the kinetic diameters of H_2 and iso-butane are 0.29 and 0.50 nm, respectively, the MFI-type zeolite membrane with 0.55 nm pores is iso-butane selective at room temperature in terms of both selectivity and permeance because the permeation is an adsorption-controlled process. However, at 500 °C the zeolite membrane becomes perm-selective for hydrogen over hydrocarbons (C1–C4) due to the weakening effect of adsorption at high temperatures [6]. LTA zeolite membranes demonstrate excellent separation properties for ethanol–water and other organic–water mixtures due to their strong hydrophilic nature, but do not have useful selectivity for separation of permanent gas mixtures such as O_2–N_2, H_2–CO_2, and so on. This transport mechanism, governed by adsorption and diffusion, is also called surface diffusion [7]. At high temperatures when the adsorption phenomena become negligible, surface diffusion is weak and activated gaseous diffusion will take place.

It is known that zeolite membranes essentially contain intercrystalline non-zeolitic pores (defects). This irregular nature of zeolite membranes with intercrystalline pores adds to the complexity of the transport process in addition to the contribution of a support layer to the permeation resistance. For zeolite membranes, selectivity similar to that expected for Knudsen flow generally indicates the presence of intercrystalline pores. Separation based primarily on adsorption differences, which is generally true in the separation of liquid mixtures by pervaporation, may have tolerance to the intercrystalline pores. However, in order to obtain high perm-selectivity, the zeolite membranes must have negligible amounts of intercrystalline pores and pinholes of larger than 2 nm so as to reduce the gas flux from these defects [3].

3.2.3 Permeation Flux in Zeolite Membranes

Transport in zeolites depends on the adsorption properties and is an activated phenomenon. For the surface diffusion mechanism, the diffusion flux can be expressed as [8]

$$J^s = -\varphi\rho_s q_m D^s \Gamma \cdot \frac{dq}{dz} \tag{3.1}$$

where φ is the corrective factor associated with the porous structure of the membrane; ρ_s is the zeolite density; q_m is the amount of adsorbed gas at saturation; D^s is the surface diffusion coefficient; p is the partial pressure; q is the adsorptive capacity; and Γ is the thermodynamic factor expressed by $\Gamma = \delta \ln(p)/\delta \ln(q)$.

At intermediate temperatures (298–350 K), the adsorption can be described by a Langmuir isotherm:

$$\theta = \frac{q}{q_m} = \frac{Kp}{1+Kp} \tag{3.2}$$

where θ stands for the occupation rate, $\theta = q/q_m$; K is the Langmuir constant (Pa^{-1}). Its temperature dependency can be correlated with the Van't Hoff relation:

$$K = K_0 \exp\left(\frac{Q_a}{RT}\right) = \exp\left(\frac{\Delta S_{ad}}{R} + \frac{Q_a}{RT}\right) \tag{3.3}$$

where Q_a is the heat of adsorption ($J\,mol^{-1}$); ΔS_{ad} is the adsorption entropy ($J\,mol^{-1}\,K^{-1}$); R is the ideal gas constant ($8.314\,J\,mol^{-1}\,K^{-1}$); and T is the temperature.

Integration of Eq. (3.1) over the membrane thickness δ gives the permeation flux:

$$J^s = -\frac{\varphi\rho_s q_m D^s}{\delta} \cdot \ln\left(\frac{q_m - q_u}{q_m - q_d}\right)$$

or

$$J^s = \frac{\varphi\rho_s q_m D^s}{\delta} \cdot \ln\left(\frac{1+Kp_u}{1+Kp_d}\right) \tag{3.4}$$

where p_u and p_d are the upstream and downstream pressure, respectively.

When the surface diffusivity (D^s) is independent of occupation rate, its temperature dependence can be described by the Arrhenius equation:

$$D^s = D_0^s \exp\left(-\frac{E_a^s}{RT}\right) \tag{3.5}$$

where D_0^s and E_a^s are the pre-exponential coefficient and the activation energy for surface diffusion, respectively.

In the case of low concentrations, Eq. (3.4) then reduces to

$$J^s = \varphi\rho_s q_m K_0 D_0^s \exp\left(\frac{Q_a - E_a^s}{RT}\right)\cdot\frac{p_u - p_d}{\delta} \tag{3.6}$$

This indicates that the permeation flux is dependent on the difference between the adsorption heat and the diffusion activation energy.

At high temperatures, activated gaseous diffusion or translational diffusion becomes the dominating transport mechanism. Then the permeation flux can be expressed by

$$J^g = -\varphi\frac{D^g}{RT}\cdot\frac{dp}{dz} \tag{3.7}$$

where the gas diffusion coefficient can be given as

$$D^g = d\sqrt{\frac{8RT}{\pi M}}\cdot\exp\left(-\frac{E_a^g}{RT}\right) \tag{3.8}$$

with d the average pore diameter or diffusional length and M the gas molecular weight. Considering gas permeability through the zeolite membrane in a large temperature range, the total permeation flux is a combination of the low-temperature (surface) and high-temperature (activated gaseous) flux. Combining Eqs (3.1), (3.2) and (3.7) gives the total flux for one component:

$$J = J^s + J^g = -\varphi\rho_s q_m\frac{D^s}{1-\theta}\cdot\frac{d\theta}{dz} - \varphi\frac{D^g}{RT}\cdot\frac{dp}{dz} \tag{3.9}$$

Knudsen diffusion can be applied to describe transport in the intercrystalline non-zeolitic pores or defects [9]. If the intercrystalline non-zeolitic pores also contribute to the permeation, the total permeation flux through the membrane is then

$$J = J^s + J^g + J^K \tag{3.10}$$

For the multi-component permeation system in zeolite membranes, the mass transfer can be described using the general Maxwell–Stefan equations [10, 11]

$$-\rho_c \frac{\theta_i}{RT} \nabla \mu_i = \sum_{j=1, j \neq i}^{n} \frac{q_j J_i - q_i J_j}{q_{mi} q_{mj} D_{ij}^s} + \frac{J_i}{q_{mi} D_i^s} \tag{3.11}$$

where the chemical potential gradient can be related to the gradient in surface coverage by a matrix of thermodynamic factors:

$$\frac{\theta_i}{RT} \nabla \mu_i = \sum_{j=1}^{n} \Gamma_{ij} \nabla \theta_j; \qquad \Gamma_{ij} = \frac{\theta_i}{p_i} \frac{\partial p_i}{\partial \theta_j} \tag{3.12}$$

The second term on the right-hand side of Eq. (3.11) stands for the interactions of the molecules with the pore wall.

3.3 PREPARATION OF ZEOLITE MEMBRANES

Zeolite membranes generally have a thickness of 2–50 μm, formed on porous materials such as Al_2O_3 and stainless steel tubes or disks owing to their greater structural stability and reduced mass-transfer resistance. The quality of zeolite membranes in terms of permeability and selectivity is determined by their intercrystalline porosity (defects), the crystal orientation relative to the membrane layer, the size of zeolite crystals, and the thickness and uniformity of the zeolite layer.

There are basically two approaches to synthesize supported zeolite membranes: liquid-phase synthesis and vapor-phase transport (or dry-gel conversion) methods [2, 12]. The liquid-phase-synthesis approach is to bring the surface of a porous support in contact with a zeolite synthesis solution (sol or gel) and keep the system under controlled conditions so that the zeolite can nucleate and grow to a continuous film on the support surface.

3.3.1 In-Situ Crystallization Method

Conventional hydrothermal synthesis is the most common method for zeolite membrane preparation. In the in-situ crystallization method, a porous support is immersed in a zeolite synthesis solution. A membrane is formed on the surface of the support by direct crystallization. Two critical stages can be distinguished during the formation of zeolite membranes, namely nucleation on the support followed by crystal growth to form a continuous zeolite film covering the support. The zeolite nuclei are either formed directly on the support by heterogeneous nucleation or deposited as embryonic seed crystals from the solution (homogeneous nucleation). Figure 3.1 shows a flow diagram of the in-situ crystallization process for membrane preparation.

The in-situ crystallization synthesis is easy to operate, but the separation properties of the resultant membranes cannot be controlled with ease, because the formation of zeolite film and microstructure depends significantly not only on the chemical and structural nature of the support surface but also on the synthesis conditions – such as synthesis solution composition, pH, temperature, presence of impurities, and even nutrient sources. Furthermore, this synthesis method also needs a relatively long crystallization time – from a few hours to a few days – leading to the formation of impure zeolites. Moreover, the zeolite crystals

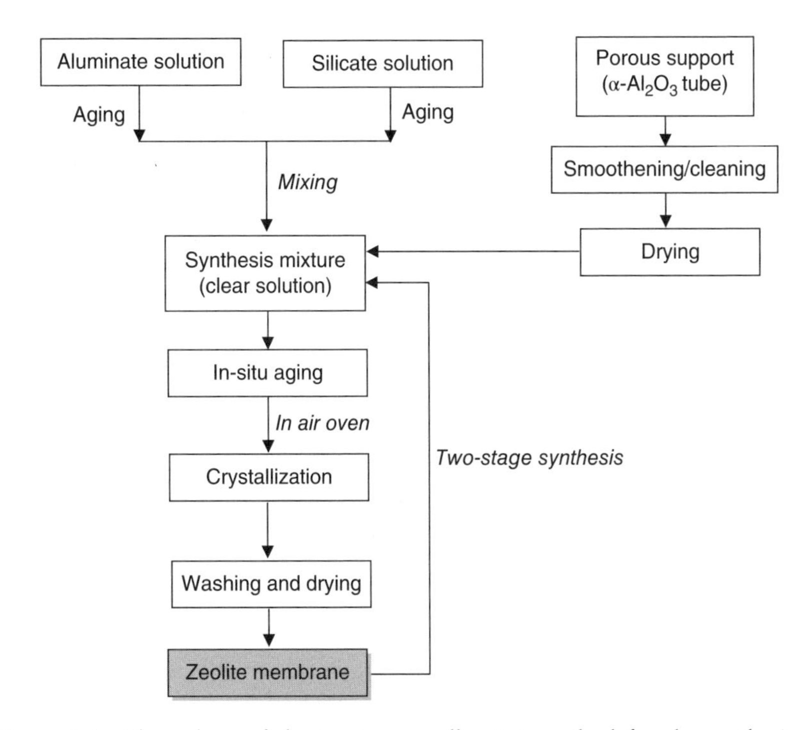

Figure 3.1 Flow chart of the in-situ crystallization method for the synthesis of zeolite membranes.

formed are usually not uniform in size since the zeolite nuclei do not form on the support surface simultaneously due to the low heating rate and inhomogeneous heating [2].

3.3.2 Secondary Growth Method

The secondary growth method is currently recognized as one of the most attractive and flexible methods to synthesize high-quality zeolite membranes. The name refers to the fact that a layer of zeolite seeds attached on the support surface in advance will re-grow to form a continuous dense layer during hydrothermal treatment. Figure 3.2 shows the process of the secondary growth method. Several steps are essentially included: making a colloidal suspension of zeolite nanocrystals, deposition of a seed layer onto a substrate, and seed growth into the zeolite film under hydrothermal crystallization conditions.

Since the location and density of nucleation sites can be well controlled in the secondary growth method, the nature of the support is less important for membrane formation than in the in-situ crystallization synthesis method. Furthermore, the elimination of the in-situ nucleation

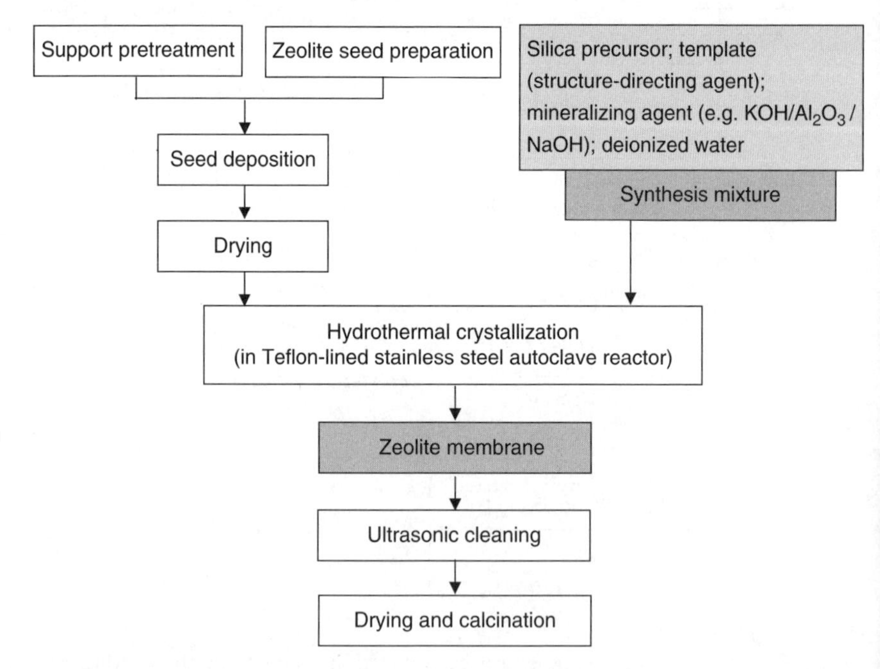

Figure 3.2 Secondary growth method for the synthesis of zeolite membranes.

step provides improved flexibility for crystal growth, better control of film microstructure (thickness, orientation), and enhanced reproducibility and scalability as well. However, the operation process becomes complicated as it involves multi-step synthesis. In some cases the use of a binder might affect the zeolite layer properties.

In order to obtain high-quality zeolite membranes, a variety of techniques have been developed to facilitate nucleation and crystal growth on the substrate and tailor the interactions between the zeolite layer and the substrate [1].

1. It has been well established that the quality of the substrate determines, to a high degree, the quality of the zeolite membranes. The substrate surface must be smooth enough for the formation of thin zeolite films on it. This can be achieved simply by common mechanical polishing. A more feasible way to create smooth porous surfaces is to deposit a thin meso-/microporous film on the support [13]. For example, a γ-Al_2O_3 layer with pore diameter of 5 nm is often coated on the α-Al_2O_3 substrate with pore diameter of 200 nm. This layer has the additional benefit of serving as a diffusion barrier, preventing the zeolite layer from penetrating the support, which would result in a longer effective diffusion path, whilst also protecting the support against possible leaching.
2. Good adherence of the zeolite seed crystals as well as the zeolite film to the support is important to guarantee the mechanical, thermal, and chemical stability of the composite membrane. The adherence of the zeolite crystals to the support surface is to a large extent determined by the hydrophilicity of the support surface. Treatment of the support with NaOH may increase the number of surface hydroxyl groups, and therefore will impart the support with more nucleation points as well as sites where crystals could adhere by means of Van der Waals interactions and H-bonding.
3. The application of a thin, uniform, and continuous layer of seed crystals on the support determines the successful synthesis of a thin, defect-free zeolite film. The simplest and most often used method is to apply seed crystals to the substrate with mechanical rubbing [14]. Slip-coating [15] and dip-coating [16, 17] the substrate in a suspension of zeolite particles, followed by drying and calcination, are also used to seed the support surface, but the process often has to be repeated a few times in order to ensure a sufficiently high coverage of the support with zeolite seed crystals. Electrostatic deposition involves charge modification of the substrate surface by

adsorption of an anionic or cationic polymer, which will then bring about electrostatic attraction of colloidal zeolite particles in suspension to the surface of the support. This process generally achieves a high coverage of the substrate with well-adhered particles and is one of the most effective techniques developed to date for seed crystal deposition [18].

4. Oriented thin zeolite membranes are attractive for both their high permeability and selectivity [5]. Much effort has been made in recent years to prepare zeolite films with controlled orientation, including by (i) controlling conditions to promote growth rates along certain crystallographic directions [19]; (ii) deposition of an oriented seed layer and proceeding with the seeded growth in the absence of further nucleation [20]; and (iii) changing the direction of the highest nutrient concentration from the out-of-plane to the in-plane direction.

Growing a layer of membrane on the inner side of the support is a challenging task due to the low accessibility to the lumen of tubular supports. This can be achieved by using a continuous flow system in which the synthesis gel is circulated continuously in the lumen of the tubular support under the action of gravity [21]. A rotating synthesis device was developed by Tiscareño-Lechuga *et al.* to prepare inner-side zeolite membranes, as shown in Figure 3.3 [15]. Zeolite seeds are deposited on the inside surface of the tubular support by slip-coating prior to hydrothermal synthesis. With the system in operation, the synthesis solution is circulated under centrifugal force.

3.3.3 Vapor-Phase Transport Method

The vapor-phase transfer (VPT) synthesis method consists of two steps: covering the support with a synthesis gel and crystallization of the dried gel in saturated steam at a similar temperature or a mixture of steam and a structure-directing agent (SDA) under autogenous pressure [22–24]. Since one of the nutrient sources is limited to the support surface, zeolite membrane formation will be limited to the support surface. With this method, the membrane thickness can be well controlled since the amount of nutrient for growing zeolite is controlled directly by the amount of gel applied. However, the crystals produced are most likely randomly oriented. Furthermore, since the density of gel is much lower than that of zeolite crystals, the crystallization process

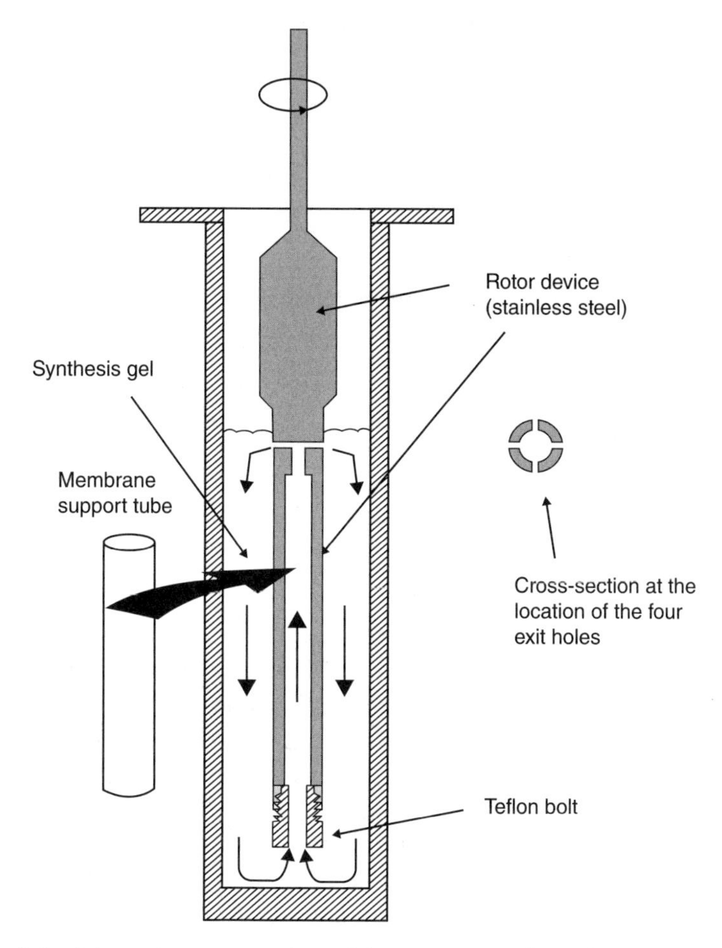

Figure 3.3 Schematic representation of the rotating synthesis device. Reproduced from [15]. With permission from Elsevier.

may produce a large number of intercrystalline voids that must be filled through a dissolution–recrystallization process.

3.3.4 Microwave Synthesis Method

Microwave-assisted crystallization provides a very efficient tool for preparing a variety of zeolite materials with small zeolite particle size, narrow particle size distribution, and high purity within a short time [25]. During microwave processing, energy is supplied by an electromagnetic field directly to the material; thus it is more efficient in transferring thermal energy to a volume of material than conventional hydrothermal

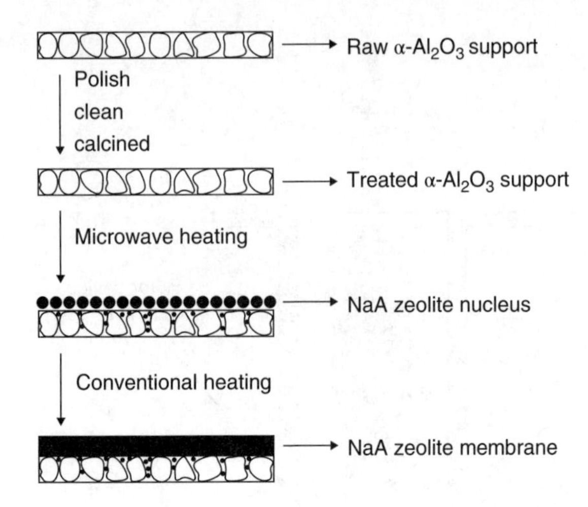

Figure 3.4 Schematic diagram for synthesis of NaA zeolite membrane together with microwave heating and conventional heating method. Reproduced from [26]. With permission from Elsevier.

processing, which transports heat through the surface of the material by convection, conduction, and radiation. Microwave synthesis of zeolite membranes starts with nucleation on a porous support under microwave heating, then growing to a continuous dense zeolite layer under hydrothermal synthesis conditions. Figure 3.4 shows a schematic diagram of the microwave synthesis method [26]. Compared with conventional hydrothermal synthesis, there is fast homogeneous heating throughout a reaction vessel and more simultaneous nucleation on the support surface, with uniform small zeolite crystals and a resulting thin membrane, as well as a reduction in processing time and energy cost compared with conventional heating. This may prove to be an effective route for the preparation of high-quality zeolite membranes on a large scale.

3.4 CONFIGURATION OF ZEOLITE MEMBRANE REACTORS

The zeolite membranes in MRs normally consist of a thin zeolite film on a porous support, typically α-Al_2O_3, stainless steel, or carbon. Generally speaking, it is more difficult to prepare defect-free zeolite membranes on tubular supports than on plate-shape supports due to the development of mechanical tensions during the drying/calcination steps that often lead to defects in the zeolite layer. However, most of the zeolite MRs are in a tubular configuration because of their higher specific area and

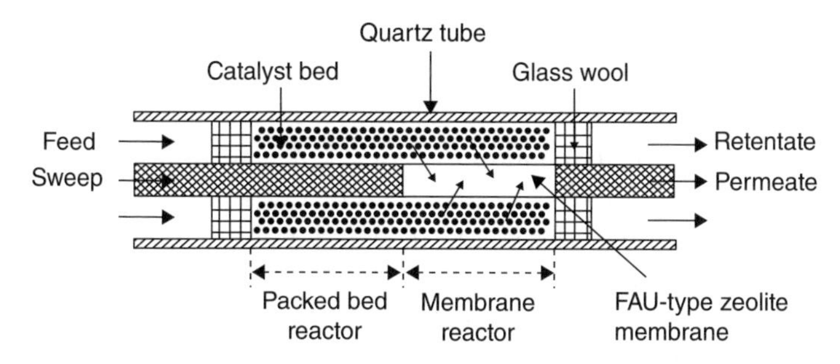

Figure 3.5 Schematic representation of a zeolite PBMR. Reproduced from [29]. With permission from Elsevier.

relatively easier integration into membrane reactors. The common application modes of zeolite MRs include PBMRs, CMRs, pervaporation membrane reactors (PVMRs), and MMRs.

3.4.1 Packed Bed Membrane Reactor

The majority of zeolite MR applications reported in the literature to date fall into the category of PBMRs. The reactor consists of a zeolite membrane with a conventional catalyst present in the form of a packed bed of pellets. The reaction takes place in the catalyst bed while the zeolite membrane serves mainly as a product separator (for H_2 or H_2O separation) [27] or a reactant distributor (for O_2 distribution) [28]. Figure 3.5 illustrates a FAU-type zeolite PBMR combined with a packed bed reactor for dehydrogenation of cyclohexane [29]. Half of the catalyst is packed in the area upstream of the permeation portion to enhance the conversion, otherwise cyclohexane will preferentially permeate at the front end of the zeolite membrane, resulting in a decrease in conversion.

3.4.2 Catalytic Membrane Reactor

In CMRs, no separate catalyst is used and reactions take place on the membrane. Zeolite membranes can be intrinsically catalytic due to the presence of catalytic sites (Brönsted acid sites, Lewis acid sites, metal ions in cationic positions, transition metal ions in zeolite lattice positions, extra-lattice transition metal compounds in channels and cavities of a zeolite, metal particles in zeolite cavities) and their high internal surface area.

The intrinsic catalytic activity of zeolites, coupled with the possibility of tuning adsorptive properties, makes them good candidates for bringing about simultaneous reaction and separation. The catalytic membranes exhibit improved selectivity and resistance to deactivation compared with the same catalyst employed in a packed bed configuration [28].

In an FTMR configuration, the reactant mixture flows through the membrane, and thereafter they react with each other on the catalytically active sites. The zeolite membrane can be non-perm-selective but acts as a reaction site facilitating the access of reactants to the catalyst.

3.4.3 Pervaporation Membrane Reactor

Figure 3.6 shows schematically a PVMR for liquid reactions, such as condensation between alcohol and carboxylic acid over acidic catalysts [30]. The water produced is removed selectively from the reaction system through a zeolite membrane by vacuum pervaporation. The zeolite membrane is required to have a high selectivity for water over alcohols and good stability to acid solutions as well.

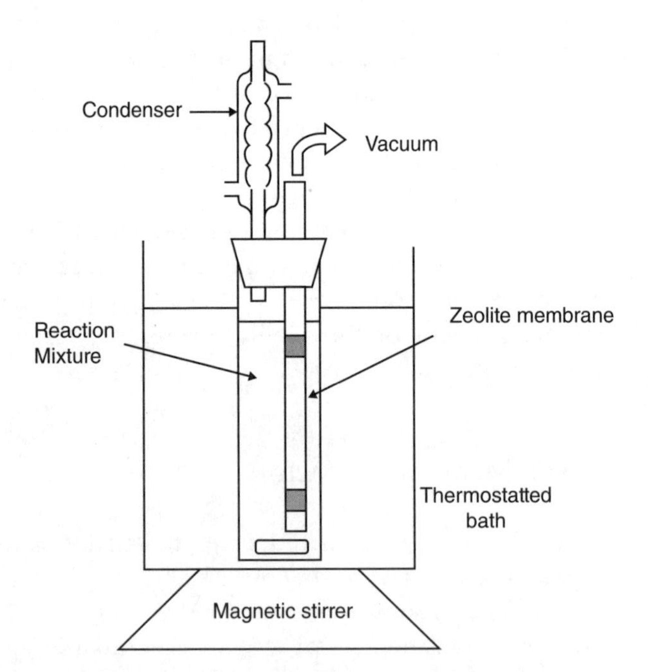

Figure 3.6 Schematic diagram of VPMR for esterification reactions. Reproduced from [30]. With permission from Elsevier.

3.4.4 Membrane Microreactor

Zeolite membranes can be prepared on a porous substrate with precisely designed microchannels to form an MMR, as shown in Figure 3.7 [31]. This configuration promises to provide rapid mass and heat transfer rates, high efficiency, and safe operation. A more detailed description of MMRs can be found in Chapter 8.

Increasing attention is being paid to particle-level membrane reactors (PLMRs), which consist of catalytic particles coated with a zeolite membrane layer allowing the selective addition of reactants to the reaction zone or the selective removal of products from the reaction zone when one product's diffusivity is much higher than that of the other products [32, 33]. The main benefit arising from this configuration is the increased membrane area per unit reactor volume compared with that of conventional membrane reactors. This is highly advantageous

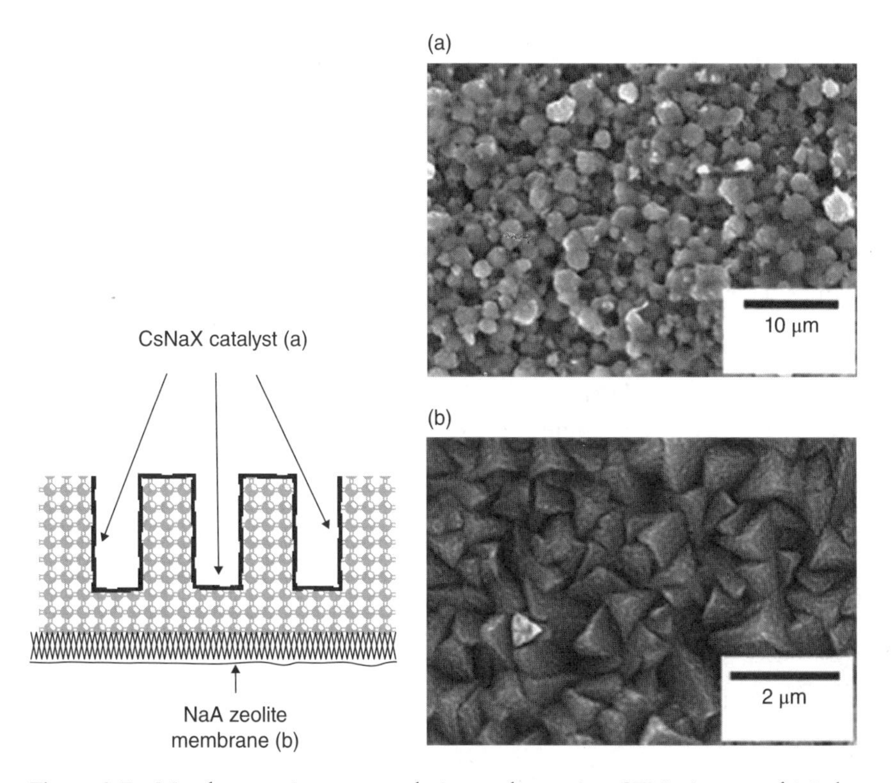

Figure 3.7 Membrane microreactor design and top view SEM pictures of (a) the 3 μm thick CsNaX catalyst deposited on the microchannel wall and (b) the 6 μm thick NaA grown on the back of the stainless steel plate. Reproduced from [5]. With permission from Elsevier.

considering the difficulty in achieving a large membrane area with the absence of defects.

3.5 APPLICATIONS OF ZEOLITE MEMBRANE REACTORS

Zeolite MRs have found extensive applications including dehydrogenation, dehydration, isomerization, and oxidation reactions, as summarized in Table 3.2. In most cases, MFI-type (silicalite/ZSM-5) zeolite membranes are applied because these high-silica zeolite membranes show good shape-selective separation behavior and are relatively easy to synthesize. As the alumina content increases, the mesopores in the MFI membranes will increase, leading to higher permeation flux but lower selectivity. For vapor separation, hydrophilic zeolite membranes such as LTA- and T-type zeolites have mostly been used because of their high selectivity for water over organics and high thermal stability.

3.5.1 Dehydrogenation Reactions

Zeolite membranes have been applied to a variety of dehydrogenation reactions, where the membrane mainly functions for H_2 separation but not for catalysis. Very high conversions and yields above the equilibrium values have been achieved for all dehydrogenation reactions due to the selective removal of produced hydrogen. The limiting parameter can be either the membrane or the catalyst, depending on the sweeping configuration that is used [52]. Since zeolite membranes have pore diameters larger than hydrogen molecules, separation is not based on molecular exclusion but on a combination of competitive adsorption and the different mobilities of molecules, arising from differences in molecular shape. Small-pore, defect-free zeolite membranes with pore sizes between the size of hydrogen (0.29 nm) and alkanes (0.45 nm) are desired to improve reactor performance.

3.5.2 Dehydration Reactions

The condensation reactions between alcohol and carboxylic acid over acidic catalysts are thermodynamically reversible with water formed as a by-product:

$$R_1COOH + HO - R_2 \Leftrightarrow R_1COOR_2 + H_2O$$

Table 3.2 Applications of zeolite MRs

Reaction type	Reaction/reactants	Membrane/ support	Catalyst	Reactor configuration	Operating conditions	Advantages	Ref.
Dehydrogenation	$C_6H_{12} \Leftrightarrow C_6H_6 + 3H_2$	NaY/Al_2O_3-tube	Pt(1 wt%)-Al_2O_3	PBMR	$P = 0.3$ MPa; $T = 423–523$ K; Ar as sweep gas	$X = 72.1\%$ above equilibrium 32.2% at 473 K	[29]
	$C_6H_5–C_2H_5 \Leftrightarrow C_6H_5–C_2H_3 + H_2$	Silicalite-1/PSS tube	Fe_2O_3	PBMR	$P = 0.8$ atm; $T = 580–640$ °C; N_2 as sweep gas	X increased by 7% from TR	[34]
	Iso-butane dehydrogenation $C_4H_{10} \Leftrightarrow C_4H_8 + H_2$	MFI/α-Al_2O_3 tube	PtIn-MFI	PBMR	$P = 100–170$ kPa; $T = 730$ K; N_2 as sweep gas	Y increased 4 times higher than TR	[35]
		DD3R/α-Al_2O_3 tube, 1–2 μm	$Cr_2O_3–Al_2O_3$	PBMR	$P = 101$ kPa; $T = 712–762$ K; N_2 as sweep gas	Y increased by 50% above equilibrium yield	[36]
Water gas shift reaction	$CO + H_2O \Leftrightarrow CO_2 + H_2$	MFI/α-Al_2O_3 disk	CuO–ZnO–Al_2O_3	PBMR	$P = 6$ bar; $T = 220–290$ °C; N_2 as sweep gas	$X = 95.4\%$ above equilibrium conversion	[37]
Esterification reactions $RCOOH + R'OH \Leftrightarrow RCOOR' + H_2O$	acetic acid + ethanol	HZSM-5/α-Al_2O_3 or PSS tube	HZSM-5	CMR	$T = 333–363$ K	Higher conversion than TR	[38]
		zeolite T/α-Al_2O_3 tube	Cation exchange resin	PVMR	$T = 343$ K; vacuum	X reached up to 100%	[30]
	palmitic acid + isopropanol	NaA/mullite tube	*p*-toluene sulfonic acid	PVMR	$T = 368–395$ K; vacuum	X dramatically enhanced	[39]
	lactic acid + ethanol	NaA/C–TiO_2–ZrO_2 tube	*p*-toluene sulfonic acid	PVMR	$T = 70$ °C; vacuum	Y increased from 40% to >90%	[40]

(*Continued*)

Table 3.2 (*Cont'd*)

Reaction type	Reaction/reactants	Membrane/support	Catalyst	Reactor configuration	Operating conditions	Advantages	Ref.
Synthesis reactions	TBA + MeOH ⇔ MTBE + H_2O	Mordenite or NaA/Al_2O_3 or PSS tube, 6–8 μm	Cation exchange resin	PBMR	T = 334 K; N_2 sweep gas	Y = 67.6% above equilibrium 67.6%	[41]
	$CO_2 + 3H_2 \Leftrightarrow CH_3OH + H_2O$	LTA/α-Al_2O_3 tube	Cu-ZnO/Al_2O_3	PBMR	T = 210–250 °C; P = 24 bar; N_2 as sweep gas	Y = 8.7% higher than 2.4% in TR	[42]
Oxidation reactions	Propane → propylene $C_3H_8 + 1/2O_2 \rightarrow C_3H_6 + H_2O$	MFI and V-MFI/α-Al_2O_3 tube	/	CMR	T = 550–650 °C; P = 1.5–4 bar	Y = 8%; S = 40–50%	[43]
		Silicalite-1 α-Al_2O_3	V-MgO or Ni-based	CMR PBMR	T = 300–600 °C	Y increased in PBMR but no effect in CMR	[28]
	CO selective oxidation $CO + O_2 \Leftrightarrow CO_2$	Y-type zeolite/α-Al_2O_3 tube	Pt-zeolite	FTCMR	T = 200–250 °C; P = 4–6 bar	X_{CO} up to 100%	[44, 45]
	Combustion of VOCs (*n*-hexane)	ZSM-5 zeolite/γ- and α-Al_2O_3 tube	Pt-zeolite	FTCMR	T = 125–250 °C	Complete combustion above 210 °C	[46]
Isomerization reactions	Xylene isomerization 2 m-X ⇔ o-X + p-X	HZSM-5/PSS disk, 7 μm	HZSM-5	PBMR	T = 300–400 °C; N_2 as sweep gas	Y increased by 10%	[47, 48]
		MFI/Al_2O_3 tube	Pt-HZSM-5	CMR	T = 577 K; N_2 as sweep gas	S up to 100% over 58% in TR	[49, 50]
	Metathesis of propene $2C_3H_6 \Leftrightarrow C_2H_4 + C_4H_8$	silicalite-1/PSS plate	Re_2O_7/γ-Al_2O_3	PBMR	T = 296 K; He as sweep gas	X increased by 13% above equilibrium conversion	[51]

X = conversion; Y = yield; S = selectivity; TR = traditional reactor.

Using a hydrophilic zeolite membrane such as LTA and zeolite T to remove the water from the reaction system by sweep gas or vacuum (pervaporation), the esterification reaction would drive toward the product side with the result of enhanced yields. Beneficial aspects of PVMRs also include low energy consumption and the possibility of carrying out esterification at a selected temperature.

The esterification reaction can also be carried out in CMRs. The catalytic membranes are prepared by ion exchange modification. CMRs may outperform PVMRs in conversion with the same loading of catalyst dispersed in the liquid bulk [38].

Fischer–Tropsch (FT) synthesis enables the production of liquid synthetic hydrocarbons from various feedstocks such as coal, natural gas, and biomass:

$$\text{FT synthesis}: CO + 2H_2 \rightarrow (CH_2) + H_2O,\ \Delta H_r = -158\,kJ\,mol^{-1}$$

The selective removal of the by-product H_2O using zeolite membranes offers threefold benefits: (i) to boost the per pass conversion and reactor productivity; (ii) to reduce the H_2O-promoted catalyst deactivation; and (iii) to displace the WGS equilibrium to enhance the conversion of CO to hydrocarbons. The MR performance is mainly determined by the membrane selectivity toward H_2O [53, 54].

Other reactions associated with water formation as a by-product, such as methanol synthesis by CO_2 hydrogenation, can also be enhanced using zeolite MRs [42]:

$$CO_2 + 3H_2 \Leftrightarrow CH_3OH + H_2O,\ \Delta H^0_{r298} = -49.4\,kJ\,mol^{-1}$$

3.5.3 Oxidative Reactions

Zeolite membrane reactors applied to the oxidative dehydrogenation of alkanes are potentially useful for (i) controlling the oxygen feed, thereby limiting highly exothermic, total combustion and (ii) improving the contact between the reactant and the catalyst. Immobilizing transition metal ions in zeolites by ion exchange or by incorporation into the lattice leads to stable isolated and well-defined redox active catalytic sites.

CO selective oxidation is the primary method used for the purification of H_2-rich gas streams. A selective catalyst, typically Pt-supported, is used to avoid H_2 consumption:

$$2CO + O_2 \rightarrow 2CO_2$$

$$2H_2 + O_2 \rightarrow 2H_2O$$

In the flow-through CMRs, noble metal nanoparticles are entrapped in a thin zeolite membrane a few micrometers thick. The CO permeates through the zeolite catalytic membrane and reacts with oxygen. Such catalytic zeolite membranes demonstrate considerable advantages, including: (i) efficient reactant/catalytic phase contact, reducing the bypassing and misdistribution generally shown in a packed bed; (ii) short contact-time; and (iii) hot-spot control that is beneficial since a local temperature increase can lead to promotion of reverse WGS reactions.

3.5.4 Isomerization Reactions

Xylene isomerization is an important reaction in the petrochemical industry, by which more PX may be recovered from the less-used MX and OX isomers. During the past decade, continuous efforts have been directed toward improving the PX yield and selectivity in xylene isomerization and reducing the costs of PX separation from xylene isomers. The MR process integrates xylene isomerization and PX separation in a single MFI membrane reactor based on the shape selectivity of the zeolite membrane. By immediate and selective removal of PX from the reaction system, the reaction can be enhanced significantly toward high selectivity and yield. High-flux zeolite membranes are required to improve the efficiency of membrane reactors for xylene isomerization [47]. For successful implementation of this type of membrane reactor, it is necessary to synthesize b-oriented MFI zeolite membranes and introduce catalytically active sites (functional groups and subsequent oxidation to acid sites) into the membrane [2].

For the propene metathesis reaction in MRs, the membrane functions mainly as an extractor for the main products (iso-butene) to avoid further side reactions [51].

3.6 PROSPECTS AND CHALLENGES

Zeolite membranes represent a new branch of inorganic membranes. Significant progress has been made so far in the synthesis of thin, high-flux, defect-free zeolite membranes with new techniques of preparation and modification. In fact, the hydrophilic LTA zeolite membranes have already been applied in industry for dehydration from organic solutions

[5, 55]. However, few reports on the practical application of zeolite membranes in high-temperature CMRs exist. Commercial applications of zeolite membranes in reactions face a number of challenges, as follows [1].

1. *Synthesis of membranes with high permeability and selectivity, that is, oriented and thin zeolite membranes.* Optimal MR operation requires the membrane flux to be in balance with the reaction rate. A large number of factors – such as the support, organic additives, temperature, and profile – have a significant influence on the microstructure and overall quality of the membrane. However, the precise correlation between the synthesis procedure and conditions and the properties of the resultant zeolite membranes is not clear. In contrast, the majority of membranes synthesized so far are MFI-type zeolite membranes that have pore diameters ~5 Å, which are still too big to separate selectively small gaseous molecules. Zeolite membranes with pores in the ~3 Å range should be developed for membrane reactors, to separate small gas molecules on the basis of size exclusion. In addition, a method to produce zeolite membranes without non-zeolite pores or defects has to be found.
2. *Reproducibility and long-term stability of membrane performance.* Reliability and durability are key criteria in the process industry and one of the main obstacles for industrial zeolite membrane reactors. To date, the preparation of a desired zeolite membrane is often largely based on trial and error, which limits significantly the reproducibility of membrane preparation. Coke deposition, triggered to a large extent by the acidic nature of the zeolite framework, necessitates the possibility of repeated membrane regeneration and a technology for repairing defective membranes.
3. *Sealing at high temperature (>250 °C) and pressure.* It is generally acknowledged that the high-temperature application of zeolitic membranes is limited strongly by the high-temperature-resistant sealing of the module. Studies with the emphasis on high-temperature seals themselves are seldom reported in the literature.
4. *Up-scaling.* Up-scaling is faced with a number of challenges, such as the synthesis of large continuous membranes. In addition to the elucidation of the membrane formation mechanism, a closer look at temperature profiles and control in large synthesis reactors is necessary for large-scale membrane production. Membrane reactor modules have to be designed in order to support large-surface-area

membranes. The ease and effectiveness of sealing inside the module also need attention.

5. *Theoretical development.* Transport in zeolite membranes is a complex process. The multi-component transport and separation behavior through zeolitic and non-zeolitic pathways in the membranes at industrially relevant operating conditions have to be clarified.

NOTATION

Symbol	Description
d	pore diameter (m)
D	diffusion coefficient ($m^2 s^{-1}$)
D_0	pre-exponential coefficient in Arrhenius equation ($m^2 s^{-1}$)
E_a	activation energy for diffusion ($J mol^{-1}$)
J	molar flow ($mol m^{-2} s^{-1}$)
K	Langmuir constant (Pa^{-1})
K_0	pre-exponential factor for Langmuir constant
M	molecular weight ($kg mol^{-1}$)
p	pressure (Pa)
q	amount of adsorbed gas ($mol kg^{-1}$)
q_m	saturation amount adsorbed gas ($mol kg^{-1}$)
Q_a	heat adsorption ($J mol^{-1}$)
R	ideal gas constant ($8314 J mol^{-1} K^{-1}$)
ΔS	adsorption entropy ($J mol^{-1} K^{-1}$)
T	temperature (K)
z	distance coordinate (m)
Γ	thermodynamic factor
θ	occupancy ($\theta = q/q_m$)
ρ_s	density of zeolite ($kg m^{-3}$)
φ	geometric constant of pore structure
δ	membrane thickness (m)

Superscripts and subscripts

d	downstream
g	gas phase
0	pre-exponential factor
u	upstream
s	surface
K	Knudsen diffusion
i, j	i or j component

REFERENCES

[1] McLeary, E.E., Jansen, J.C. and Kapteijn, F. (2006) Zeolite based films, membranes and membrane reactors: Progress and prospects. *Microporous and Mesoporous Materials*, 90, 198–220.

[2] Fong, Y., Abdullah, A., Ahmad, A. and Bhatia, S. (2008) Development of functionalized zeolite membrane and its potential role as reactor combined separator for para-xylene production from xylene isomers. *Chemical Engineering Journal*, 139, 172–193.

[3] Lin, Y.S. (2001) Microporous and dense inorganic membranes: Current status and prospective. *Separation and Purification Technology*, 25, 39–55.

[4] Caro, J., Noack, M., Kolsch, P. and Schafer, R. (2000) Zeolite membranes – state of their development and perspective. *Microporous and Mesoporous Materials*, 38, 3–24.

[5] Caro, J. and Noack, M. (2008) Zeolite membranes – recent developments and progress. *Microporous and Mesoporous Materials*, 115, 215–233.

[6] Caro, J., Noack, M. and Kolsch, P. (2005) Zeolite membranes: From the laboratory scale to technical applications. *Adsorption*, 11, 215–227.

[7] Xiao, J. and Wei, J. (1992) Diffusion mechanism of hydrocarbons in zeolites – I. Theory, *Chemical Engineering Science*, 47, 1123–1141.

[8] Romero, J., Gijiu, C., Sanchez, J. and Rios, G.M. (2004) A unified approach of gas, liquid and supercritical solvent transport through microporous membranes. *Chemical Engineering Science*, 59, 1569–1576.

[9] Wirawan, S.K., Creaser, D., Lindmark, J., Hedlund, J., Bendiyasa, I.M. and Sediawan, W.B. (2011) H_2/CO_2 permeation through a silicalite-1 composite membrane. *Journal of Membrane Science*, 375, 313–322.

[10] Sommer, S., Melin, T., Falconer, J.L. and Noble, R.D. (2003) Transport of C_6 isomers through ZSM-5 zeolite membranes. *Journal of Membrane Science*, 224, 51–67.

[11] Hanebuth, M., Dittmeyer, R., Mabande, G.T.P. and Schwieger, W. (2005) On the combination of different transport mechanisms for the simulation of steady-state mass transfer through composite systems using permeation through stainless steel supported silicalite-1 membranes as a model system. *Catalysis Today*, 104, 352–359.

[12] Wee, S.-L., Tye, C.-T. and Bhatia, S. (2008) Membrane separation process – pervaporation through zeolite membrane. *Separation and Purification Technology*, 63, 500–516.

[13] Tsai, C.-Y., Tam, S.-Y., Lu, Y. and Brinker, C.J. (2000) Dual-layer asymmetric microporous silica membranes. *Journal of Membrane Science*, 169, 255–268.

[14] Ahn, H., Lee, H., Lee, S.-B. and Lee, Y. (2006) Pervaporation of an aqueous ethanol solution through hydrophilic zeolite membranes. *Desalination*, 193, 244–251.

[15] Tiscareño-Lechuga, F., Téllez, C., Menéndez, M. and Santamarí, J. (2003) A novel device for preparing zeolite-A membranes under a centrifugal force field. *Journal of Membrane Science*, 212, 135–146.

[16] Xu, X., Bao, Y., Song, C., Yang, W., Liu, J. and Lin, L. (2005) Synthesis, characterization and single gas permeation properties of NaA zeolite membrane. *Journal of Membrane Science*, 249, 51–64.

[17] Sato, K. and Nakane, T. (2007) A high reproducible fabrication method for industrial production of high flux NaA zeolite membrane. *Journal of Membrane Science*, 301, 151–161.

[18] Boudreau, L.C., Kuck, J.A. and Tsapatsis, M. (1999) Deposition of oriented zeolite A films: In situ and secondary growth. *Journal of Membrane Science*, 152, 41–59.

[19] Jeong, H.K., Krohn, J., Sujaoti, K. and Tsapatsis, M. (2002) Oriented molecular sieve membranes by heteroepitaxial growth. *Journal of the American Chemical Society*, 124, 12966–12968.

[20] Mabande, G.T.P., Ghosh, S., Lai, Z., Schwieger, W. and Tsapatsis, M. (2005) Preparation of b-oriented MFI films on porous stainless steel substrates. *Industrial and Engineering Chemistry Research*, 44, 9086–9095.

[21] Pera-Titus, M., Bausach, M., Llorens, J. and Cunill, F. (2008) Preparation of inner-side tubular zeolite NaA membranes in a continuous flow system. *Separation and Purification Technology*, 59, 141–150.

[22] Matsufuji, T., Nishiyama, N., Ueyama, K. and Matsukata, M. (2000) Permeation characteristics of butane isomers through MFI-type zeolitic membranes. *Catalysis Today*, 56, 265–273.

[23] Alfaro, S., Arruebo, M., Coronas, J. and Menéndez, M. (2001) J. Santamaría, Preparation of MFI type tubular membranes by steam-assisted crystallization. *Microporous and Mesoporous Materials*, 50, 195–200.

[24] Matsufuji, T., Nishiyama, N., Matsukata, M. and Ueyama, K. (2000) Separation of butane and xylene isomers with MFI-type zeolitic membrane synthesized by a vapor-phase transport method. *Journal of Membrane Science*, 178, 25–34.

[25] Li, Y., Chen, H., Liu, J. and Yang, W. (2006) Microwave synthesis of LTA zeolite membranes without seeding. *Journal of Membrane Science*, 277, 230–239.

[26] Huang, A. and Yang, W. (2007) Hydrothermal synthesis of NaA zeolite membrane together with microwave heating and conventional heating. *Materials Letters*, 61, 5129–5132.

[27] Illgen, U., Schäfer, R., Noack, M., Kölsch, P., Kühnle, A. and Caro, J. (2001) Membrane supported catalytic dehydrogenation of iso-butane using an MFI zeolite membrane reactor. *Catalysis Communications*, 2, 339–345.

[28] Pantazidis, A., Dalmon, J.A. and Mirodatos, C. (1995) Oxidative dehydrogenation of propane on catalytic membrane reactors. *Catalysis Today*, 25, 403–408.

[29] Jeong, B.-H., Sotowa, K.-I. and Kusakabe, K. (2003) Catalytic dehydrogenation of cyclohexane in an FAU-type zeolite membrane reactor. *Journal of Membrane Science*, 224, 151–158.

[30] Tanaka, K., Yoshikawa, R., Ying, C., Kita, H. and Okamoto, K. (2001) Application of zeolite membranes to esterification reactions. *Catalysis Today*, 67, 121–125.

[31] Lai, S.M., Ng, C.P., Martin-Aranda, R. and Yeung, K.L. (2003) Knoevenagel condensation reaction in zeolite membrane microreactor. *Microporous and Mesoporous Materials*, 66, 239–252.

[32] Nishiyama, N., Miyamoto, M., Egashira, Y. and Ueyama, K. (2001) Zeolite membrane on catalyst particle for selective formation of *p*-xylene in disproportionation of toluene. *Chemical Communications*, 18, 1746–1747.

[33] Nishiyama, N., Ichioka, K., Park, D.-H. *et al.* (2004) Reactant selective hydrogenation over composite silicalite-1-coated Pt/TiO_2 particles. *Industrial and Engineering Chemistry Research*, 43, 1211–1215.

[34] Kong, C., Lu, J., Yang, J. and Wang, J. (2007) Catalytic dehydrogenation of ethylbenzene to styrene in a zeolite silicalite-1 membrane reactor. *Journal of Membrane Science*, 306, 29–35.

[35] Ciavarella, P., Casanave, D., Moueddeb, H., Miachon, S., Fiaty, K. and Dalmon, J.A. (2001) Isobutane dehydrogenation in a membrane reactor – influence of the operating conditions on the performance. *Catalysis Today*, 67, 177–184.
[36] van den Bergh, J., Gücüyener, C., Gascon, J. and Kapteijn, F. (2011) Isobutane dehydrogenation in a DD3R zeolite membrane reactor. *Chemical Engineering Journal*, 166, 368–377.
[37] Zhang, Y., Wu, Z., Hong, Z., Gu, X. and Xu, N. (2012) Hydrogen-selective zeolite membrane reactor for low temperature water gas shift reaction. *Chemical Engineering Journal*, 197, 314–321.
[38] Bernal, M.P., Coronas, J., Menendez, M. and Santamaria, J. (2002) Coupling of reaction and separation at the microscopic level: Esterification processes in a H-ZSM-5 membrane reactor. *Chemical Engineering Science*, 57, 1557–1562.
[39] Peng, X., Lei, Q., Lv, G. and Zhang, X. (2012) Zeolite membrane-assisted preparation of high fatty esters. *Separation and Purification Technology*, 89, 84–90.
[40] Jafar, J.J., Budd, P.M. and Hughes, R. (2002) Enhancement of esterification reaction yield using zeolite – a vapour permeation membrane. *Journal of Membrane Science*, 199, 117–123.
[41] Salomon, M.A., Coronas, J., Menendez, M. and Santamaria, J. (2000) Synthesis of MTBE in zeolite membrane reactors. *Applied Catalysis A: General*, 200, 201–210.
[42] Gallucci, F., Paturzo, L. and Basile, A. (2004) An experimental study of CO_2 hydrogenation into methanol involving a zeolite membrane reactor. *Chemical Engineering and Processing: Process Intensification*, 43, 1029–1036.
[43] Julbe, A., Farrusseng, D., Jalibert, J.C., Mirodatos, C. and Guizard, C. (2000) Characteristics and performance in the oxidative dehydrogenation of propane of MFI and V-MFI zeolite membranes. *Catalysis Today*, 56, 199–209.
[44] Bernardo, P., Algieri, C., Barbieri, G. and Drioli, E. (2006) Catalytic zeolite membrane reactors for the selective CO oxidation. *Desalination*, 200, 702–704.
[45] Bernardo, P., Algieri, C., Barbieri, G. and Drioli, E. (2008) Hydrogen purification from carbon monoxide by means of selective oxidation using zeolite catalytic membranes. *Separation and Purification Technology*, 62, 629–635.
[46] Aguado, S., Coronas, J. and Santamaría, J. (2005) Use of zeolite membrane reactors for the combustion of VOCs present in air at low concentrations. *Chemical Engineering Research and Design*, 83, 295–301.
[47] Zhang, C., Hong, Z., Gu, X.H., Zhong, Z.X., Jin, W.Q. and Xu, N.P. (2009) Silicalite-1 zeolite membrane reactor packed with HZSM-5 catalyst for meta-xylene isomerization. *Industrial and Engineering Chemistry Research*, 48, 4293–4299.
[48] Haag, S., Hanebuth, M., Mabande, G.T.P., Avhale, A., Schwieger, W. and Dittmeyer, R. (2006) On the use of a catalytic H-ZSM-5 membrane for xylene isomerization. *Microporous and Mesoporous Materials*, 96, 168–176.
[49] Dyk, L.v., Lorenzen, L., Miachon, S. and Dalmon, J.-A. (2005) Xylene isomerization in an extractor type catalytic membrane reactor. *Catalysis Today*, 104, 274–280.
[50] Daramola, M.O., Burger, A.J., Giroir-Fendler, A., Miachon, S. and Lorenzen, L. (2010) Extractor-type catalytic membrane reactor with nanocomposite MFI-alumina membrane tube as separation unit: Prospect for ultra-pure para-xylene production from m-xylene isomerization over Pt-HZSM-5 catalyst. *Applied Catalysis A: General*, 386, 109–115.

[51] van de Graaf, J.M., Zwiep, M., Kapteijn, F. and Moulijn, J.A. (1999) Application of a zeolite membrane reactor in the metathesis of propene. *Chemical Engineering Science*, 54, 1441–1445.

[52] Casanave, D., Ciavarella, P., Fiaty, K. and Dalmon, J.A. (1999) Zeolite membrane reactor for isobutane dehydrogenation: Experimental results and theoretical modelling. *Chemical Engineering Science*, 54, 2807–2815.

[53] Rohde, M.P., Unruh, D. and Schaub, G. (2005) Membrane application in Fischer–Tropsch synthesis reactors – overview of concepts. *Catalysis Today*, 106, 143–148.

[54] Espinoza, R.L., du Toit, E., Santamaria, J., Menendez, M., Coronas, J. and Irusta, S. (2000) Use of membranes in Fischer–Tropsch reactors, in Studies in Surface Science and Catalysis (eds F.V.M.S.M. Avelino Corma and G.F. José Luis), Elsevier, Amsterdam, pp. 389–394.

[55] Morigami, Y., Kondo, M., Abe, J., Kita, H. and Okamoto, K. (2001) The first large-scale pervaporation plant using tubular-type module with zeolite NaA membrane. *Separation and Purification Technology*, 25, 251–260.

4

Dense Metallic Membrane Reactors

4.1 INTRODUCTION

Dense metallic membranes are typified by palladium (Pd) membrane for hydrogen permeation and silver (Ag) membrane for oxygen permeation. Pd-based membranes possess very reasonable hydrogen permeability, but the oxygen permeability of Ag-based membranes is orders of magnitude lower than the hydrogen permeability of Pd-based membranes. In fact, crystalline Pd-alloy such as Pd/Ag (23–25% Ag) membranes are commercially available by virtue of their high permeability and stability, and have been used for decades to provide ultrapure hydrogen, for example, in the semiconductors industry and for the operation of fuel cells. In recent years, there has been growing interest in the application of MRs for various reactions toward higher conversion and yield improvement. Considerable effort has been made to improve membrane performance (high hydrogen permeability and perm-selectivity combined with good mechanical/thermal and long-term stability) and lower membrane costs.

This chapter focuses mainly on Pd-based MRs with respect to the gas permeation mechanism, membrane preparation, MR construction and operation, as well as applications in a variety of chemical reactions. In addition to a general description of Pd membranes and MRs, recent progress and critical issues in the dense metal membrane area will also be presented at the end of this chapter.

Inorganic Membrane Reactors: Fundamentals and Applications, First Edition. Xiaoyao Tan and Kang Li.

4.2 GAS PERMEATION IN DENSE METALLIC MEMBRANES

4.2.1 Types of Dense Metallic Membranes

Metals exhibit a large spread of hydrogen permeability values in the temperature range of interest for MR applications (200–700°C) [1]. Refractory metals such as Nb, Ta, and V have high permeability values, but their use is limited by their high hydrogen embrittlement and high reactivity with gases, forming dense inorganic layers (oxides, nitrides, etc.) over their surface and reducing considerably the hydrogen permeation. In practice, Pd and its alloys are mostly used to form self-supporting or supported composite dense metal membranes for hydrogen permeation [2]. Alloying Pd with Ag, Cu, Au, Y, and so on reduces the critical temperature for hydrogen embrittlement and leads to an increase in hydrogen permeability. The maximal value of hydrogen flow is reached for an alloy with approximately 23 wt% silver.

4.2.1.1 Self-Supporting Palladium-Based Membranes

Early work focused on the application of unsupported dense Pd membranes prepared by conventional metallurgical processes. These self-supporting membranes are in the form of tubes or foils, and possess wall thicknesses greater than 50–100 μm to maintain sufficient mechanical strength. This leads to low hydrogen flux but a high cost of the membrane (precious Pd and Pd alloys), which limits their industrial applications.

4.2.1.2 Supported Palladium-Based Composite Membranes

In order to meet the membrane performance targets of H_2 flux exceeding 1 mol m^{-2} s^{-1} and H_2 permeate purity of at least 99.99%, with a membrane cost target of $150/ft^2 for commercial applications, the membrane thickness should be reduced to about 1–10 μm without the creation of cracks and pinholes [3]. This can be realized through supported composite membranes prepared by various techniques. The thin Pd membrane provides high hydrogen flux, while the porous support provides satisfactory mechanical strength.

Palladium-based composite membranes consist of a thin dense layer of palladium or a palladium alloy on a porous support. Tubes or disks of porous glass, porous ceramics, or PSS can all be employed as supports.

Usually, an intermediate layer of reduced porosity is used to improve the metallic membrane quality as well as the adhesion between the metal and the support. When metal supports (stainless steel, nickel, etc.) are used, the thermal mismatching between the Pd alloy films and the porous substrate can be reduced.

4.2.1.3 *Metal Membranes with Low or No Pd Contents*

Amorphous Ni-based alloy membranes are composed of inexpensive Ni and early transition metals (ETMs) such as Ta, Nb, Hf, Zr, Ti with approximate compositions of Ni(60–70%)–ETM(40–30%). Such membranes demonstrate a significant performance/cost saving over Pd-based alloys, although Pd-alloy catalyst layers at both surfaces (typically 500 nm or less) are still required to promote the dissociation of hydrogen and prevent inhibition by CO, H_2S, and steam, and thus of particular interest for the large-scale production of hydrogen from carbon-based fuels. The membrane performance may be degraded by two thermally dependent processes: (i) crystallization, leading to a reduction in hydrogen permeability and mechanical strength and (ii) diffusion of Pd catalyst into bulk amorphous alloy, resulting in a decrease in rate of hydrogen dissociation and potential inhibition of the membrane surface. The hydrogen permeability of these alloy materials is generally inversely proportional to their thermal stability. Therefore, the design of such alloy membranes is a compromise between hydrogen production rate and durability. Ni–Nb–Zr amorphous alloys demonstrate prospective materials for use due to their noticeable hydrogen permeability, $2.0–2.5 \times 10^{-9}\,mol\,m^{-1}\,s^{-1}\,Pa^{-0.5}$ at 400°C, as well as good stability [4]. Metal molybdenum (Mo) and Mo–carbon membranes exhibit high catalytic activity toward dehydrogenation propane in addition to the selective hydrogen permeation [5]. Although Ni-based amorphous alloy membranes are attractive due to their low cost, satisfactory hydrogen perm-selectivity, and high catalytic activities, they can be used only below their crystallization temperature to maintain good separation performance.

4.2.2 Hydrogen Permeation Mechanism in Pd-Based Membranes

Hydrogen permeation through Pd-based membranes is based on the dissolution of H_2 in metal Pd to form a Pd–H alloy and the diffusion of the H atoms across the membrane, which is described by the solution/diffusion

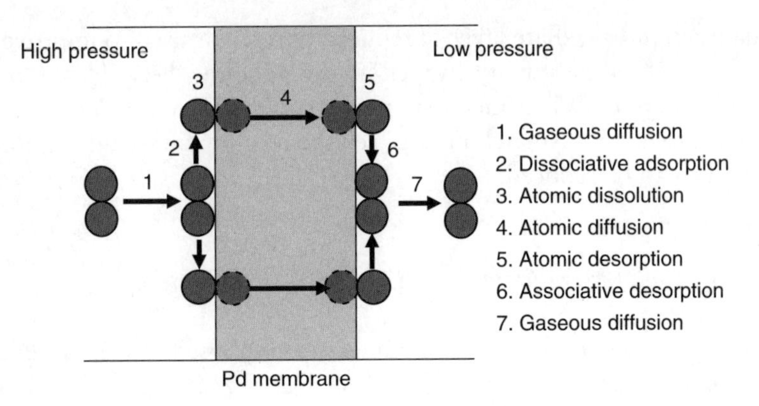

Figure 4.1 Schematic mechanism of hydrogen permeation in dense Pd-based membranes.

mechanism. As depicted schematically in Figure 4.1, the hydrogen permeation process consists of the following steps:

1. Hydrogen diffusion from the bulk gas to the membrane surface.
2. H_2 chemically adsorbed on the membrane surface and dissociated into atomic H.
3. Migration of atomic H on the membrane surface into the membrane bulk (dissolution).
4. Diffusion of atomic H through the Pd lattice (diffusion).
5. Transition of atomic H from the bulk metal to the membrane surface on the permeate side.
6. Associative desorption of atomic H into H_2 molecules.
7. Diffusion of H_2 from the membrane surface into the bulk gas.

In general, the external mass transfer resistance is much smaller than the others, and steps 1 and 7 can be negligible. At temperatures higher than 150°C (achieved in practical applications), the dissociative adsorption (steps 2–3) and associative desorption (steps 5–6) may be approximately at thermodynamic equilibrium. When the membrane thickness is larger than 1 μm, H-atom diffusion through the bulk of the Pd (step 4) may be the rate-limiting step for hydrogen permeation. But at low temperatures or if the Pd film thickness is very small (<1 μm), H_2 desorption may become the rate-limiting step.

When the diffusion of H atoms through the membrane bulk is the rate-limiting step, Fick's first law can be used to describe the H_2 permeation flux:

$$J_H = -D_H \frac{dc_H}{dz}$$

or

$$J_{H_2} = -\frac{D_H}{2}\frac{dc_H}{dz} \tag{4.1}$$

where J_{H_2} is the H_2 flux (mol m^{-2} s^{-1}), D_H the diffusivity of the H-atom in the membrane (m^2 s^{-1}), and c_H the H concentration in the membrane bulk (mol m^{-3}).

At elevated temperatures (>150°C), an instant thermodynamic equilibrium exists on the membrane surface between the H_2 in gas phase and the atomic H dissolved in the membrane bulk:

$$H_2 \xleftrightarrow{K_e} 2H \tag{4.2}$$

At low H_2 pressures, the relation between the dissolved H concentration on the membrane surface and the H_2 pressure in the gas phase can be given by

$$K_e = \frac{c_H^2}{p_{H_2}} \tag{4.3}$$

where K_e is the equilibrium constant and p is the hydrogen partial pressure in the gas phase (Pa).

Therefore, the boundary conditions for the diffusion equation are given by

$$z=0,\ c_{H,up} = \sqrt{K_e p_{H_2,up}^{0.5}};\ z = \delta,\ c_{H,down} = \sqrt{K_e p_{H_2,down}^{0.5}} \tag{4.4}$$

Integration of Eq. (4.1) over the membrane thickness gives the hydrogen permeation flux as

$$J_{H_2} = \frac{P}{\delta}\cdot\left(p_{H_2,up}^{0.5} - p_{H_2,down}^{0.5}\right) \tag{4.5}$$

Equation (4.5) is Sieverts' equation, where δ is the membrane thickness (m), P is the hydrogen permeability of the membrane (mol m^{-1} s^{-1} Pa$^{-0.5}$) expressed by $P = D_H\sqrt{K_e}/2$, $p_{H_2,up}$ and $p_{H_2,down}$ are the partial pressures of hydrogen in the upstream and downstream (Pa), respectively.

The permeability can be expressed by the product of diffusivity and hydrogen solubility in the Pd lattice:

$$P = D_H\sqrt{K_e}/2 = D_H \cdot S_{H_2} \tag{4.6}$$

in which

$$D_H = D_{H,0} \cdot \exp\left(\frac{-E_d}{RT}\right) \tag{4.7}$$

$$S_{H_2} = S_{H_2,0} \cdot \exp\left(\frac{\Delta H_a}{RT}\right), \tag{4.8}$$

where E_d is the activation energy for the diffusion hydrogen atoms (J mol^{-1}), R is the universal gas constant (J mol^{-1}K^{-1}), T is the absolute temperature (K), and ΔH_a is the enthalpy of hydrogen absorption in the Pd membrane (J mol^{-1}).

As can be seen, the hydrogen permeation flux is inversely proportional to the membrane thickness. A thin film of Pd membrane is always preferred in practical applications. Since only hydrogen is allowed to dissolve in the membrane, the Pd-based membranes may exhibit a very high perm-selectivity to hydrogen.

It is noteworthy that Sieverts' law – Eq. (4.5) – is valid only at low H_2 pressures, otherwise deviations may occur. In addition, if the Pd membrane is defective and allows for the passage of other gases, or the Pd membrane surface has low activity due to the adsorption of contaminants, the permeation behavior of the membrane also deviates from Sieverts' law. In a general form, the hydrogen permeation flux in Pd-based membranes can be expressed by Richardson's equation:

$$J_{H_2} = \frac{P_0 \exp(-E_a / RT)}{\delta}\left(p^n_{H_2,up} - p^n_{H_2,down}\right) \tag{4.9}$$

where $P_0 = D_{H,0} \cdot S_{H_2,0}$ is the pre-exponential factor of permeability, $E_a = E_{diff} - \Delta H_a$ is the activation energy of H_2 permeation in Pd (J mol^{-1}), and n is the pressure factor depending on the operating pressure and the membrane surface properties. In general, the values of n are between 0.5 and 1. When operated at low pressure, n is equal to 0.5 and Eq. (4.9) becomes Sieverts' law.

Values of the permeation parameters – that is, the pre-exponential factor P_0, activation energy E_a, and pressure factor n – may be quite different depending on the membrane properties and measuring conditions. They are obtained by the permeation measurement carried out at different transmembrane pressure gradients and temperatures. Figure 4.2 shows the experimental setup for permeation measurement. The pressure of the H_2 feed is controlled using a pressure regulator, while N_2 is used as the sweep gas. The membrane tube is glazed except

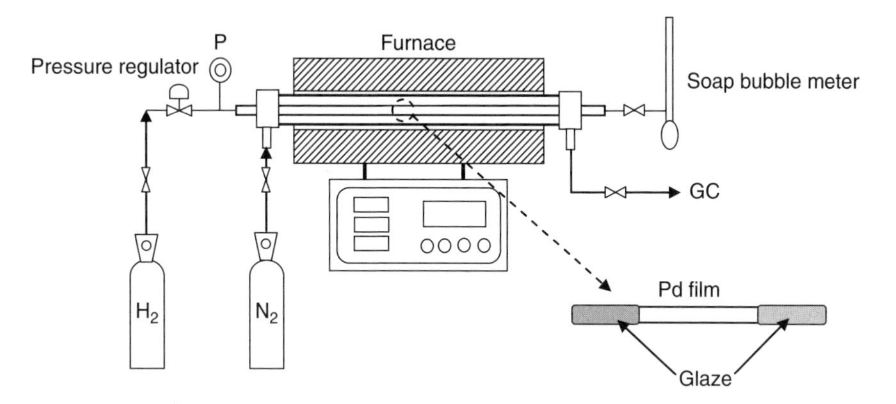

Figure 4.2 Experimental setup for hydrogen permeation measurement.

Table 4.1 Literature values of the H_2 permeation kinetic parameters of Pd membranes

Membrane	Thickness (μm)	Temperature (°C)	P_0 (mmol m^{-2} s^{-1} Pa^{-n})	E_a (kJ mol^{-1})	n	Ref.
Pd–Ag; self-supported tube	60	200–300	7.2×10^{-6}	2.90	0.5	[6]
Pd–Ag; self-supported tube	50	400	3.61×10^{-8}	8.58	0.5	[7]
Pd–Ag(25 wt%) tube	50	200–300	5.44×10^{-8}	10.72	0.5	[8]
Pd–Au/porous Ni disk	4	350–390	5.75×10^{-9}	14.62	0.65	[9]
Pd–Ag/α-Al_2O_3	3.9	225–300	7.76×10^{-8}	15.70	0.5	[10]
Pd/γ-Al_2O_3–α-Al_2O_3	15	350–650	1.31×10^{-11}	10.0	0.61	[2]
Pd–Cu/α-Al_2O_3	11	450	2.55×10^{-10}	—	1.0	[11]

for the central section, which has a constant heating temperature so that the membrane area for permeation can be calculated accurately. Some literature values of the permeation parameters are listed in Table 4.1.

As can be seen from Table 4.1, the permeation parameters are quite different for various membranes and depend on the measurement conditions [12]. For example, both the pre-exponential factor and the activation energy are increased noticeably under reaction conditions compared with pure hydrogen permeation [6]. The factors affecting the membrane permeation parameters include:

1. *Membrane thickness.* As the membrane thickness decreases, surface processes like hydrogen adsorption and dissociation/association

will play a more important role in permeation, resulting in deviation of the hydrogen pressure exponent from the Sieverts' value of 0.5 [13]. It has been claimed that the surface processes control the permeance of very thin Pd films, and for ultrathin Pd layers the values of n are close to 1.0 [2].

2. *Membrane surface morphology.* The rough surface of Pd membranes can lead to a noticeable increase in surface area and thus in hydrogen permeability. For example, H_2 permeability through the Pd–Ru 6 wt% tubular membrane can be enhanced by a factor of 2 due to the formation of an active porous layer after boiling the membrane in CCl_4, rinsing with HCl, and annealing in H_2 at 700–800°C [2].
3. *Surface poisoning.* Surface poisoning, which is caused by the adsorption of contaminants such as C, CO, CO_2, or hydrocarbons on the membrane surface, results in a decrease in the adsorption rate of H_2. If the poisoning is too severe, the dissociative reactions at the surface may dominate the H_2 permeation.
4. *Presence of leaks.* The presence of defects in Pd membranes may result in Knudsen viscous flow, leading to an n factor higher than 0.5. Moreover, the Knudsen viscous mechanism is characterized by the activation energy of gas diffusion, which is lower than that of solution diffusion. As a result, the activation energy for H_2 permeation in the presence of leakage will be lower than normal values.
5. *Support resistance and high operating pressure.* For supported Pd membranes where the porous support provides a prominent resistance to H_2 permeation, the pressure factor n is higher than 0.5 and the activation energy for H_2 permeation is decreased.

It should be noted that the pressure factor is mostly an empirical parameter because it is determined from measured fluxes by regression with a prescribed pressure exponent of 0.5 or 1, but the sensitivity of the flux to the pressure exponent is weak.

4.2.3 Effect of Substrate on H_2 Permeation

Pd membranes are usually formed on a porous substrate. The porous substrate mainly provides mechanical strength for the Pd composite membrane, but also influences the hydrogen permeation [14]. Figure 4.3 shows schematically the hydrogen permeation through a composite Pd

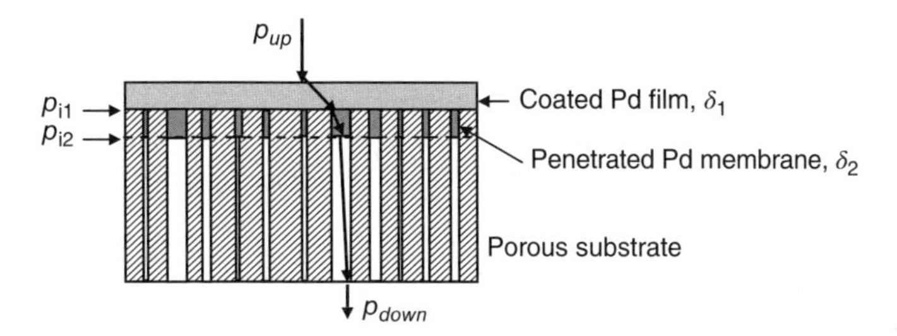

Figure 4.3 Scheme of hydrogen permeation through a composite Pd membrane.

membrane where three layers can be distinguished: the Pd thin film, the penetrated layer, and the porous support. Formation of the penetrated layer is due to the entry of Pd into a depth of the pores during the membrane preparation.

Therefore, the hydrogen permeation flux through the composite membrane may be given by

$$J_{H_2} = \frac{P_{H_2}}{\delta_1}\left(p_{up}^n - p_{i,1}^n\right) = \frac{P_{H_2}\varepsilon_p}{\delta_2}\left(p_{i,1}^n - p_{i,2}^n\right) = Pe_{sub}\left(p_{i,2} - p_{down}\right) \quad (4.10)$$

where $p_{i,1}$ and $p_{i,2}$ are the hydrogen partial pressures on the interfaces; δ_1 and δ_2 are the thicknesses of the surface-coated Pd layer and the penetrated Pd depth, respectively; ε_p is the surface porosity of the substrate for Pd penetration; and Pe_{sub} is the H_2 permeance in the porous substrate, which may be calculated based on the microstructure.

The effect of the substrate on the hydrogen permeation through the composite membrane can be evaluated by the substrate influence factor, defined by the ratio of the pressure drop across the substrate to the total pressure difference across the composite membrane:

$$\eta = \frac{p_{i,1} - p_{down}}{p_{up} - p_{down}} \quad (4.11)$$

When $\eta \to 0$, the influence of the substrate can be negligible, whereas the permeation is controlled by the substrate when $\eta \to 1$.

Figure 4.4 shows the hydrogen permeation flux as a function of temperature for composite membranes with different effective porosities [15]. When the effective porosity is low, for example, $\varepsilon_p/L_p = 20\ \text{m}^{-1}$, the hydrogen permeation flux decreases as the permeation temperature is increased. This implies that for this effective porosity the permeation

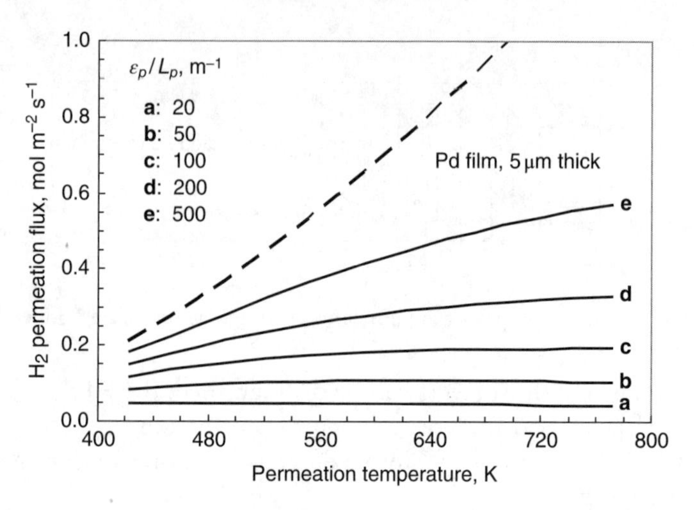

Figure 4.4 Hydrogen permeation flux through Pd/Al_2O_3 composite hollow fiber membranes as a function of temperature for substrates with different effective porosities. Reproduced from [15]. With permission from Elsevier.

rate is dominated by the permeation in the substrate. With the effective porosity increased to above 100 m^{-1}, the hydrogen permeation flux increases with increasing permeation temperature, suggesting that the permeation in the Pd film also plays an important role in the process. However, as the temperature increases, the role of the permeation in the Pd film becomes less dominant. This can be seen from Figure 4.4, where the solid lines deviate from the dashed line more and more as the permeation temperature is increased.

4.3 PREPARATION OF DENSE METALLIC MEMBRANES

4.3.1 Cold-Rolling and Diffusion Welding Method

The cold-rolling and diffusion welding method is used to produce self-supporting metal membranes [1]. The rolled Pd/Ag (23–25 wt% Ag) foils are welded by operating a thermal treatment at 1000°C for 1–2 h in a furnace under vacuum or inert gas (Ar, He) atmosphere. The high diffusion of silver atoms into the metal lattice allows the metal parts to be joined. Figure 4.5 shows the welding device for a thermomechanical press to join thin-wall Pd/Ag permeators. It consists of stainless steel sheets closed by a threatened bar made of a metal alloy having a negligible thermal expansion coefficient. Metal foils of 50–60 μm thickness are wrapped around an alumina bar and then the limbs of the foil to be

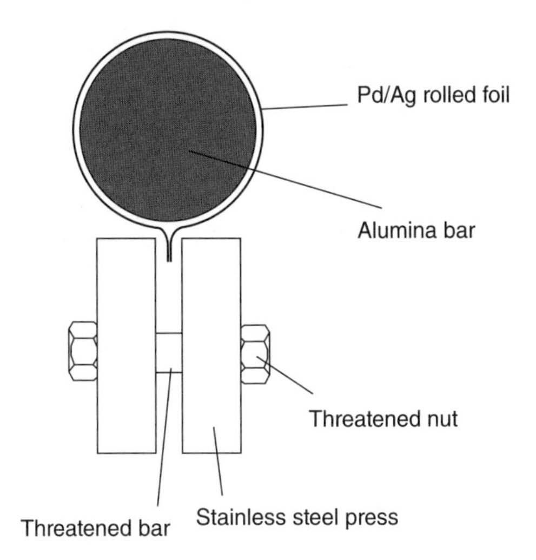

Figure 4.5 Thermomechanical press for joining thin-walled Pd–Ag tubes via diffusion welding. Reproduced from [1]. With permission from Elsevier.

joined are constrained through a thermomechanical press. By increasing the temperature, all materials expand except for the threatened bar which applies the pressure needed to ensure the welding of the Pd/Ag foil. Such a membrane wall thickness permits operation at 300–400°C with a transmembrane pressure of 200–330 kPa. Moreover, the resultant metal membrane is dense and defect-free, thus exhibiting very high hydrogen selectivity in addition to significant hydrogen permeation fluxes.

4.3.2 Electroless Plating Method

The electroless plating (ELP) method is the most applied method for the preparation of Pd-based membranes since it can form a uniform deposition on complex shapes and large substrate areas with simple equipment. It is based on the autocatalytic reduction of a metastable metallic salt complex on an activated substrate surface. The details of the ELP process and the typical operating conditions are illustrated in Figure 4.6. Prior to plating, the specified surface of the substrate has to be cleaned by ultrasonic rinsing with a detergent solution composed of ammonia, isopropyl alcohol, and deionized water for a ceramic substrate. In case of PSS tubes, the cleaning sequence should be extended by ultrasonic rinsing in acetone to remove adherent contaminants like oil or grease. The substrate is then activated by immersion into an acidic $SnCl_2$ sensitization solution followed by the Pd nuclei seeding in an acidic $PdCl_2$

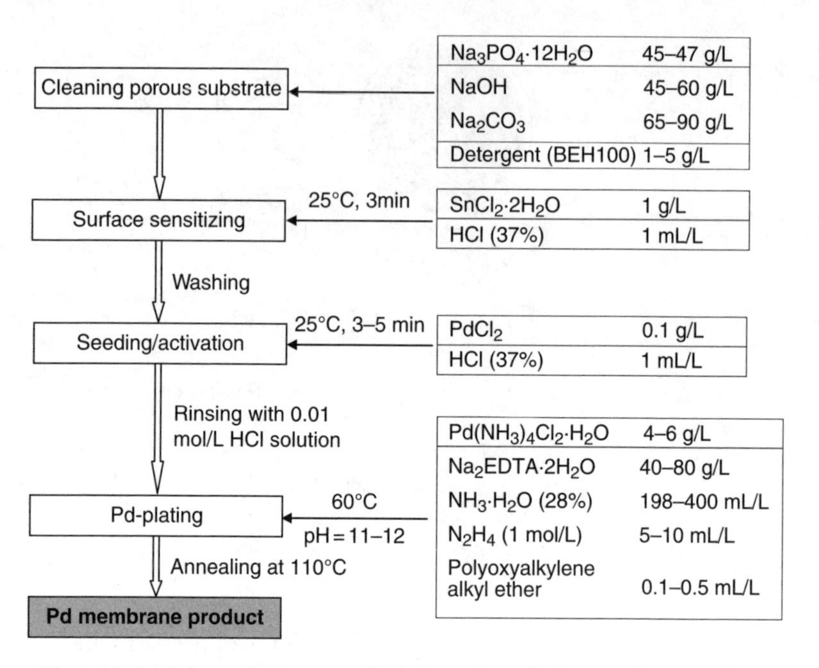

Figure 4.6 Schematic process of Pd electroless plating on the PSS tube.

solution. This activation process is repeated until the surface turns dark brown. A copious amount of deionized water is used to rinse the substrates between immersions. The ELP is carried out at around 60°C in the plating bath, with the main component $PdCl_2$ as palladium source and hydrazine as reducing agent. For tubular substrates, they have to be turned upside down after each plating step so as to avoid the generation of defects at the top end caused by ascending bubbles in the upper part of the tubes. Drying of the plated membranes at 110°C, at least after every second plating step, is essential for a successful coating, as it greatly improves the adhesive strength of the palladium layer. Figure 4.7 shows the morphology of the Pd/Al_2O_3 membrane by the ELP method.

Although ELP exhibits the advantages of ease to scale up and flexibility to coat metal film on supports of different geometry, the composition of the deposited membranes is difficult to control. Furthermore, to make defect-free membranes with submicrometer thickness depends strongly on the perfection of the porous supports. Any imperfections – like particulates on the substrate, non-uniformities in pore size, and so on – would result in failure of formation of defect-free membranes. Therefore, the surface of commercially available supports should generally be modified by very precise techniques such as sol-gel to form an intermediate layer with reduced pore size in advance of coating. For high selectivity, the ELP-derived Pd membranes usually have a large thickness (>5 μm).

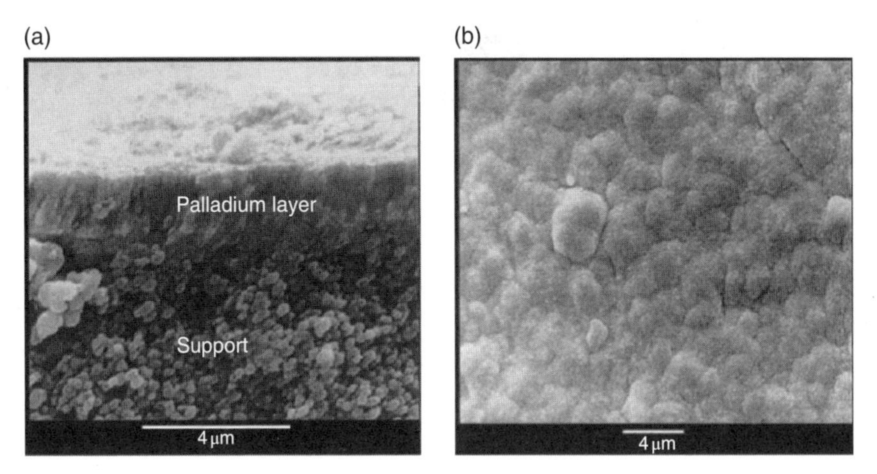

Figure 4.7 SEM images of a composite Pd–ceramic membrane prepared by the electroless plating method: (a) cross-section; (b) palladium surface. Reproduced from [16]. With permission from Elsevier.

In order to improve the adhesion between the metal and the support in addition to good flux and perm-selectivity, a modified ELP method has been developed to form "Pd nanopore" composite membranes [3]. The synthesis process includes the addition of Pd nucleation sites onto a γ-Al_2O_3 layer, followed by another coating of a γ-Al_2O_3 layer for Pd deposition inside, as depicted in Figure 4.8. This sequence of steps ensures that the pores are filled completely, irrespective of the pore size of the substrate. Depending on the duration of the ELP step, Pd film growth may proceed beyond the pore mouths of the top alumina layer.

In order to reduce the thermal mismatch between the Pd alloy film and the porous substrate, metal porous supports (stainless steel, nickel, etc.) are used with an intermediate layer to reduce the intermetallic diffusion (with consequent poisoning) between the metal porous support and the Pd/Ag layer. The intermediate layer can be a ceramic or a porous Pd/Ag layer prepared by the bi-metal multi-layer deposition technique [17]. The resulting membranes demonstrate high operating temperature (over 500°C) and long-term durability.

4.3.3 Electroplating Method

The electroplating (EP) is generally applied for the deposition of thin Pd membranes on metal substrates such as PSS. Cleaning, sensitizing, and activating steps are essential before the EP process. The electroplating is performed in an acidic Pd chloride solution with supplemented

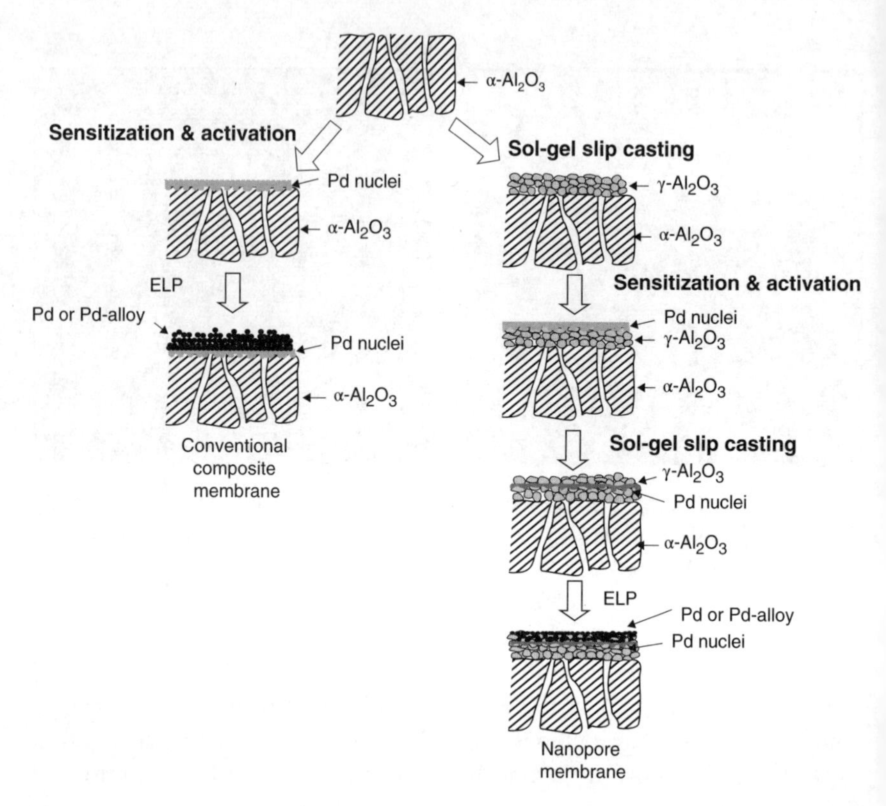

Figure 4.8 Comparison of synthesis methods for conventional top-layer and nanopore Pd membranes. Reproduced from [3]. With permission from Elsevier.

ammonia, usually at 50–60°C and 0.1–0.5 mA cm^{-2} current density. Higher current densities and temperatures would reduce the adhesive strength and durability of the Pd layer. After plating, the membranes are rinsed in deionized water and dried at 110°C for several hours. Sometimes a vacuum is applied downstream of the substrate (Vacuum electrodeposition) [18].

4.3.4 Chemical Vapor Deposition Method

The CVD method usually employs pyrolysis or hydrogen reduction of a suitable chemical, such as palladium hexafluoroactylacetonate $Pd(C_5HF_6O_2)_2$ as the palladium precursor. The precursor is vaporized at around 60°C and fed to the CVD reactor using nitrogen as carrier gas. Then, it is decomposed at a higher temperature (~200°C) and Pd is deposited on the cleaned substrate. This method has the advantages of fast deposition rate, easy control of the membrane thickness, uniform deposition on complex-shaped surfaces, high hardness, no entry of impurities, and a lower tendency toward hydrogen embrittlement. However, it relies heavily on the availability of suitable materials for deposition. Moreover, it also requires a high purity of constituents and strict process conditions.

4.3.5 High-Velocity Oxy-Fuel Spraying Method

The high-velocity oxy-fuel spraying (HVOFS) method is a thermal spraying technique for coating supports based on powder-type spraying materials. The Pd powder is injected into the nozzle of a spray gun using nitrogen as a carrier gas. It is partially molten in a fuel/air flame at a high temperature (2600–2900°C), accelerated to high velocities (500–600 m s^{-1}), and propelled onto the surface of the substrate. On the substrate surface, the impinging particles are deformed due to their high thermal and kinetic energy. Repeated coating will finally lead to a continuous Pd layer. The HVOFS operation can tolerate relatively rough substrate surfaces and is expected to give a good adhesion of the Pd-film to the support.

4.3.6 Magnetron Sputtering Method

Magnetron sputtering can produce very thin films of alloys onto almost any porous substrate. The process involves ion bombardment of the target material to eject atoms, which then adhere to the porous substrate. By applying different sputtering targets, the membrane composition can easily be controlled. The adhesion strength of the film can be enhanced by exposing the substrate to a low-frequency radio frequency glow discharge. Very thin Pd-alloy membranes with good quality can be prepared using the sputtering method in combination with surface modification. However, the preparation procedure is quite complicated and time-consuming, because of the batch production process.

4.3.7 Summary

Each of the above-mentioned methods to prepare Pd-based membranes has its own advantages and disadvantages, which are summarized in Table 4.2. A suitable method should be fitted to the desired membrane properties – such as membrane thickness, geometry, impurities, and so on – and the economy as well. ELP is more practical in commercial

Table 4.2 Comparison of the methods to synthesize Pd-based membranes

Method	Membrane thickness	Operation	Economy	Composition control	Adhesion strength	Substrate requirement
Electroless plating	High	Easy (scale-up)	Low	Hard	Low	Strict
Electroplating	Medium	Easy	Low	Hard	Low	High (metal, only)
CVD method	Low	Complex	High	Easy	High	Low
HVOFS	High	Complicated	High	Easy	High	Low
Magnetron sputtering	Low	Complicated	High	Easy	High	High

practice due to its low cost, but CVD or sputtering can be used to prepare thinner defect-free membranes. In general, ELP is a suitable method for coating ceramic supports, while HVOFS and combined electroplating/electroless plating are more promising for the preparation of Pd/PSS composite membranes [16].

4.4 CONFIGURATIONS OF METALLIC MEMBRANE REACTORS

4.4.1 Packed Bed Membrane Reactor

In general, Pd-MRs are composed of a membrane tube fixed in a module shell made of steel or ceramic on both ends, with the catalyst preferably packed inside the membrane tube (lumen side) as shown in Figure 4.9. The feed is introduced into the catalyst bed while the hydrogen permeated through the membrane is collected on the shell side of the reactor. Different operating modes can be applied to collect the permeated hydrogen. To measure membrane rector performance, an inert gas such as Ar or N_2 is usually used as the sweep gas to generate the hydrogen concentration gradient, passing by the membrane tube concurrently or countercurrently on the shell side. By using steam as the sweep gas, the hydrogen produced can easily be recovered as a product by subsequent condensation. However, the steam may affect the membrane performance by reaction with the membrane. In practice, pure hydrogen can be obtained as a product by applying a vacuum to the membrane permeate side, or by carrying out the reaction under high pressure without the use of any sweep gas. In order to prevent any membrane damage due to direct contact between

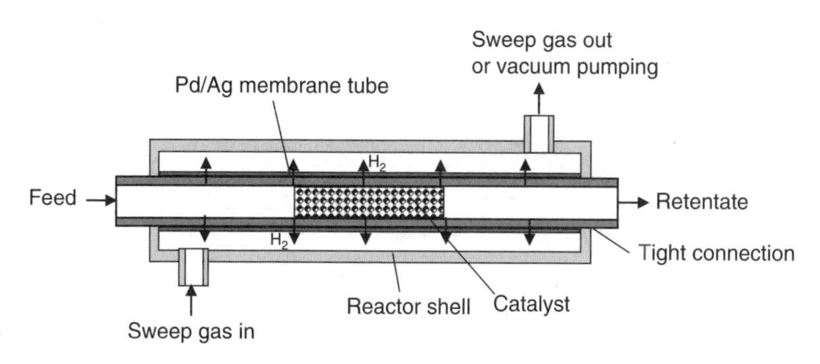

Figure 4.9 Scheme of a Pd-membrane reactor: Catalyst packed in lumen.

the Pd membrane and catalyst pellets, the catalyst is usually located on the bare side of the substrate (opposite to the Pd film).

The Pd-MR performance can be enhanced by using air or CO as the sweep gas. This enables a more effective removal of hydrogen due to a reaction with the permeated hydrogen. In addition, the heat produced due to this reaction can compensate for the endothermic dehydrogenation reaction and maintain a heat balance in the system. However, an oxide layer may form on the Pd/Ag membrane surface, leading to a decrease in membrane performance. In order to suppress any potential oxide formation, the oxygen or CO concentrations in the sweep gas should be lower than 5% or 2%, respectively. The potential formation of metallic oxide on the membrane surface can be eliminated by a high-temperature reduction with H_2 prior to the reaction experiments [19].

In MRs, the Pd-based membrane has to be connected tightly to the module in order to guarantee selective hydrogen separation. However, such a connection has to avoid any mechanical stress to the membrane which could be damaged, especially when thin Pd layers are used. The interaction between hydrogen and Pd produces significant modification of the metal lattice, thus producing the expansion (contraction) of the membranes as a consequence of the hydrogen uploading (desorption). The hydrogenated Pd-based tube under thermal cycling elongates or reduces much more than the membrane module shell does. Consequently, a combined compressive and bending stress may be applied to the membrane tube. A finger-like configuration has been developed to alleviate this stress [20]. As shown in Figure 4.10, the membrane tube with a closed end is gas-tightly connected to the module by means of a graphite gasket; in this way, the lumen and shell side are separated. The gaseous mixture is fed inside the membrane tube via a smaller stainless steel tube placed inside the permeator. Since the membrane tube is fixed to the module only at one end, its free elongation/contraction is allowed – hence avoiding

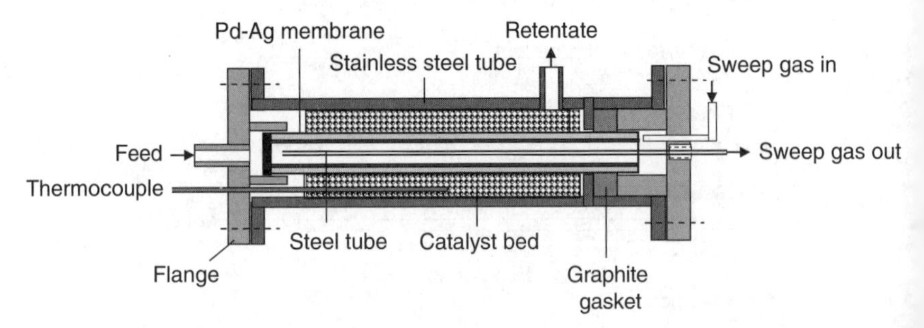

Figure 4.10 Scheme of a finger-like membrane reactor: Catalyst on the shell side.

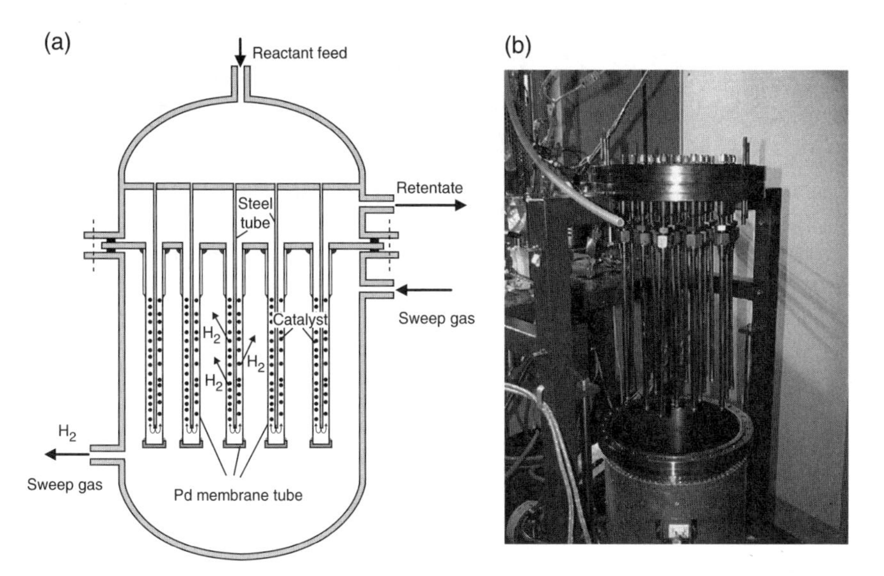

Figure 4.11 Scheme (a) and view (b) of the multi-tube membrane reactor. Reproduced from [20]. With permission from Elsevier.

mechanical stress and ensuring a long lifetime. Furthermore, concurrent or countercurrent modes can be operated by easily changing the feed and retentate flows, as well as changing the inlet and outlet of the sweep gas streams.

The finger-like permeator tube assembly has also been applied for the manufacture of multi-tube MRs, as shown in Figure 4.11, in which the catalyst is packed in the lumen of the membrane tubes. Of course, the reactor can also be designed to locate the catalyst outside the membrane tube.

In the MR, back-permeation of hydrogen to the reaction side may take place in part of the membrane area due to the non-uniform distribution of the hydrogen partial pressure gradient. The length of the back-permeation sector depends on the kinetics: for fast kinetic reactions (e.g., methane steam reforming) it is very short, but for reactions characterized by slow kinetics (e.g., WGS reactions) it is significantly higher. Moreover, maintaining a certain partial pressure of hydrogen on the reaction side of the MR is favorable to reduce the deactivation rate of catalyst and membrane by carbon deposition [21]. A combined configuration of a traditional section with a typical MR section is proposed to allow good exploitation of the whole membrane area for permeation. As shown in Figure 4.12, the membrane is located only in the second part of the catalyst bed. This configuration is of particular importance in all the reactions characterized by slow kinetics. The catalyst packing

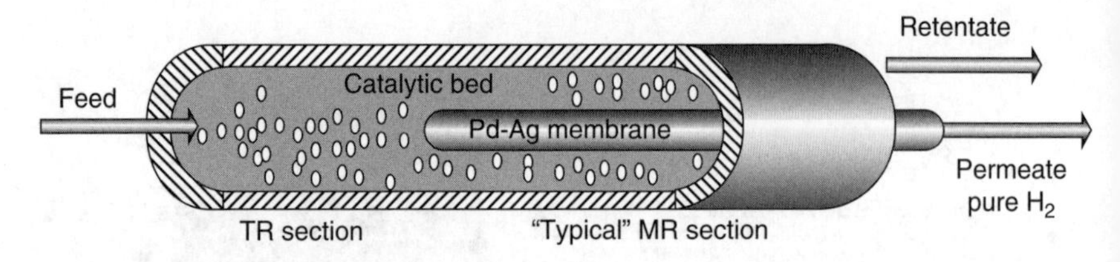

Figure 4.12 Scheme of the membrane reactor with two reaction zones. Reproduced from [6]. With permission from Elsevier.

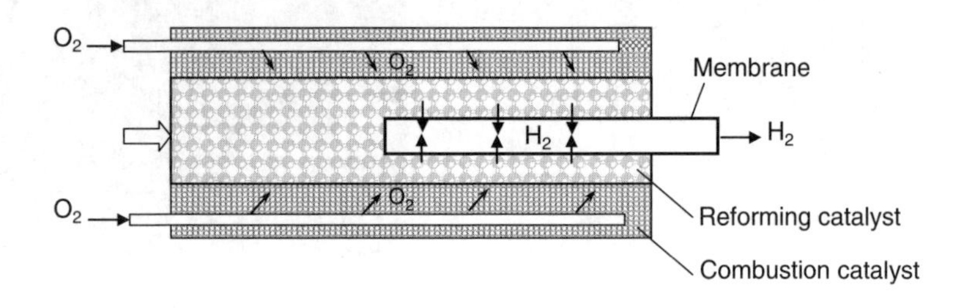

Figure 4.13 Autothermal reforming membrane reactor.

in front of the membrane zone enhances the MR efficiency [22]. Sometimes, hydrogen is premixed into the feed so as to obtain stable conversions, whereas the excess hydrogen is removed by membrane permeation – leading to improved conversions [23].

Figure 4.13 shows an autothermal membrane reactor (ATMR) consisting of a reaction tube containing the reforming and combustion catalyst beds and a perm-selective membrane tube closed at one end. The hydrogen produced by MSR permeates through the membrane, while the heat needed for the endothermic reaction is given by the combustion of methane. Oxygen (or air) is fed and dosed through small tubes with ports along the flow direction of the reaction so as to control the methane combustion and ensure an optimum temperature profile. Temperature peaks which could damage the membrane can be avoided in such an MR configuration.

In addition to this finger-like configuration, harmful mechanical stresses can be avoided by using two pre-tensioned stainless steel metal bellows inserted between the Pd/Ag tube ends and module, as shown in Figure 4.14. The reactor joined to the metal bellows is shorter than the module by about 7.5 mm, which is the assessed maximum elongation of the Pd/Ag tube under hydrogenation. Then the permeator is connected to the module by applying a traction force which expands the bellows. Although the membrane tube is fixed tightly to the module at both ends,

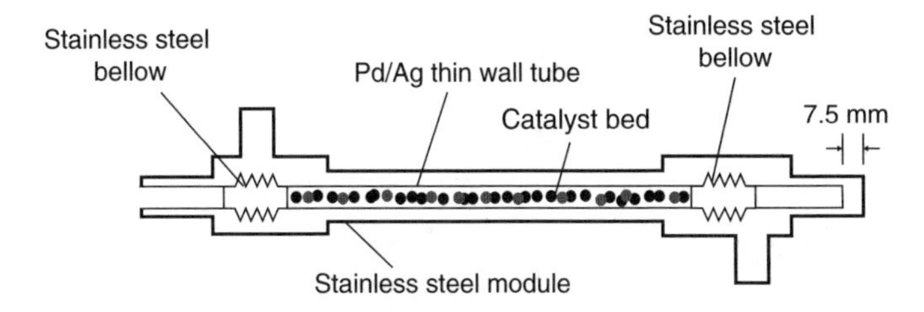

Figure 4.14 Scheme of the membrane reactor with two metal bellows. Reproduced from [1]. With permission from Elsevier.

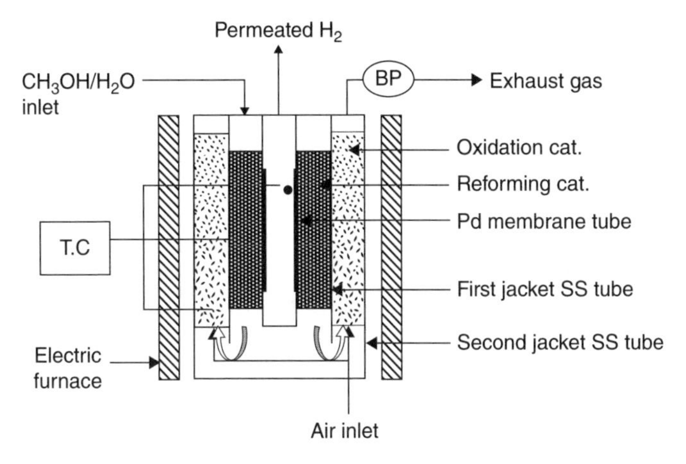

Figure 4.15 Schematic diagram of the double-jacketed membrane reactor. Reproduced from [24]. With permission from Elsevier.

its elongation during operation is compensated by the bellows and no compressive stress is applied to the thin wall tube.

A double-jacketed MR configuration is developed to combine the dehydrogenation reactions with a combustion reaction to generate high-purity hydrogen as shown in Figure 4.15. The concentric module includes a supported Pd-membrane tube located at the center position and two stainless steel tubes assembled separately as double outer jackets. The steam reforming catalyst is loaded adjacent to the outer surface of the membrane tube, while an oxidation catalyst is packed at the outer surface of the first jacket. The mixture of reactants is fed into the reforming catalyst bed to produce H_2 and CO_x. Under the transmembrane pressure difference, the instantaneously produced H_2 penetrates through the Pd-membrane tube and is removed directly as high-purity hydrogen. The rejected gases – H_2, CO, CO_2 – and/or a trace amount of unreacted reactants then enter the oxidation catalyst bed and are converted into

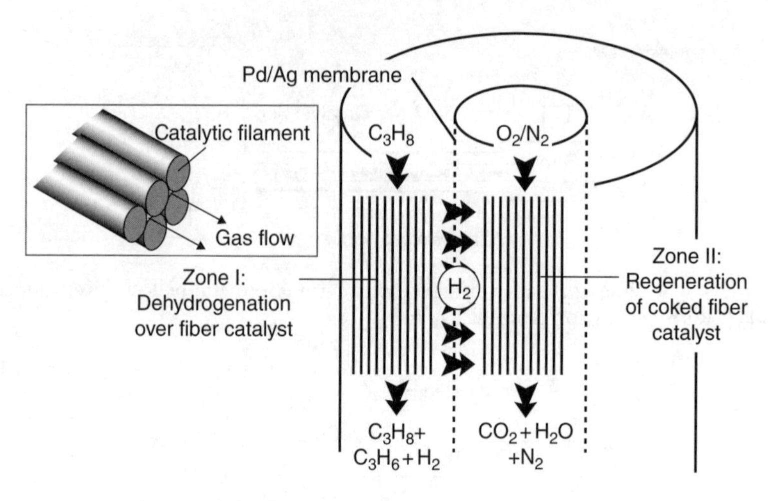

Figure 4.16 Scheme of the membrane reactor with a two-zone filamentous catalytic bed. Reproduced from [25]. With permission from Elsevier.

CO_2 and H_2O as a result of total oxidation reaction with air, while the heat generated is provided for the endothermic reforming reaction. This reactor design can be used in on-board hydrogen generation for fuel cell vehicles, with the use of liquid fuel such as gasoline or methanol [24].

It is acknowledged that the packed catalyst bed in MRs should provide a low pressure drop and a narrow residence time distribution in order to obtain high selectivity. This can be achieved using filamentous catalysts placed in parallel in a tubular MR as shown in Figure 4.16 [25]. In this reactor, the dehydrogenation reaction coupled with the hydrogen combustion reaction represents an ideal design of Pd-MR. As shown in the figure, two zones containing structured filamentous catalyst beds are included in the reactor. Therefore, the reaction and regeneration can be carried out in a cyclic mode by switching the reaction mixture from one zone to the other.

4.4.2 Membrane Microreactor

The Pd-membrane microreactor is shown schematically in Figure 4.17. It consists of a stainless steel feed housing parallel channels. The Pd/Ag film is placed between the channel housing and a stainless steel plate with apertures corresponding to the channel geometry. Catalysts may be coated on the surface of the channels. Both the feed housing and the perforated steel plate are highly polished before mounting. On the permeate

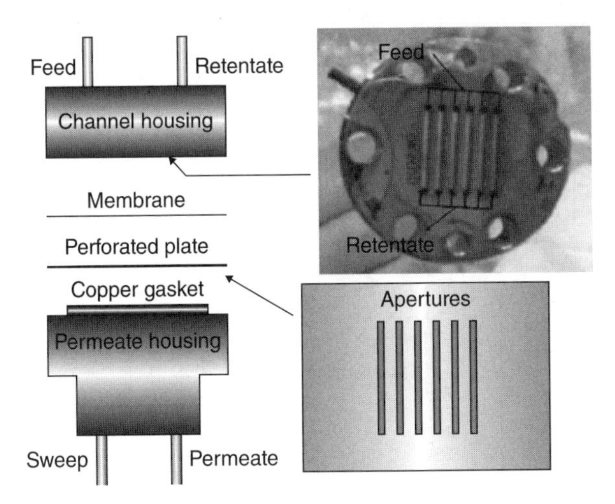

Figure 4.17 Sketch of the microchannel membrane reactor. Reproduced from [26]. With permission from Elsevier.

side an open housing is employed, and a copper gasket is placed between the perforated steel plate and the permeate housing for sealing. For this design, 1–3 μm sputtered Pd/Ag films are sufficiently strong to sustain the pressure differences that enable hydrogen separation from gas mixtures without the use of a sweep gas [26, 27]. Furthermore, the boundary layer resistances can be neglected in this microchannel configuration. A more detailed description of membrane microreactors will be given in Chapter 8.

4.5 APPLICATIONS OF DENSE METALLIC MEMBRANE REACTORS

4.5.1 Dehydrogenation Reactions

Owing to the excellent properties of hydrogen perm-selectivity, the Pd-based membranes so far have been experimented on for a variety of dehydrogenation reactions – that is, dehydrogenation of paraffin into olefin, or dehydrogenation of alcohol into aldehyde or ketone. Recent studies on the dehydrogenation reactions in Pd-MRs are summarized in Table 4.3. As can be seen, the tubular PBMR configuration with sweep gas is applied mostly in experiments. All the experimental results have demonstrated the increase in conversion due to hydrogen separation.

Table 4.3 Dehydrogenation reactions in the Pd-MRs

Reaction	Membrane	Catalyst/Reactor configuration	Operation	Ref.
$C_2H_6 \leftrightarrow C_2H_4 + H_2$	Pd on Ta/Nb alloy tube, 1–3 μm	5% Re/HZSM-5; PBMR	$T = 773$–858 K; vacuum pumping	[28]
$C_3H_8 \leftrightarrow C_3H_6 + H_2$ $\Delta H^0_{298} = 124\,kJ\,mol^{-1}$	Pd on γ-/α-Al_2O_3 hollow fiber, ~6 μm	Pt (0.5 wt%)/SBA-15; CMR	$T = 773$ and 823 K; Ar as sweep gas	[29]
n-butane → n-butene	Pd–Ag on vycor glass tube, 6 μm	Pd–$A1_2O_3$; PBMR	$T = 670$ K; N_2, air, and CO as sweep gas	[19]
$iC_4H_{10} \leftrightarrow iC_4H_8 + H_2$ $\Delta H^0_{873} = 118\,kJ\,mol^{-1}$	Pd foil, 25/75 μm	Pt/α-Al_2O_3, Rh/α-Al_2O_3, Pt/γ-Al_2O_3; PBMR	$T = 400$–700°C; N_2 as sweep gas	[30]
	Pd–Ag(20%) on PSS tube, 10 μm	Pt/Al_2O_3; PBMR	$T = 450$–520°C; N_2 as sweep gas	[31]
	Pd–Ag(23%) on α-Al_2O_3 tube, 8.6 μm	Cr_2O_3/Al_2O_3; PBMR	$T = 450$–520°C; N_2 as sweep gas	[21]
Cyclohexane → benzene with H_2 recovery	Pd on α-alumina tube, 4 μm	0.5 wt% Pt/Al_2O_3; PBMR	$T = 250$–300°C; $P = 2$–4 bar; no sweep gas	[32]
Ethylbenzene → styrene	Pd on α-Al_2O_3 tube, 10 μm	Iron oxide-cat; PBMR	$T = 560$–580°C; N_2 as sweep gas	[33]
	Pd on PSS tube, 0.5 μm	Fe_2O_3–K_2O; PBMR	$T = 450$–625°C; 1.2 atm; N_2 as sweep gas	[34]

$CH_3CH_2OH \leftrightarrow CH_3CHO + H_2$, $\Delta H^0_{f,298} = 68.45$ kJ mol^{-1}	Pd–Ag(23 wt%) on α-Al_2O_3 tube, 2.2 μm	14.5 wt% Cu/SiO_2; PBMR	$T = 250$–410°C; N_2 as sweep gas	[35]
	Pd on PSS tube, 5.88–29 μm	Cu–Zn/Al_2O_3; PBMR	$T = 623$–723 K; $P = 2$–11 bar; N_2 as sweep gas	[36]
2-butanol → methyl ethyl ketone	Pd–Ag(23 wt%) on α-Al_2O_3 tube, 2.2 μm	14.4 wt% Cu/SiO_2; PBMR	$T = 190$–240°C; N_2 as sweep gas	[37]
$C_6H_{11}OH \leftrightarrow C_6H_{10}O + H_2$	Pd on porous glass disk, 9.7 μm	Pd(0.5%)/$A1_2O_3$; PBMR	$T = 513$ K; N_2 or O_2 as sweep gas	[38]
Dehydrogenation of cis-3-hexen-1-ol	Pd–Ag(23 wt%)	CuO–ZnO; PBMR	$T = 443$–503 K; Ar as sweep gas	[39]

The level of yield improvement in Pd-MRs depends on both the catalyst activity (reaction kinetics) and the membrane permeability (hydrogen removal). In case the process is limited by the slow kinetics of the dehydrogenation reaction, a small amount of oxygen can be added in the feed stream to suppress the kinetic limitation, while the removal of H_2 through the Pd membrane can alleviate the equilibrium limitation [30]. Most of the heat for dehydrogenation reactions can be provided by the exothermicity of the oxidative reaction. Furthermore, the addition of oxygen can balance the increasing carbonaceous coverage through the formation of carbon dioxide [40]. However, if the process is controlled mainly by hydrogen removal, any measures to increase the withdrawal rate of hydrogen from the membrane may be taken. For example, in the dehydrogenation of ethanol, the conversion is increased with increasing flow rates of sweep gas and decreasing transmembrane pressures. This effect of sweep-gas flow rates on enhancing H_2 flux is more remarkable at higher temperatures.

Generally, the hydrogen deficiency in the reaction zone due to the membrane separation also results in an increased coke formation, leading to catalyst and membrane deactivation. This suggests that more effort needs to be deployed on the development of new dehydrogenation catalysts with high activity per unit volume and high resistance to deactivation.

4.5.2 Reforming Reactions for H_2 Production

The majority of the dense Pd or its alloy membrane is used in hydrogen production because of its infinite perm-selectivity. It has been demonstrated that the H_2 recovery and purity levels in Pd-MRs can reach as high as 99% and 99.9999%, respectively without the use of a sweep gas. Table 4.4 summarizes the reactions for H_2 production in the Pd-MRs.

4.5.2.1 Water Gas Shift Reaction

The WGS reaction is exothermic and limited by thermodynamic constraints:

$$\text{WGS: } CO + H_2O \Leftrightarrow CO_2 + H_2,\ \Delta H^0_{298} = -41.1\,\text{kJ}\,\text{mol}^{-1}$$

MRs have been utilized to increase CO conversion by shifting the equilibrium conversion toward the products (CO_2 and H_2). In order to reduce the reactor volume, the WGS reaction is usually conducted under high

Table 4.4 Applications of Pd-MRs for hydrogen production

Reaction	Membrane	Catalyst	Reactor configuration	Operation	Ref.
WGS: $CO + H_2O \Leftrightarrow H_2 + CO_2$	60 μm Pd–Ag tube	$CuO/CeO_2/Al_2O_3$	PBMR	$T = 275–325°C$; $P = 2–6$ bar; no sweep gas	[41]
	3 μm Pd/α-Al_2O_3 tube	$Cu–Zn/Al_2O_3$	PBMR	$T = 300°C$; $P = 4.46$ bar; steam as sweep gas	[42]
	20 μm Pd/PSS tube	Fe–Cr-cat	PBMR	$T = 390°C$; $P = 7–11$ bar; no sweep gas	[43]
	6 μm Pd–Ag/α-Al_2O_3 hollow fiber	30% CuO/CeO_2	CMR	$T = 300–500°C$; $P = 1$ bar; Ar as sweep gas	[44]
	5 μm Pd–Ag/α-Al_2O_3 hollow fiber	Pt/CeO_2	PBMR	$T = 375–525°C$; Ar as sweep gas	[45]
MSR: $CH_4 + 2H_2O \leftrightarrow CO_2 + 4H_2$	50 μm Pd–Ag (23 wt%) tube	$Ni–ZrO_2$; $Ni–Al_2O_3$	PBMR	$T = 450°C$; N_2 as sweep gas	[46]
	24.3 μm Pd/Ag on PSS	$Ni/MgO/Al_2O_3$	PBMR	$T = 35–470°C$; $P = 3–7$ atm	[47]
MDR: $CH_4 + CO_2 \leftrightarrow 2CO + 2H_2$	50 μm Pd–Ag (23 wt%) tube	$Rh/La_2O_3–SiO_2$	PBMR	$T = 550°C$; $P = 1$ bar; Ar as sweep gas	[48]
HSR: steam reforming of liquid hydrocarbons	4.5 μm Pd on α-Al_2O_3 tube	$Ni–La_2O_3–Al_2O_3$	PBMR	$T = 723–823$ K; $P = 2–9$ bar; Ar as sweep gas	[49]
	Pd on α-Al_2O_3 tube	2.0% Ru/α-Al_2O_3	PBMR	$T = 500°C$; $P = 1$ atm; Ar as sweep gas	[50]

(*Continued*)

Table 4.4 (*Cont'd*)

Reaction	Membrane	Catalyst	Reactor configuration	Operation	Ref.
Methanol-SR: $CH_3OH + H_2O \leftrightarrow CO_2 + 3H_2$	50 μm Pd–Ag tube	Cu/Zn/Mg-cat		$T = 300°C$; no sweep gas	[51]
	3.9 μm Pd–Ag/ α-Al_2O_3 hollow fiber	Cu–ZnO–Al_2O_3	PBMR	$T = 250–300°C$; $P = 3–5$ bar; N_2 as sweep gas	[10]
ESR: $C_2H_5OH + 3H_2O \leftrightarrow 2CO_2 + 6H_2$, $\Delta H^0_{r298} = 157\ kJ\,mol^{-1}$	150 μm Pd–Ag tube	5 wt% Ru–Al_2O_3	PBMR	$T = 400–450°C$; $P = 1–8$ bar; N_2 as sweep gas	[52]
	30 μm Pd–Ag/ PSS tube	$Co_3(Si_2O_5)_2(OH)_2$	PBMR	$T = 598–673\ K$; $P = 5–15$ bar; No sweep gas	[53]
	20 μm Pd/PSS tube	Ni/ZrO_2 and Co/ Al_2O	PBMR	$T = 400°C$; $P = 8–12$ bar; No sweep gas	[54]
	2 μm Pd–Cu/ α-Al_2O_3 hollow fiber	Co–Na/ZnO	PBMR	$T = 553–633\ K$; $P = 1$ atm; Ar as sweep gas	[55]
OESR: $C_2H_5OH + 3H_2O + 1/2O_2 \leftrightarrow 2CO_2 + 5H_2$	60 μm Pd–Ag tube	0.5% Pt/Al_2O_3	PBMR	$T = 400, 450°C$; $P = 1–2$ bar; N_2 as sweep gas	[56]
	20 μm Pd–Ag/ PSS	Zn–Cu/Al_2O_3	PBMR	$T = 593–723\ K$; $P = 3–10$ atm	[57]

Selective oxidation of ethanol: $C_2H_5OH + 1/2O_2 \leftrightarrow 2CO_2 + 3H_2$	50 μm Pd–Ag tube	Rh/Al_2O_3	PBMR	$T = 450°C$; $P = 1.0–3.0$ bar	[58]
ASR: $CH_3COOH + 2H_2O \leftrightarrow 2CO_2 + 4H_2$	50 μm Pd–Ag tube	$Ni–Al_2O_3$; $Ru–Al_2O_3$	PBMR	$T = 400–450°C$; $P = 1.5–2.5$ bar; N_2 as sweep gas	[59]
GSR: $C_3H_8O_3 + 3H_2O \leftrightarrow 3CO_2 + 7H_2$	25 μm Pd–Ag/PSS tube	$Ni–CeO_2/Al_2O_3$	PBMR	$T = 450°C$; $P = 1–5$ atm	[60]
	50 μm Pd–Ag tube	0.5 wt% Ru/Al_2O_3	PBMR	$T = 400°C$; $P = 1–5$ atm; N_2 as sweep gas	[61]
$2NH_3 \leftrightarrow N_2 + 3H_2$	4 μm Pd on γ-/α-Al_2O_3 hollow fiber	70% $Ni–Al_2O_3$	PBMR	$T = 500, 600°C$; $P = 1–5$ atm; vacuum pumping	[3]
	40 μm Pd/PSS tube	Ru–C	PBMR	$T = 500, 600°C$; $P = 1–5$ atm; He as sweep gas	[62]
	3 μm Pd/α-Al_2O_3 tube	$Ni/La–Al_2O_3$	PBMR	$T = 698–773$ K; $P = 1–5$ atm; N_2 as sweep gas	[63]

pressure with steam as the sweep gas or without the use of a sweep gas. An appropriate choice of the feed pressure, sweep gas flow rate, and temperature has beneficial effects on the conversion and temperature profile, mitigating eventual dangerous hot spots and avoiding thermal runaway [64]. For thin membranes ($<5\,\mu m$) with high permeance, the mass transfer resistance due to boundary layer formation may become rate limiting at a low Reynolds number. This gas boundary resistance can be reduced by increasing the velocity of the reaction mixture, but it will adversely affect the H_2 recovery due to decreased residence time. The membrane separation efficiency can be impaired due to the reversible surface inhibition by CO, CO_2, or H_2O and the irreversible surface inhibition by coke formation. They can be suppressed by increasing the temperature and the presence of H_2O or CO_2, respectively [65].

4.5.2.2 *Steam Reforming of Methane*

Hydrogen production from methane includes two reactions: MSR into syngas and WGS reactions, as described in Chapter 2. With a Pd-MR, high-purity hydrogen can be obtained with improved methane conversion and decreased reaction temperature as low as 400–500°C. Addition of some oxygen in the reactant feed induces partial oxidation of methane, providing the MSR reaction with the required endothermic heat. This is thus called autothermal reforming [47]. However, the removal of hydrogen increases the tendency to poisoning from H_2S and carbon deposition. These effects can be reduced by operating at higher temperatures when H_2S poisoning is dominant or by increased pressure to lessen the extent of carbon deposition [66].

4.5.2.3 *Dry Reforming of Methane*

The use of Pd-MRs for MDR helps to reduce the operating temperature, leading to suppression of catalyst deactivation due to coke deposition. By adding H_2O vapor in the reactant feed, MSR will take place together with MDR. As a result, carbon formation can be reduced due to the oxidation of the carbon precursors – such as partially hydrogenated CH_x species – leading to improved catalyst stability. A desirable H_2/CO ratio can be obtained conveniently via adjustment of the CH_4/H_2O ratio in the feed [67]. Furthermore, methane conversion and hydrogen yield can be increased, whereas CO_2 conversion and CO yield will be decreased.

4.5.2.4 *Steam Reforming of High Hydrocarbons*

Reforming of liquid hydrocarbon fuels in a Pd-alloy MR represents a promising method to provide pure hydrogen product stream without any additional purification steps for PEM fuel cells [68]. In the steam reforming process, the high hydrocarbons are converted irreversibly into CO and H_2, followed by fast methanation and WGS reactions [49]:

$$C_nH_m + nH_2O = nCO + (n + m/2)H_2$$
$$\text{(for iso-octane, } n = 8,\ \Delta H^0_{298} = 1273\,\text{kJ mol}^{-1}\text{)}$$

The coke formation can be suppressed through a pre-reforming process to lower the concentration of olefins, aromatics, and unreacted dodecane in the feed. This will prevent deactivation of both the catalyst and the membrane, with higher hydrogen yield achieved [50].

4.5.2.5 *Methanol Steam Reforming and Ethanol Steam Reforming*

The Pd-MR system for methanol-SR and ESR provides a very attractive device for on-board hydrogen generation in fuel-cell electric vehicles and on-site generation of power electricity. The countercurrent operation is better than the concurrent mode in terms of higher hydrogen yield and CO-free hydrogen recovery, even at low reaction pressure [51]. The water–ethanol mixture from the bioprocess can be used directly as the feed, which can provide a higher energy efficiency (up to 50%) than the distilled bio-ethanol [69].

4.5.2.6 *Oxidative Ethanol Steam Reforming*

Oxidative ethanol steam reforming (OESR), also called autothermal reforming of ethanol, takes place in the presence of oxygen in feed:

$$\text{OESR}: C_2H_5OH + 3H_2O + 1/2O_2 \leftrightarrow 2CO_2 + 5H_2,\ \Delta H^0_{298} = -50\,\text{kJ mol}^{-1}$$

This is an exothermic reaction and represents a good compromise between the hydrogen yield and the energy balance [56]. The addition of oxygen in feed can reduce catalyst deactivation due to the carbon deposition that is thermodynamically promoted above 400°C. In particular, the oxygen addition can prevent ethylene and ethane formation caused by dehydration of ethanol [70]. The amount of oxygen influences the reaction process significantly. The steam reforming reaction prevails with deficient oxygen feed, whereas high O_2 addition will shift the reaction scenario to be partial oxidation dominating with the CO_2 selectivity increased. Increasing pressure favors the OESR reaction to produce high-purity hydrogen [71]. The hydrogen recovery reduces slightly with increasing oxygen input due to the fast oxidation reaction that consumes hydrogen before the onset of the steam reforming reaction.

4.5.2.7 *Acetic Acid Steam Reforming*

Acetic acid is renewable and can easily be obtained from biomass by fermentation. It is also a safe hydrogen carrier due to its non-inflammability. Acetic acid can be converted to hydrogen with high selectivity at low temperature over effective catalysts such as Ni–Co and Ru:

$$CH_3COOH + 2H_2O \leftrightarrow 2CO_2 + 4H_2$$

Using a Pd-MR can give a rather high acetic acid conversion, with 30–35% hydrogen recovery. The overall reaction system can be optimized by tuning the right amount of Ru-based catalyst and Ni-based catalyst [59].

4.5.2.8 *Glycerol Steam Reforming*

Glycerol as a by-product of biodiesel production represents a renewable energy source. Conversion to hydrogen by steam reforming is an effective way to utilize the redundant low-value glycerol:

$$\text{GSR}: C_3H_8O_3 + 3H_2O \leftrightarrow 3CO_2 + 7H_2,\ \Delta H^0_{r298} = 127.67\,\text{kJ}\,\text{mol}^{-1}$$

The main problem for the process is carbon formation on the acidic catalyst support [61]. Using a Pd-MR enables us to lower the reaction temperature.

4.5.2.9 *Decomposition of NH_3*

NH_3 is a good hydrogen carrier due to the high H-content, mild liquefied condition, and environmental benignity (free of carbon). The ammonia decomposition is a mildly endothermic and thermodynamically limited process:

$$2NH_3 \leftrightarrow N_2 + 3H_2, \Delta H^0_{298} = 46.2\,\text{kJ}\,\text{mol}^{-1}$$

Based on thermodynamic data, temperatures above 673 K are needed to reach complete ammonia decomposition (>99%). Coupling NH_3 decomposition with Pd-membrane separation can be an ideal way to provide high-purity CO_x-free H_2 with improved yield at low temperatures [62]. Although NH_3 may inhibit hydrogen permeation slightly, the permeability of the membrane used can be recovered by air and subsequent H_2 treatments. However, the combination of an NH_3 cracker with a Pd-membrane separator is more preferable for on-site H_2 generation [63].

4.5.3 Direct Hydroxylation of Aromatic Compounds

Direct conversion of benzene to phenol is an attractive alternative to the currently used cumene process, which is a multi-step reaction system with the disadvantages of high energy consumption and a large amount of acetone by-product [72]. Since the reactivity of phenol is higher than that of benzene over general heterogeneous catalysts, consecutive oxidation of benzene would take place – leading to low phenol selectivity and yields. The principle of direct hydroxylation of aromatic compounds in Pd-MRs is illustrated in Figure 4.18. Hydrogen and oxygen are supplied separately on opposite sides of the membrane. The active hydrogen species formed by the permeation from one side of the Pd

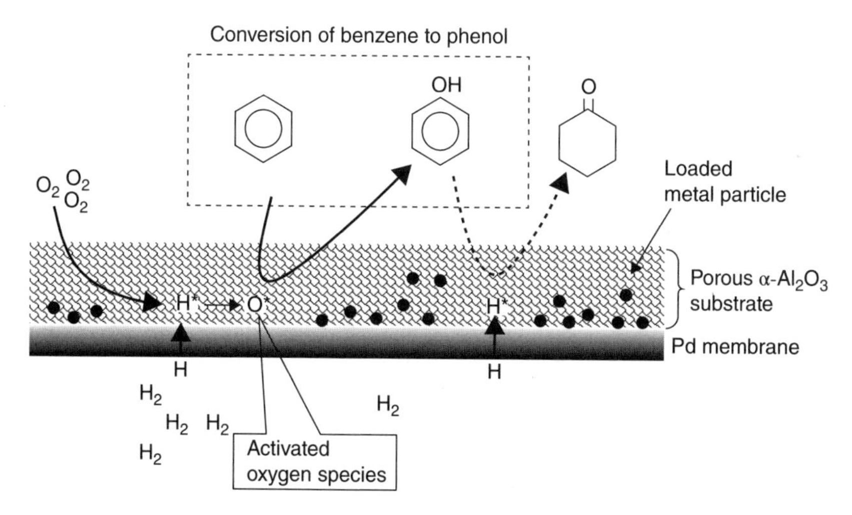

Figure 4.18 Conceptual illustration of direct hydroxylation of benzene to phenol by using a Pd membrane reactor. Reproduced from [73]. With permission from Elsevier.

membrane produces active oxygen species such as HO^* or HOO^* on the opposite side by reacting with oxygen gas. These active oxygen species then react directly with the adsorbed benzene on Pd into phenol. It is more effective to produce active oxygen species than to mix a supply of oxygen and hydrogen from the gas phase to the Pd surface. No expensive oxidizing reagents are needed. Furthermore, the risk of explosion can be eliminated because the supplies of oxygen and hydrogen are physically separated. When Cu is loaded on the porous α-Al_2O_3 substrate, the hydroxylation activity can be enhanced while the activity of side reactions – that is, complete oxidation and hydrogenation – can be suppressed [73]. Another route for higher productivity is to promote mass transfer through minimization of the MR size [74].

4.5.4 Direct Synthesis of Hydrogen Peroxide

Direct synthesis of hydrogen peroxide (H_2O_2) is an economical, simple, and environmentally friendly route for H_2O_2 production. Figure 4.19 depicts the H_2O_2 synthesis from H_2 and O_2 in a Pd-MR. The MR design also provides a promising approach to overcoming the problem of safety [75]. Both the yield and selectivity of H_2O_2 increases with H_2 feed pressure. The H_2O_2 yield also increases with increasing temperature, although the H_2O_2 selectivity may decline [76].

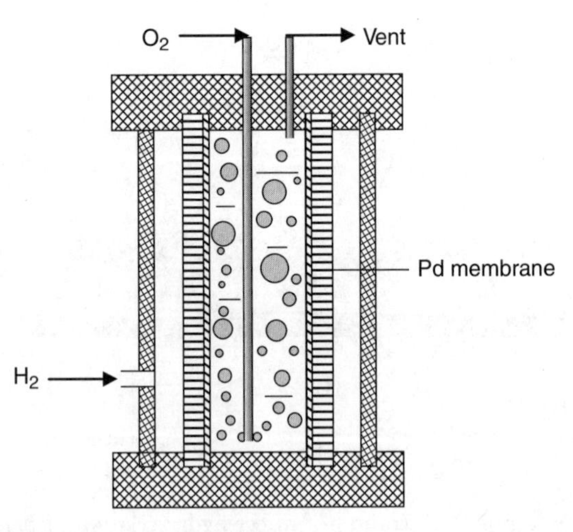

Figure 4.19 Schematic synthesis of H_2O_2 in Pd-membrane reactor. Reproduced from [75]. With permission from Elsevier.

4.6 CHALLENGES AND PROSPECTS

Small-scale Pd-alloy MRs became available in the 1980s for CH_4/CH_3OH reforming to produce H_2 for stationary fuel-cell-type applications. However, large-scale industrial applications have not yet materialized because of a number of concerns, as follows.

Membrane cost. Pd is an expensive metal. The self-supporting, mechanically resistant Pd or Pd-alloy dense membranes must typically be a few hundred micrometers thick, thus the cost is too high for affordable large-scale applications. Many studies have been aimed at reducing the use of precious metals or using alternative alloys that are cheaper. Another effective way to reduce the cost is to develop composite metal membranes whereby a thin (1–10 μm) Pd-alloy film is deposited on (or in the pores of) an underlying inorganic porous support, such as PSS or an asymmetric ceramic composite tube. However, the fabrication of thin Pd-based films has to take into consideration the hydrogenation of Pd, which elongates significantly with respect to the support, thus involving loss of adhesion and selectivity of the membrane. In addition, the combination of membrane with the porous support should be compatible with no delamination of the membrane layer occurring upon repetitive cycling of temperatures.

On the contrary, once the Pd and Pd-alloy membrane thickness is thin, the porous substrate may become the main factor contributing to the cost of the membranes. It is generally acknowledged that the minimum thickness of a dense Pd-based membrane is approximately three times the size of the largest pore present in the substrate surface. This implies that ultra-thin Pd-based membranes with high H_2 permeance and low cost can only be deposited on defect-free surfaces with a small and narrow pore size distribution. However, the fabrication of asymmetric ceramic composite tubes also presents high costs because multiple fabrication steps are required to obtain gradually reduced porosity.

Membrane stability. Pure Pd membranes become brittle upon thermal cycling in a H_2 atmosphere due to the phase transition between different Pd hydride phases (α and β) with distinctly different crystal lattice parameters. Alloying Pd with various other metals (e.g., Ag, Cu, Au, Ru, Rh) tends to lower the phase transition temperature and under some circumstances improves the hydrogen permeability. In addition to the phase-transition problem, mechanical and thermal stability problems also have to be faced. Owing to the significant differences in thermal expansion coefficients between the metal film and the porous support, and the metal atom/ion counter diffusion, the membrane may lose its mechanical stability in long-term operation. Furthermore, thermal cycling of hydrogenated Pd alloys involves significant elongation/contraction of the membrane tubes, which may influence strongly the durability of membrane modules.

Poisoning and carbon deposition. Poisoning by sulfur and chloride containing gas-phase impurities and deactivation due to carbon deposition are matters of concern with Pd-based membranes. The presence of small amounts of these chemicals may reduce the membrane permeability significantly, or even result in complete membrane failure. In contrast carbon deposition, which often results from exposure to hydrocarbons and has a tendency to dissolve in the bulk metal, may reduce hydrogen solubility and permeability of the membrane and will eventually result in membrane mechanical failure. Re-exposure of the poisoned Pd/Ag to hydrogen atmosphere can remove the poisoning agent but cannot restore the mechanical integrity of the Pd/Ag membrane that is destroyed by the poisoning. The poisoning effect is more significant for thin membranes. For example, it has been proved that only membranes with thickness above 7 μm can be stable at high temperatures [77]. In addition, since Pd-based membranes generally operate best at high pressures (to provide the driving force for separation), the MR needs to be able to operate at high feed pressures.

Important R&D studies concerning the large-scale application of dense metallic MRs in recent and future years include: (i) development of generic fabrication techniques for simple and large-scale production of very thin, flawless membranes over large, complex supports; (ii) design of Pd-based MRs aimed at optimizing the temperature control and reactants dosing and maximizing the hydrogen permeation, especially for small mobile applications with reduced sizes and costs.

NOTATION

Symbol	Description
C	concentration (mol m^{-3})
D	diffusion coefficient (m^2 s^{-1})
E_d, E_a	activation energy for diffusion or permeation (J mol^{-1})
ΔH_a	enthalpy of hydrogen absorption (J mol^{-1})
J	molar flux (mol m^{-2} s^{-1})
K_e	equilibrium constant for dissociation reaction (mol^2 m^{-6} Pa^{-1})

L_p	pore length (m)
M	molecular weight (g mol^{-1})
n	pressure factor in Richardson's equation
p	pressure (Pa)
P	permeability (mol m $m^{-2}\,s^{-1}\,Pa^{-n}$)
Pe	permeance (mol $m^{-2}\,s^{-1}\,Pa^{-1}$)
R	ideal gas constant, 8.314 J $mol^{-1}\,K^{-1}$
S	solubility (mol $m^{-3}\,Pa^{-0.5}$)
T	temperature (K)
z	diffusion length variable (m)
δ	membrane thickness (m)
ε_p	porosity of substrate
η	substrate influence factor defined in Eq. (4.11)

Subscripts and superscripts

up	upstream
down	downstream
H	atomic hydrogen in metal lattice
0	pre-exponential factor

REFERENCES

[1] Tosti, S. (2010) Overview of Pd-based membranes for producing pure hydrogen and state of art at ENEA laboratories. *International Journal of Hydrogen Energy*, 35, 12650–12659.

[2] Dittmeyer, R., Hollein, V. and Daub, K. (2001) Membrane reactors for hydrogenation and dehydrogenation processes based on supported palladium. *Journal of Molecular Catalysis A: Chemical*, 173, 135–184.

[3] Israni, S.H., Nair, B.K.R. and Harold, M.P. (2009) Hydrogen generation and purification in a composite Pd hollow fiber membrane reactor: Experiments and modeling. *Catalysis Today*, 139, 299–311.

[4] Dolan, M., Dave, N., Morpeth, L. *et al.* (2009) Ni-based amorphous alloy membranes for hydrogen separation at 400°C. *Journal of Membrane Science*, 326, 549–555.

[5] Bobrov, V.S., Digurov, N.G. and Skudin, V.V. (2005) Propane dehydrogenation using catalytic membrane. *Journal of Membrane Science*, 253, 233–242.

[6] Barbieri, G., Brunetti, A., Tricoli, G. and Drioli, E. (2008) An innovative configuration of a Pd-based membrane reactor for the production of pure hydrogen. *Journal of Power Sources*, 182, 160–167.

[7] Iulianelli, A. and Basile, A. (2010) An experimental study on bio-ethanol steam reforming in a catalytic membrane reactor. Part I: Temperature and sweep-gas flow configuration effects. *International Journal of Hydrogen Energy*, 35, 3170–3177.

[8] Mendes, D., Chibante, V., Zheng, J.-M. *et al.* (2010) Enhancing the production of hydrogen via water–gas shift reaction using Pd-based membrane reactors. *International Journal of Hydrogen Energy*, 35, 12596–12608.

[9] Hwang, K.-R., Lee, C.-B., Ryi, S.-K. and Park, J.-S. (2012) Hydrogen production and carbon dioxide enrichment using a catalytic membrane reactor with Ni metal catalyst and Pd-based membrane. *International Journal of Hydrogen Energy*, 37, 6626–6634.

[10] Israni, S.H. and Harold, M.P. (2011) Methanol steam reforming in single-fiber packed bed Pd–Ag membrane reactor: Experiments and modeling. *Journal of Membrane Science*, 369, 375–387.

[11] Roa, F., Way, J.D., McCormick, R.L. and Paglieri, S.N. (2003) Preparation and characterization of Pd–Cu composite membranes for hydrogen separation. *Chemical Engineering Journal*, 93, 11–22.

[12] Guazzone, F., Engwall, E.E. and Ma, Y.H. (2006) Effects of surface activity, defects and mass transfer on hydrogen permeance and n-value in composite palladium–porous stainless steel membranes. *Catalysis Today*, 118, 24–31.

[13] Deveau, N.D., Ma, Y. and Datta, R. (2013) Beyond Sieverts' law: A comprehensive microkinetic model of hydrogen permeation in dense metal membranes. *Journal of Membrane Science*, 437, 298–311.

[14] Gabitto, J. and Tsouris, C. (2008) Hydrogen transport in composite inorganic membranes. *Journal of Membrane Science*, 312, 132–142.

[15] Hatim, M.D.I., Tan, X., Wu, Z. and Li, K. (2011) Pd/Al_2O_3 composite hollow fibre membranes: Effect of substrate resistances on H2 permeation properties. *Chemical Engineering Science*, 66, 1150–1158.

[16] Quicker, P., Hollein, V. and Dittmeyer, R. (2000) Catalytic dehydrogenation of hydrocarbons in palladium composite membrane reactors. *Catalysis Today*, 56, 21–34.

[17] Ayturk, M.E., Mardilovich, I.P., Engwall, E.E. and Ma, Y.H. (2006) Synthesis of composite Pd–porous stainless steel (PSS) membranes with a Pd/Ag intermetallic diffusion barrier. *Journal of Membrane Science*, 285, 385–394.

[18] Nam, S.E. and Lee, K.H. (2000) A study on the palladium/nickel composite membrane by vacuum electrodeposition. *Journal of Membrane Science*, 170, 91–99.

[19] Gobina, E. and Hughes, R. (1996) Reaction assisted hydrogen transport during catalytic dehydrogenation in a membrane reactor. *Applied Catalysis A: General*, 137, 119–127.

[20] Tosti, S., Basile, A., Bettinali, L., Borgognoni, F., Gallucci, F. and Rizzello, C. (2008) Design and process study of Pd membrane reactors. *International Journal of Hydrogen Energy*, 33, 5098–5105.

[21] Guo, Y., Lu, G., Wang, Y. and Wang, R. (2003) Preparation and characterization of Pd–Ag/ceramic composite membrane and application to enhancement of catalytic dehydrogenation of isobutane. *Separation and Purification Technology*, 32, 271–279.

[22] Kikuchi, E., Nemoto, Y., Kajiwara, M., Uemiya, S. and Kojima, T. (2000) Steam reforming of methane in membrane reactors: comparison of electroless-plating and CVD membranes and catalyst packing modes. *Catalysis Today*, 56, 75–81.

[23] Itoh, N., Xu, W.C., Hara, S., Kakehida, K., Kaneko, Y. and Igarashi, A. (2003) Effects of hydrogen removal on the catalytic reforming of *n*-hexane in a palladium membrane reactor. *Industrial and Engineering Chemistry Research*, 42, 6576–6581.
[24] Lin, Y.M. and Rei, M.H. (2000) Process development for generating high purity hydrogen by using supported palladium membrane reactor as steam reformer. *International Journal of Hydrogen Energy*, 25, 211–219.
[25] Wolfrath, O., Kiwi-Minsker, L. and Renken, A. (2001) Novel membrane reactor with filamentous catalytic bed for propane dehydrogenation. *Industrial and Engineering Chemistry Research*, 40, 5234–5239.
[26] Mejdell, A.L., Jøndahl, M., Peters, T.A., Bredesen, R. and Venvik, H.J. (2009) Experimental investigation of a microchannel membrane configuration with a 1.4 μm Pd/Ag23 wt.% membrane – effects of flow and pressure. *Journal of Membrane Science*, 327, 6–10.
[27] Mejdell, A.L., Peters, T.A., Stange, M., Venvik, H.J. and Bredesen, R. (2009) Performance and application of thin Pd-alloy hydrogen separation membranes in different configurations. *Journal of the Taiwan Institute of Chemical Engineers*, 40, 253–259.
[28] Wang, L., Murata, K. and Inaba, M. (2003) Production of pure hydrogen and more valuable hydrocarbons from ethane on a novel highly active catalyst system with a Pd-based membrane reactor. *Catalysis Today*, 82, 99–104.
[29] Gbenedio, E., Wu, Z., Hatim, I., Kingsbury, B.F.K. and Li, K. (2010) A multifunctional Pd/alumina hollow fibre membrane reactor for propane dehydrogenation. *Catalysis Today*, 156, 93–99.
[30] Raybold, T.M. and Huff, M.C. (2000) Oxidation of isobutane over supported noble metal catalysts in a palladium membrane reactor. *Catalysis Today*, 56, 35–44.
[31] Liang, W. and Hughes, R. (2005) The catalytic dehydrogenation of isobutane to isobutene in a palladium/silver composite membrane reactor. *Catalysis Today*, 104, 238–243.
[32] Itoh, N., Tamura, E., Hara, S. *et al.* (2003) Hydrogen recovery from cyclohexane as a chemical hydrogen carrier using a palladium membrane reactor. *Catalysis Today*, 82, 119–125.
[33] Yu, C. and Xu, H. (2011) An efficient palladium membrane reactor to increase the yield of styrene in ethylbenzene dehydrogenation. *Separation and Purification Technology*, 78, 249–252.
[34] She, Y., Han, J. and Ma, Y.H. (2001) Palladium membrane reactor for the dehydrogenation of ethylbenzene to styrene. *Catalysis Today*, 67, 43–53.
[35] Keuler, J.N. and Lorenzen, L. (2002) Comparing and modeling the dehydrogenation of ethanol in a plug-flow reactor and a Pd–Ag membrane reactor. *Industrial and Engineering Chemistry Research*, 41, 1960–1966.
[36] Lin, W.-H. and Chang, H.-F. (2004) A study of ethanol dehydrogenation reaction in a palladium membrane reactor. *Catalysis Today*, 97, 181–188.
[37] Keuler, J.N. and Lorenzen, L. (2002) The dehydrogenation of 2-butanol in a Pd-Ag membrane reactor. *Journal of Membrane Science*, 202, 17–26.
[38] Schramm, O. and Seidel-Morgenstern, A. (1999) Comparing porous and dense membranes for the application in membrane reactors. *Chemical Engineering Science*, 54, 1447–1453.
[39] Sato, T., Yokoyama, H., Miki, H. and Itoh, N. (2007) Selective dehydrogenation of unsaturated alcohols and hydrogen separation with a palladium membrane reactor. *Journal of Membrane Science*, 289, 97–105.

[40] Amandusson, H., Ekedahl, L.G. and Dannetun, H. (1999) Methanol-induced hydrogen permeation through a palladium membrane. *Surface Science*, 442, 199–205.
[41] Brunetti, A., Barbieri, G. and Drioli, E. (2009) Upgrading of a syngas mixture for pure hydrogen production in a Pd–Ag membrane reactor. *Chemical Engineering Science*, 64, 3448–3454.
[42] Abdollahi, M., Yu, J., Liu, P.K.T., Ciora, R., Sahimi, M. and Tsotsis, T.T. (2012) Ultra-pure hydrogen production from reformate mixtures using a palladium membrane reactor system. *Journal of Membrane Science*, 390/391, 32–42.
[43] Liguori, S., Pinacci, P., Seelam, P.K. *et al.* (2012) Performance of a Pd/PSS membrane reactor to produce high purity hydrogen via WGS reaction. *Catalysis Today*, 193, 87–94.
[44] García-García, F.R., Rahman, M.A., Kingsbury, B.F.K. and Li, K. (2010) A novel catalytic membrane microreactor for CO_x-free H_2 production. *Catalysis Communications*, 12, 161–164.
[45] García-García, F.R., Torrente-Murciano, L., Chadwick, D. and Li, K. (2012) Hollow fibre membrane reactors for high H_2 yields in the WGS reaction. *Journal of Membrane Science*, 405/406, 30–37.
[46] Basile, A., Campanari, S., Manzolini, G. *et al.* (2011) Methane steam reforming in a Pd–Ag membrane reformer: An experimental study on reaction pressure influence at middle temperature. *International Journal of Hydrogen Energy*, 36, 1531–1539.
[47] Chang, H.-F., Pai, W.-J., Chen, Y.-J. and Lin, W.-H. (2010) Autothermal reforming of methane for producing high-purity hydrogen in a Pd/Ag membrane reactor. *International Journal of Hydrogen Energy*, 35, 12986–12992.
[48] Coronel, L., Múnera, J.F., Lombardo, E.A. and Cornaglia, L.M. (2011) Pd based membrane reactor for ultra pure hydrogen production through the dry reforming of methane. Experimental and modeling studies. *Applied Catalysis A: General*, 400, 185–194.
[49] Chen, Y., Xu, H., Wang, Y. and Xiong, G. (2006) Hydrogen production from the steam reforming of liquid hydrocarbons in membrane reactor. *Catalysis Today*, 118, 136–143.
[50] Miyamoto, M., Hayakawa, C., Kamata, K., Arakawa, M. and Uemiya, S. (2011) Influence of the pre-reformer in steam reforming of dodecane using a Pd alloy membrane reactor. *International Journal of Hydrogen Energy*, 36, 7771–7775.
[51] Iulianelli, A., Longo, T. and Basile, A. (2008) Methanol steam reforming reaction in a Pd–Ag membrane reactor for CO-free hydrogen production. *International Journal of Hydrogen Energy*, 33, 5583–5588.
[52] Tosti, S., Fabbricino, M., Moriani, A. *et al.* (2011) Pressure effect in ethanol steam reforming via dense Pd-based membranes. *Journal of Membrane Science*, 377, 65–74.
[53] Domínguez, M., Taboada, E., Molins, E. and Llorca, J. (2012) Ethanol steam reforming at very low temperature over cobalt talc in a membrane reactor. *Catalysis Today*, 193, 101–106.
[54] Seelam, P.K., Liguori, S., Iulianelli, A. *et al.* (2012) Hydrogen production from bioethanol steam reforming reaction in a Pd/PSS membrane reactor. *Catalysis Today*, 193, 42–48.
[55] Yun, S., Lim, H. and Ted Oyama, S. (2012) Experimental and kinetic studies of the ethanol steam reforming reaction equipped with ultrathin Pd and Pd–Cu membranes for improved conversion and hydrogen yield. *Journal of Membrane Science*, 409/410, 222–231.

[56] Tosti, S., Zerbo, M., Basile, A., Calabrò, V., Borgognoni, F. and Santucci, A. (2013) Pd-based membrane reactors for producing ultra pure hydrogen: Oxidative reforming of bio-ethanol. *International Journal of Hydrogen Energy*, 38, 701–707.

[57] Lin, W.-H., Hsiao, C.-S. and Chang, H.-F. (2008) Effect of oxygen addition on the hydrogen production from ethanol steam reforming in a Pd–Ag membrane reactor. *Journal of Membrane Science*, 322, 360–367.

[58] Iulianelli, A., Liguori, S., Calabrò, V., Pinacci, P. and Basile, A. (2010) Partial oxidation of ethanol in a membrane reactor for high purity hydrogen production. *International Journal of Hydrogen Energy*, 35, 12626–12634.

[59] Basile, A., Gallucci, F., Iulianelli, A., Borgognoni, F. and Tosti, S. (2008) Acetic acid steam reforming in a Pd–Ag membrane reactor: The effect of the catalytic bed pattern. *Journal of Membrane Science*, 311, 46–52.

[60] Chang, A.C.C., Lin, W.-H., Lin, K.-H., Hsiao, C.-H., Chen, H.-H. and Chang, H.-F. (2012) Reforming of glycerol for producing hydrogen in a Pd/Ag membrane reactor. *International Journal of Hydrogen Energy*, 37, 13110–13117.

[61] Iulianelli, A., Seelam, P.K., Liguori, S. *et al.* (2011) Hydrogen production for PEM fuel cell by gas phase reforming of glycerol as byproduct of bio-diesel. The use of a Pd–Ag membrane reactor at middle reaction temperature. *International Journal of Hydrogen Energy*, 36, 3827–3834.

[62] García-García, F.R., Ma, Y.H., Rodríguez-Ramos, I. and Guerrero-Ruiz, A. (2008) High purity hydrogen production by low temperature catalytic ammonia decomposition in a multifunctional membrane reactor. *Catalysis Communications*, 9, 482–486.

[63] Zhang, J., Xu, H. and Li, W. (2006) High-purity CO_x-free H_2 generation from NH_3 via the ultra permeable and highly selective Pd membranes. *Journal of Membrane Science*, 277, 85–93.

[64] Chiappetta, G., Clarizia, G. and Drioli, E. (2008) Theoretical analysis of the effect of catalyst mass distribution and operation parameters on the performance of a Pd-based membrane reactor for water–gas shift reaction. *Chemical Engineering Journal*, 136, 373–382.

[65] Augustine, A.S., Ma, Y.H. and Kazantzis, N.K. (2011) High pressure palladium membrane reactor for the high temperature water–gas shift reaction. *International Journal of Hydrogen Energy*, 36, 5350–5360.

[66] Hou, K., Fowles, M. and Hughes, R. (1999) Potential catalyst deactivation due to hydrogen removal in a membrane reactor used for methane steam reforming. *Chemical Engineering Science*, 54, 3783–3791.

[67] Soria, M.A., Mateos-Pedrero, C., Guerrero-Ruiz, A. and Rodríguez-Ramos, I. (2011) Thermodynamic and experimental study of combined dry and steam reforming of methane on Ru/ZrO_2–La_2O_3 catalyst at low temperature. *International Journal of Hydrogen Energy*, 36, 15212–15220.

[68] Damle, A.S. (2009) Hydrogen production by reforming of liquid hydrocarbons in a membrane reactor for portable power generation–experimental studies. *Journal of Power Sources*, 186, 167–177.

[69] Manzolini, G. and Tosti, S. (2008) Hydrogen production from ethanol steam reforming: Energy efficiency analysis of traditional and membrane processes. *International Journal of Hydrogen Energy*, 33, 5571–5582.

[70] Iulianelli, A., Longo, T., Liguori, S., Seelam, P.K., Keiski, R.L. and Basile, A. (2009) Oxidative steam reforming of ethanol over Ru–Al_2O_3 catalyst in a dense Pd–Ag

membrane reactor to produce hydrogen for PEM fuel cells. *International Journal of Hydrogen Energy*, 34, 8558–8565.

[71] Lin, W.-H., Liu, Y.-C. and Chang, H.-F. (2010) Autothermal reforming of ethanol in a Pd–Ag/Ni composite membrane reactor. *International Journal of Hydrogen Energy*, 35, 12961–12969.

[72] Sato, K., Hanaoka, T.-a., Niwa, S.-i., Stefan, C., Namba, T. and Mizukami, F. (2005) Direct hydroxylation of aromatic compounds by a palladium membrane reactor. *Catalysis Today*, 104, 260–266.

[73] Sato, K., S.H., Natsui, M., Nishioka, M., Inoue, T. and Mizukami, F. (2010) Palladium-based bifunctional membrane reactor for one-step conversion of benzene to phenol and cyclohexanone. *Catalysis Today*, 156, 276–281.

[74] Wang, X.B., Zhang, X.F., Liu, H.O., Qiu, J.S., Han, W. and Yeung, K.L. (2012) Investigation of Pd membrane reactors for one-step hydroxylation of benzene to phenol. *Catalysis Today*, 193, 151–157.

[75] Wang, L., Bao, S., Yi, J., He, F. and Mi, Z. (2008) Preparation and properties of Pd/Ag composite membrane for direct synthesis of hydrogen peroxide from hydrogen and oxygen. *Applied Catalysis B: Environmental*, 79, 157–162.

[76] Shi, L., Goldbach, A., Zeng, G. and Xu, H. (2010) Direct H_2O_2 synthesis over Pd membranes at elevated temperatures. *Journal of Membrane Science*, 348, 160–166.

[77] Armor, J.N. (1998) Applications of catalytic inorganic membrane reactors to refinery products. *Journal of Membrane Science*, 147, 217–233.

5

Dense Ceramic Oxygen-Permeable Membrane Reactors

5.1 INTRODUCTION

Dense ceramic membranes are made from composite oxides usually having a perovskite or fluorite crystalline structure. A large number of oxygen vacancies are present in the membrane, mostly generated by the doping strategy and leading to noticeable oxygen ionic conductivity at elevated temperatures (>700°C). Under an electrochemical potential gradient, oxygen gas can permeate through the membrane in a dissociated or ionized form (hence such membranes are also called ionic transport membranes (ITMs)) rather than the conventional molecular diffusion, thus an extremely high oxygen perm-selectivity can be achieved. Depending on the mediator of the electronic flux, the dense ceramic membranes can be divided into three categories:

1. A single material membrane based on a material exhibiting mixed ionic and electronic conductivity (MIEC membrane).
2. A dual-phase composite consisting of percolating phases of an ionic conductor and an electronic conductor (dual-phase membrane).
3. A pure ionic conductor with suitable electrodes connected to an external circuit for the electronic current (electrolyte membrane).

Figure 5.1 shows schematically the oxygen permeation through dense ceramic membranes. For the electrolyte membrane, an external circuit is

Inorganic Membrane Reactors: Fundamentals and Applications, First Edition. Xiaoyao Tan and Kang Li.

required for electron transport with two electrodes attached to the membrane surfaces to distribute/collect electrons. When an external power source is applied, oxygen can be pumped electrochemically from the negative side to the positive side of the membrane regardless of the oxygen partial pressures on each side, as shown in Figure 5.1(a). When an oxygen partial pressure gradient is present across the membrane, the oxygen can also be permeated through the electrolyte membrane with the aid of the external circuit to transport electrons, as shown in Figure 5.1(b). As a result, electrical power can be co-generated along with the oxygen permeation. For the MIEC and dual-phase membranes, oxygen can permeate through the membranes under an oxygen partial pressure gradient at high temperatures without the need for electrodes

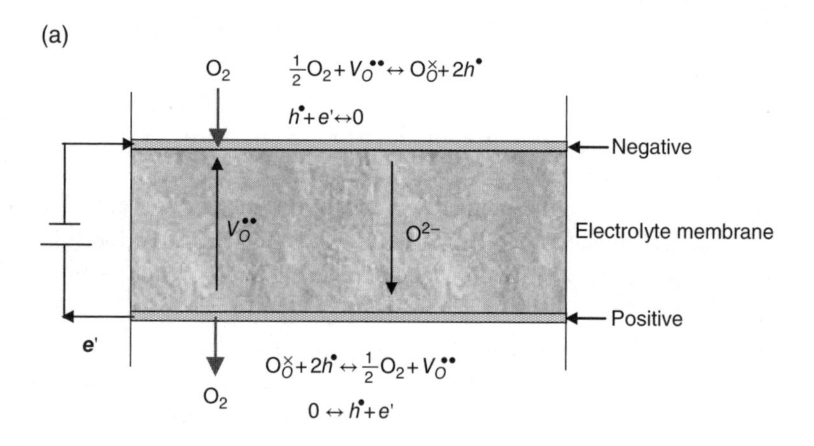

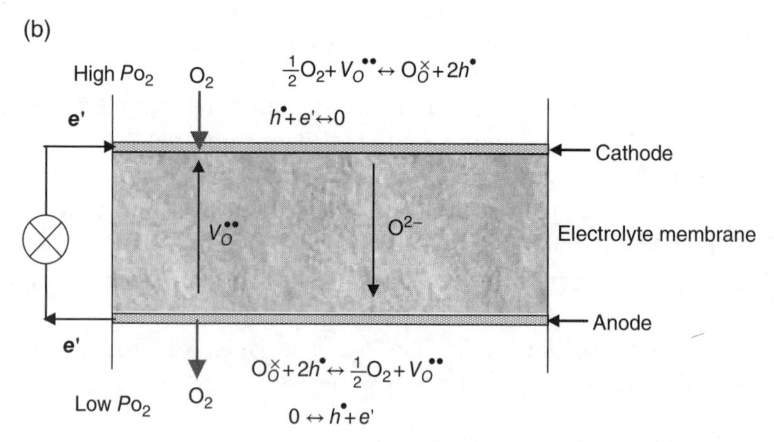

Figure 5.1 Oxygen permeation in dense ceramic membranes: (a) electrochemical oxygen pump; (b) oxygen permeation together with production of electricity;

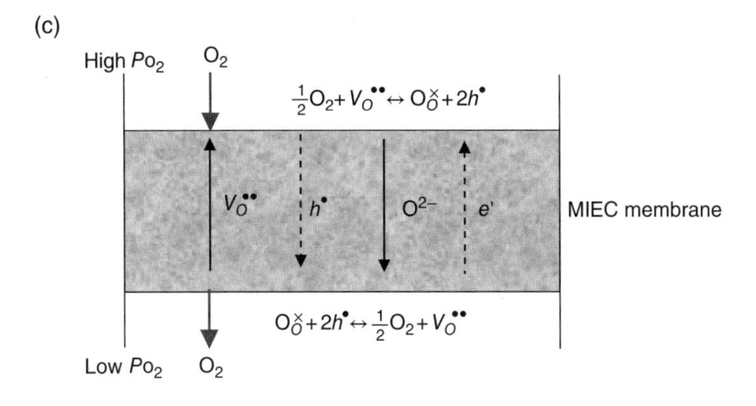

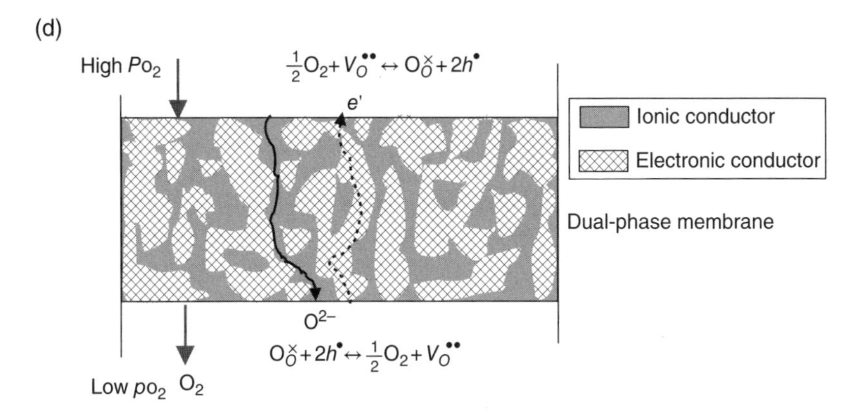

Figure 5.1 (*Continued*) (c) oxygen permeation through MIEC membrane; (d) oxygen permeation through dual-phase membrane.

and external circuits, as shown in Figure 5.1(c,d), respectively. This makes the oxygen permeation through the membrane much simpler and accordingly the operational cost can be reduced remarkably. Owing to their infinite oxygen perm-selectivity as well as their high thermal resistance, dense ceramic membranes can potentially be applied to a variety of oxygen-related reactions such as partial oxidation or oxidative dehydrogenation, where the membrane functions as either an oxygen distributor or an extractor [1].

This chapter will describe extensively the dense ceramic oxygen permeable membrane reactors from their principles, fabrication, and design to their applications. The prospects and critical issues of the dense ceramic membrane reactors in commercial applications will also be presented and discussed at the end of the chapter.

5.2 OXYGEN PERMEATION IN DENSE CERAMIC MEMBRANES

5.2.1 Membrane Materials

The materials of oxygen permeable membranes are required to have both high ionic and electronic conductivity as well as high thermal and chemical stability when exposed to large oxygen partial pressure gradients. So far, a large number of oxygen-permeable membrane materials have been developed [2–6]. These materials can be classified into two main groups based on crystalline structure: fluorites and perovskites. In recent years, dual-phase membranes have also been developed in an attempt to obtain both high oxygen permeability and good stability.

- **Fluorite oxides**
  Fluorite oxides for oxygen-permeable membranes are typically represented by ZrO_2, CeO_2, and Bi_2O_3 due to their high oxygen ion mobility. These oxides have different crystal phases depending on temperature that exhibit different oxygen ionic conductivity. Doping with some other oxides – such as CaO, MgO, Y_2O_3, Er_2O_3 – helps to maintain the crystal phase with the highest oxygen ionic conductivity in a wide temperature range.
  Among these oxides, ZrO_2 and Bi_2O_3 have low electronic conductivity, and thus can only be used commonly as the electrolyte membrane for oxygen permeation, while CeO_2 has noticeable electronic conductivity because of the transition of Ce^{4+} to Ce^{3+}. Compared with ZrO_2, Bi_2O_3 and CeO_2 show higher oxygen ionic conductivity at intermediate temperatures (<800°C). By introducing an oxide with high electronic conductivity, the doped ZrO_2 and Bi_2O_3 can become single-phase MIEC materials. However, their electronic conductivity is still much lower than their ionic conductivity.
- **Perovskite oxides**
  Most dense ceramic membranes are made from perovskite oxides with general formula ABO_3, where the A-site cation is coordinated to 12 oxygen ions forming a cuboctahedral coordination environment while the B-site cation is coordinated to six oxygen ions with

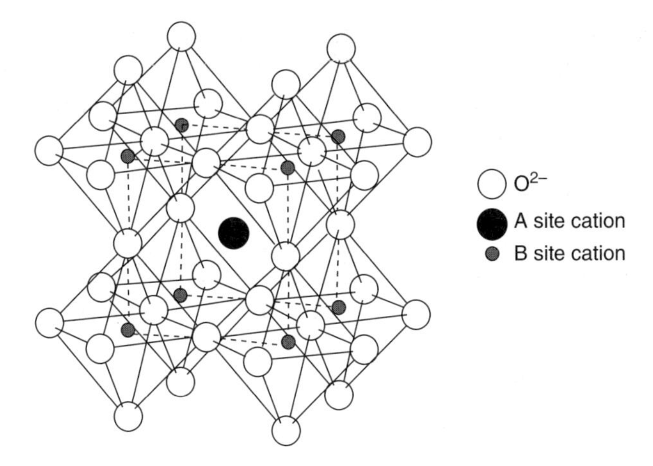

Figure 5.2 Structure of ABO_3 perovskite oxides, emphasizing the coordination number of the A cation at the body center. Large cations such as La^{3+}, Ba^{2+}, and Sr^{2+} occupy the A-sites. Smaller cations, with multiple valence essential for electron conduction, occupy the B-sites and often include ions of Ni, Fe, Co, Mn, and Cr. Reproduced from [7]. With permission from Elsevier.

an octahedral geometry. Figure 5.2 shows the cubic structure of ABO_3 perovskites, emphasizing the coordination environment around the A-site cation.

The structural stability of perovskite can be evaluated by the Goldschmidt tolerance factor

$$\eta = \frac{r_A + r_O}{\sqrt{2}\left(r_B + r_O\right)} \tag{5.1}$$

where r_O, r_A, and r_B are the radius of oxygen ion, A-site ion, and B-site ion, respectively. In general, the structural stability of a perovskite increases as η increases, with η values between 0.75 and 1.0 representing a stable perovskite.

In most cases the A site is occupied by 2+ large alkali earth metals such as Ba, La, or Sr and the B site by 4+ smaller first-row transition elements such as Ce, Co, or Fe. But the combinations $A^{1+}B^{5+}O_3$ and $A^{3+}B^{3+}O_3$ have also been found. The A or B-site cation can be substituted partially by other metal cations having different valances whereas the perovskite structure is still retained. As a result, a large number of oxygen vacancies are generated in the crystalline lattice for electrical neutrality. The oxygen ions with enough energy can move between adjacent vacancies leading to appreciable oxygen ionic conductivity. In the meantime, electronic conductivity may arise due to the presence of ions with charges deviating from the normal lattice ions, or the transition of electrons from normally filled energy levels to normally empty levels. Therefore, the doped perovskite oxides with general formula $A_{1-x}A'_xB_{1-y}B'_yO_{3-\delta}$ $(0 \le x \le 1; 0 \le y \le 1)$, where δ represents the amount of oxygen vacancies (defects), demonstrate excellent mixed ionic–electronic conduction properties, which stems from the high degree of oxygen deficiency and high mobility of the oxygen vacancies [8].

- **Dual-phase composites**
  Dual-phase composite ceramic membrane materials combine a material having high oxygen ionic conductivity with a second material having high electronic conductivity. Oxygen ion and electron transport occur in different phases. The ion- and electron-conducting phases must form a continuous and intimately intermixed network to allow simultaneous and balanced transport of oxygen ions and electrons under the electroneutrality constraint of zero net charge. The conductivity of the composite materials depends strongly on the volume ratio of two phases. Theoretically, the high-performance dual-phase membranes can be designed by selecting appropriate materials and adjusting the ratio of two phases. Nowadays, the oxygen ionic conductors for dual-phase membranes are mainly the doped fluorites such as Y_2O_3-stabilized ZrO_2, Er_2O_3-doped Bi_2O_3, Gd_2O_3-doped CeO_2, $Ce_{0.8}Sm_{0.2}O_{2-\delta}$ while the electronic conductors can be noble metals or perovskites exhibiting high electronic conductivity [9–13].

  A more detailed description of dense ceramic oxygen permeable membrane materials is beyond the scope of this book and can be found elsewhere [2–6].

5.2.2 Oxygen Permeation Flux in MIEC Membranes

The oxygen permeation in MIEC membranes takes place under an oxygen partial pressure gradient at elevated temperatures. In the presence of oxygen gas, the oxygen vacancies on the membrane surface can be filled by the oxygen atoms accompanied by consumption of electrons (n-type conductor) or formation of electron holes (p-type conductor) according to the following reactions:

$$\frac{1}{2}O_2 + V_O^{\bullet\bullet} + 2e' \xleftrightarrow{k_f/k_r} O_O^{\times} \text{ (for n-type conductor)} \qquad (5.2a)$$

$$\frac{1}{2}O_2 + V_O^{\bullet\bullet} \xleftrightarrow{k_f/k_r} O_O^{\times} + 2h^{\bullet} \text{ (for p-type conductor)} \qquad (5.2b)$$

where the charged defects are defined using the Kröger–Vink notation, that is, $O_O^{\times}$ stands for lattice oxygen, $V_O^{\bullet\bullet}$ for oxygen vacancy, and $h^{\bullet}$ for electron hole. k_f and k_r are the forward and reverse reaction rate constants for the surface exchange reaction, respectively. It should be noted that the above-surface exchange reaction may involve many sub-steps – such as oxygen adsorption, dissociation, recombination, and charge transfer [14] – which can be represented as follows:

1. $O_{2(g)} + S_{(ad)} \leftrightarrow O_{2(ad)}$ (adsorption)
2. $O_{2(ad)} + e' \leftrightarrow O'_2$ (charge transfer)
3. $O_{2(ad)} + e' + S_{(ad)} \leftrightarrow 2O'_{(ad)}$ (dissociation)
4. $O'_{(ad)} + e' \leftrightarrow O''_{(ad)}$ (charge transfer)
5. $O''_{(ad)} + V_O^{\bullet\bullet} \leftrightarrow O_O^{\times}$ (incorporation of oxygen ions into the lattice)

The equilibrium between the electrons and holes is given by

$$0 \longleftrightarrow e' + h^{\bullet}$$

These surface reactions are thus associated with transport of charges e', $h^{\bullet}$, $V_O^{\bullet\bullet}$ at or close to the surface of the oxide.

Figure 5.3 demonstrates schematically the oxygen permeation process through an MIEC membrane including the following steps in series: (i) oxygen molecular diffusion from the gas stream to the membrane surface

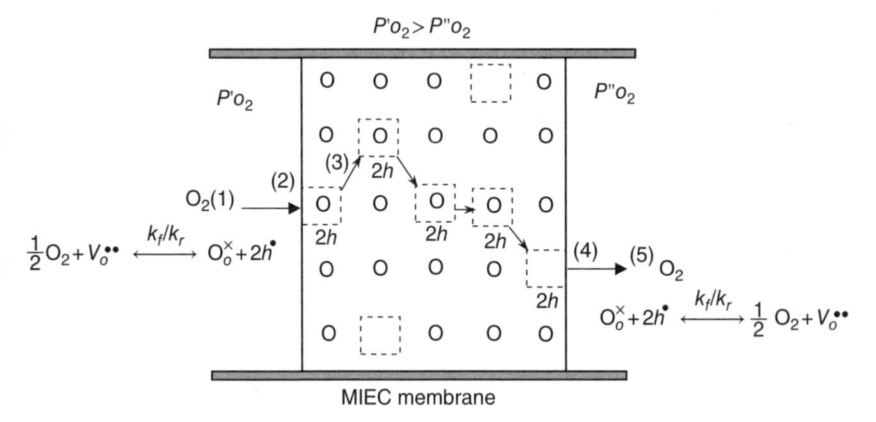

Figure 5.3 Oxygen permeation through a mixed ionic–electronic conducting membrane: □ represents oxygen vacancy.

(high-pressure side); (ii) reaction between molecular oxygen and oxygen vacancy on the membrane surface (high-pressure side); (iii) bulk diffusion of oxygen vacancy across the membrane; (iv) reaction between lattice oxygen and electron-hole on the membrane surface (low-pressure side); and (v) mass transfer of oxygen from the membrane surface to the gas stream (low-pressure side). Generally, the gaseous transfer resistances of steps 1 and 5 can be negligible, and the oxygen permeation is controlled by the diffusion of oxygen ions in the membrane and/or the surface oxygen exchange kinetics on either or both sides of the membrane.

The transport flux of charged species in electrolyte membranes can be described by the Nernst–Planck equation

$$J_i = -\frac{\sigma_i}{z_i^2 F^2} \nabla \mu_i + C_i \upsilon \tag{5.3}$$

where σ_i, μ_i, z_i, C_i are the conductivity, electrochemical potential, charge number, and concentration of species i, respectively; υ is the local velocity of inert marker; F is the Faraday constant. The electrochemical potential for each charged species consists of a chemical potential or an activity term and a local electrostatic potential term, ϕ:

$$\mu_i = \mu_i^0 + RT \ln a_i + z_i F \phi \tag{5.4}$$

where μ_i^0, a_i, ϕ, R, and T are the standard chemical potential, activity, galvanic (internal) potential, gas constant, and temperature, respectively. For the ideal state, the activity of a defect can be replaced by its concentration (with unit activity coefficient).

The conductivity of a defect can be correlated with its concentration and diffusivity, which is a measure of the random motion of species i in the lattice, using the Nernst–Einstein equation

$$\sigma_i = \frac{z_i^2 F^2}{RT} C_i D_i \tag{5.5}$$

where D_i is the diffusion coefficient of charged species i.

The relationship between the flux of mobile ions and the current density through the membrane can be expressed as

$$I = F \sum_i z_i J_i \tag{5.6}$$

For MIEC membranes without external circuit, the overall charge balance is applied or $\sum z_i J_i = 0$ and the local velocity of inert marker is negligible, $\upsilon = 0$. Accordingly, the transport flux of charged defects in the MIEC membrane at steady state can be derived (one-dimensional model) from Eqs (5.3)–(5.6) as [15]

$$J_i = -D_i C_i \left[\frac{1-t_i}{C_i} \cdot \frac{dC_i}{dx} - \sum_{j \neq i} \frac{z_i}{z_j} \cdot \frac{t_j}{C_j} \cdot \frac{dC_j}{dx} \right] \tag{5.7}$$

where t_i is the transport number of defect i:

$$t_i = \frac{\sigma_i}{\sum \sigma_j} = \frac{z_i^2 D_i C_i}{\sum_j z_j^2 D_j C_j} \tag{5.7a}$$

In the oxygen-permeable MIEC membranes, oxygen vacancies and electron holes are the primary mobile charge carriers, and the oxygen vacancy flux may be derived from Eq. (5.7) as

$$J_V = -\frac{(C_h + 4C_V) D_V D_h}{C_h D_h + 4 C_V D_V} \cdot \frac{dC_V}{dx} \tag{5.8}$$

where subscripts h and V represent hole and oxygen vacancy, respectively.

Based on the stoichoimetric relationship between oxygen and vacancy, the oxygen bulk diffusion flux in the membrane can be given by

$$J_{O_2} = -\frac{1}{2} J_V = \frac{1}{2L} \int_{(I)}^{(II)} \frac{(C_h + 4C_V) D_V D_h}{C_h D_h + 4 C_V D_V} \cdot dC_V \tag{5.9}$$

where L is the membrane thickness. It can be seen that the oxygen permeation flux through the MIEC membranes is determined by the inherently conducting properties of the membrane material and is inversely proportional to the membrane thickness. In order to obtain high oxygen permeation fluxes, the membrane has to possess high and equivalent oxide ionic and electronic conductivities.

For perovskite membranes, the electronic conductivity overwhelms the ionic conductivity, that is, $C_h D_h \gg C_V D_V$ and $C_h \gg C_V$ [16, 17]. Therefore, Eq. (5.9) reduces to

$$J_{O_2} = \frac{D_V (C_V'' - C_V')}{2L} \tag{5.10}$$

where C'_V and C''_V are the concentrations of oxygen vacancies at the high and low oxygen pressure sides of the membrane, respectively.

On the other contrary, the surface exchange reaction rates integrated with all sub-steps on the O_2-rich and O_2-lean sides can be given, respectively, by [18]

$$J_{O_2} = k_f \left(p'_{O_2}\right)^{0.5} C'_V - k_r \tag{5.11a}$$

$$J_{O_2} = k_r - k_f \left(p''_{O_2}\right)^{0.5} C''_V \tag{5.11b}$$

where k_f and k_r are, respectively, the forward and reverse reaction rate constants for the surface reactions; p'_{O_2} and p''_{O_2} are the oxygen partial pressures on the O_2-rich and O_2-lean side, respectively. It should be noted that Eq. (5.11) is based on the fact that the electron holes are essentially constant on membrane surfaces due to the overwhelming electronic conductivity in perovskites, and thus the exchange reaction rates may be pseudo zero order with respect to electron-hole concentration at steady state under isothermal conditions.

A combination of Eqs (5.10) and (5.11) gives the overall oxygen permeation flux through the perovskite membranes in terms of the partial pressure and membrane thickness as

$$J_{O_2} = \frac{k_r\left[\left(p'_{O_2}\right)^{0.5} - \left(p''_{O_2}\right)^{0.5}\right]}{\left(p''_{O_2}\right)^{0.5} + \dfrac{2Lk_f}{D_V}\cdot\left(p'_{O_2}p''_{O_2}\right)^{0.5} + \left(p'_{O_2}\right)^{0.5}} \tag{5.12}$$

In the case of bulk diffusion as the rate-limiting step, the resistance of the surface exchange reactions to oxygen permeation can be neglected. Hence, the oxygen permeation flux is then given with the assumption of equilibrium surface exchange reactions as

$$J_{O_2} = \frac{D_V k_r}{2Lk_f}\left[\left(p''_{O_2}\right)^{-0.5} - \left(p'_{O_2}\right)^{-0.5}\right] \tag{5.12a}$$

For surface exchange kinetics controlling the permeation process, the O_2 permeation flux is given by

$$J_{O_2} = \frac{k_r\left[\left(p'_{O_2}\right)^{0.5} - \left(p''_{O_2}\right)^{0.5}\right]}{\left(p''_{O_2}\right)^{0.5} + \left(p'_{O_2}\right)^{0.5}} \tag{5.12b}$$

Following the same process, the oxygen permeation flux through tubular MIEC membranes can be given by [19]

$$J_{O_2} = \frac{k_r\left[\left(p'_{O_2}\right)^{0.5} - \left(p''_{O_2}\right)^{0.5}\right]}{\frac{R_m}{R_o}\cdot\left(p''_{O_2}\right)^{0.5} + \frac{2k_f\left(R_o - R_{in}\right)}{D_V}\cdot\left(p'_{O_2}p''_{O_2}\right)^{0.5} + \frac{R_m}{R_{in}}\cdot\left(p'_{O_2}\right)^{0.5}} \tag{5.13}$$

where p'_{O_2} and p''_{O_2} are the oxygen partial pressures on the outer and inner surface of the tubular membrane, respectively; R_m is the logarithmic mean radius, $R_m = (R_o - R_{in})/\ln(R_o/R_{in})$, in which R_o and R_{in} are respectively the outer and inner radius of the tube (cm).

The permeation kinetic parameters D_V, k_f, and k_r can be determined by regressing the experimental oxygen fluxes measured at different conditions with the above permeation model. Table 5.1 lists the expressions of permeation parameters for the $La_{0.6}Sr_{0.4}Co_{0.2}Fe_{0.8}O_{3-\delta}$ perovskite membrane. The diffusion coefficients of oxygen vacancies in other perovskite membranes were also studied by the isotopic method, as summarized in Table 5.2. The activation energy for oxygen vacancy diffusion is in the range of 77 ± 21 kJ mol^{-1} [20].

Table 5.1 Pre-exponential coefficient and activation energy of D_V, k_f, and k_r for the $La_{0.6}Sr_{0.4}Co_{0.2}Fe_{0.8}O_{3-\delta}$ perovskite membrane [18]

Expression	Pre-exponential coefficients	Activation energy (kJ mol^{-1})
$D_V = D_V^0 \exp\left(-\frac{E_D}{RT}\right)$	(1.58 ± 0.01) × 10^{-2} cm^2 s^{-1}	73.6 ± 0.2
$k_f = k_f^0 \exp\left(-\frac{E_f}{RT}\right)$	(5.90 ± 0.08) × 10^{6} cm^2 atm$^{-0.5}$ s^{-1}	226.9 ± 0.2
$k_r = k_r^0 \exp\left(-\frac{E_r}{RT}\right)$	(2.07 ± 0.02) × 10^{4} mol cm^{-1} s^{-1}	241.3 ± 0.1

Table 5.2 Diffusion coefficients of oxygen vacancy in perovskite membranes

Membrane composition	D_V (cm^2 s^{-1})	Temperature (°C)	Ref.
$La_{0.9}Sr_{0.1}FeO_{3-\delta}$	6×10^{-6}	1000	[21]
$La_{0.2}Sr_{0.2}Co_{0.2}Fe_{0.8}O_{3-\delta}$	9.8×10^{-6}	850	[22]
$SrFe_{0.67}Co_{0.33}O_{3-\delta}$	4.73×10^{-6}	1000	[23]
$Sm_{0.5}Sr_{0.5}CoO_{3-\delta}$	8.6×10^{-7}	915	[24]
$SrFe_{0.2}Co_{0.8}O_{3-\delta}$	5×10^{-6}	870	[25]

5.3 PREPARATION OF DENSE CERAMIC MEMBRANES

The preparation of dense ceramic membranes usually consists of three steps: powder synthesis, shaping, and sintering. Each step plays particular roles in membrane microstructure, and thus in membrane performance. For example, powder synthesis, as the first step, plays a critical role in determining the particle size of the powder, and consequently takes effect on the microstructure of the membrane [26]. The defects or macro voids are formed mainly during the shaping process. After shaping, the membrane microstructure will be determined further by the sintering conditions including atmosphere, sintering profile, heating/cooling rate, highest sintering temperature, and dwell time. There are many routes to synthesize oxide-conducting ceramic powders, such as: solid-state reaction, spray pyrolysis, chemical co-precipitation, sol-gel process, and so on. However, membrane preparation refers mainly to the shaping process. Different fabrication methods are applied depending on the membrane configuration: disk/flat sheet, tube, or hollow fiber.

5.3.1 Isostatic Pressing

The cold isostatic pressing (CIP) method is generally applied to prepare disk/flat sheet-shaped membranes. Ceramic powders are milled into fine particles and then mixed with binder to form a uniform membrane precursor. This is then pressed into disks in a stainless steel mold under an isostatic or hydraulic pressure, followed by sintering at high temperature. Such disk-shaped membranes usually have a thickness of about 1 mm so as to provide enough mechanical strength. The disk membrane provides a very limited effective area (around a few square centimeters) for oxygen permeation and low permeation flux due to the large thickness.

CIP combined with the green machining method can be applied to fabricate tubular membranes with one dead-end [27]. The uniform membrane precursor is first pressed into green cylindrical rods. A carbide bit is used to drill the rods into dead-end geometry. The dead-end tubes are then sintered and annealed at elevated temperatures into dense membranes.

5.3.2 Extrusion

Tubular membranes are usually prepared by a plastic extrusion method [28]. Calcined and milled ceramic powder is mixed with several additives to make a slip with enough plasticity to easily be formed into tubes while retaining satisfactory strength in the green state. The additives include a solvent (butanol or xylene), a binder, and a plasticizer. After the slip is prepared, some of the solvent is allowed to evaporate to yield a plastic mass that is forced through a die at a high pressure to produce hollow tubes. The dimension of the extruded tube can be controlled by the orifice diameters of the die. The extruded tube is heated in a temperature range of 150–400°C to facilitate the removal of gaseous species formed during decomposition of the organic additives. Then, the tube is sintered at high temperature in stagnant air for a certain duration to obtain dense ceramic membrane tubes. Generally, a membrane prepared by extruding is less denser than one prepared by isostatic pressing [29].

5.3.3 Phase Inversion

The phase inversion method described in Chapter 2 has also been applied widely to fabricate dense ceramic hollow fiber membranes [30]. The spinning and sintering conditions have to be modulated to ensure the gas-tight property of the membranes. The properties of the dense ceramic membranes are strongly dependent on their morphology and microstructure (i.e., the size and shape of grains, porosity, pore size, pore size distribution, etc.) of the hollow fibers. Theoretically, the hollow fiber morphology can be tailored as expected for different applications by modulation of the suspension composition and the spinning parameters [31, 32]. Figure 5.4 shows SEM images of $La_{0.6}Sr_{0.4}Co_{0.2}Fe_{0.8}O_{3-\delta}$ hollow fiber membranes prepared using different bore liquids. However, it is still a great challenge to design precisely and control the macro- and microstructure of the ceramic hollow fiber membranes because too many factors – such as the particle size and its distribution, the shape

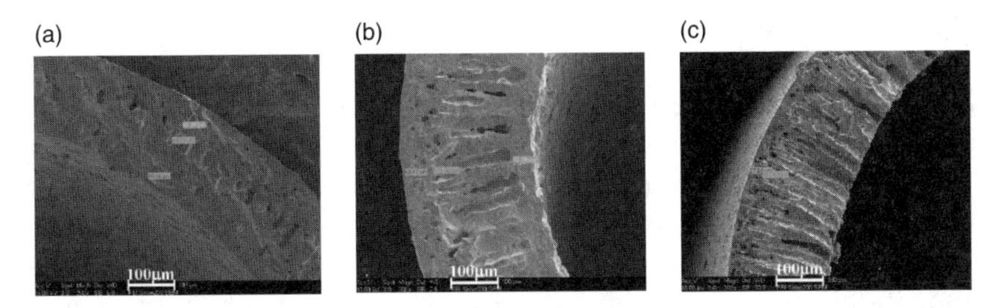

Figure 5.4 Sectional SEM images of the $La_{0.6}Sr_{0.4}Co_{0.2}Fe_{0.8}O_{3-\delta}$ hollow fiber membranes prepared using (a) 100% EtOH, (b) 50% EtOH–NMP, and (c) 10% EtOH–NMP as the bore liquid solution. Reproduced from [31, 32]. With permission from Elsevier.

and surface property of ceramic powders, the composition and viscosity of the spinning suspension, the spinning conditions (spinning rate, air gap, internal coagulant, etc.), and the sintering parameters (sintering temperature, dwell time, heating rate) – can solely or jointly affect the formation of membrane structures. Therefore, a well-designed phase inversion/sintering process coupled with an optimal spinning suspension is the key to obtaining asymmetric hollow fibers with the desired permeation characteristics as well as excellent mechanical strength.

5.3.4 Slurry Coating

The slurry coating method is used to fabricate composite ceramic membranes consisting of a thin dense ceramic film supported on porous substrates [33–35]. To start with, a green porous substrate is prepared by the CIP or plastic extrusion method. A coating slurry is then prepared by dispersing the finely pulverized ceramic powder in a water or isopropanol solution together with a dispersant, followed by being ball-milled for a few hours. The slurry is then coated on the green porous substrate by various techniques such as slip casting, spin coating, or dip coating. The green body is dried at room temperature, and finally sintered at a designed sintering profile. The thickness of the membrane can be controlled by the slurry concentration or by using multiple coating/drying/sintering cycles. In order to densify the perovskite powders coated onto the surfaces into dense films, an exact procedure should be conducted. Fissures and cracks can be minimized by pre-sintering the substrate so as to match its subsequent shrinkage with that of the coated layers.

5.3.5 Tape Casting

Flat-sheet thin, dense ceramic membranes can be fabricated by the tape casting and lamination method [36, 37]. A shear-thinning slurry is prepared from the attrition-milled ceramic powder by adding a binder (e.g., methyl methacrylate), a plasticizer (e.g., dibutyl phthalate), and a dispersing agent in an azeotropic solvent of butanone-2 and ethanol. A cohesive and flexible green tape with a controlled thickness is obtained using a doctor blade tape-casting apparatus. The green tape is then applied to a porous tape-cast support by lamination (application of pressure and heat between two rollers). Before sintering, a very slow de-bindering cycle is applied to avoid damaging the laminated structure. The laminated structure is finally co-sintered in air at an elevated temperature into dense membranes.

5.4 DENSE CERAMIC MEMBRANE REACTORS

5.4.1 Principles of Dense Ceramic Membrane Reactors

The dense ceramic membranes applied in MRs can be either MIECs or pure ionic conductors (electrolytes). In the MIEC-MRs, the membrane itself serves as the internal circuit for electron transport, whereas an external circuit for electron transport has to be provided in the pure ionic conductor MRs (EMRs). The principles of the dense ceramic MRs are illustrated schematically in Figure 5.5.

1. *Electrochemical pump membrane reactors* (*EP-MRs*). When using pure ionic conducting membranes, an external power source has to be applied to generate the electrochemical potential gradient for oxygen permeation. Two electrodes of metals or MIEC oxides are attached to the membrane surfaces to distribute/collect electrons and the electrical current passing through the cell is controlled by a potentiostat as shown in Figure 5.5(a) [38]. Oxygen is pumped electrochemically from one side to the other of the membrane, regardless of the oxygen concentration gradient. Reactions take place on the membrane surface or in a packed catalyst bed. The oxygen flux transferring across the membrane can be controlled externally by varying the electrical current.
2. *Solid oxide fuel cell membrane reactors* (*SOFC-MRs*). When an oxygen partial pressure gradient is present across the pure ionic conductor membrane, the oxygen can be permeated through the membrane with the aid of an external circuit to transport electrons, as shown in Figure 5.5(b). Gaseous oxygen is reduced on the cathode catalyst into O^{2-} and transferred through the membrane to the anode, where oxidation reactions take place. The electrons released by chemical reaction on the anode return to the cathode via an external circuit so that the energy of oxidation is converted into electrical power.
3. *MIEC membrane reactors* (*MIEC-MRs*). In an MIEC-MR, oxygen permeates through the membrane under the oxygen partial pressure gradient at high temperatures without the need for electrodes

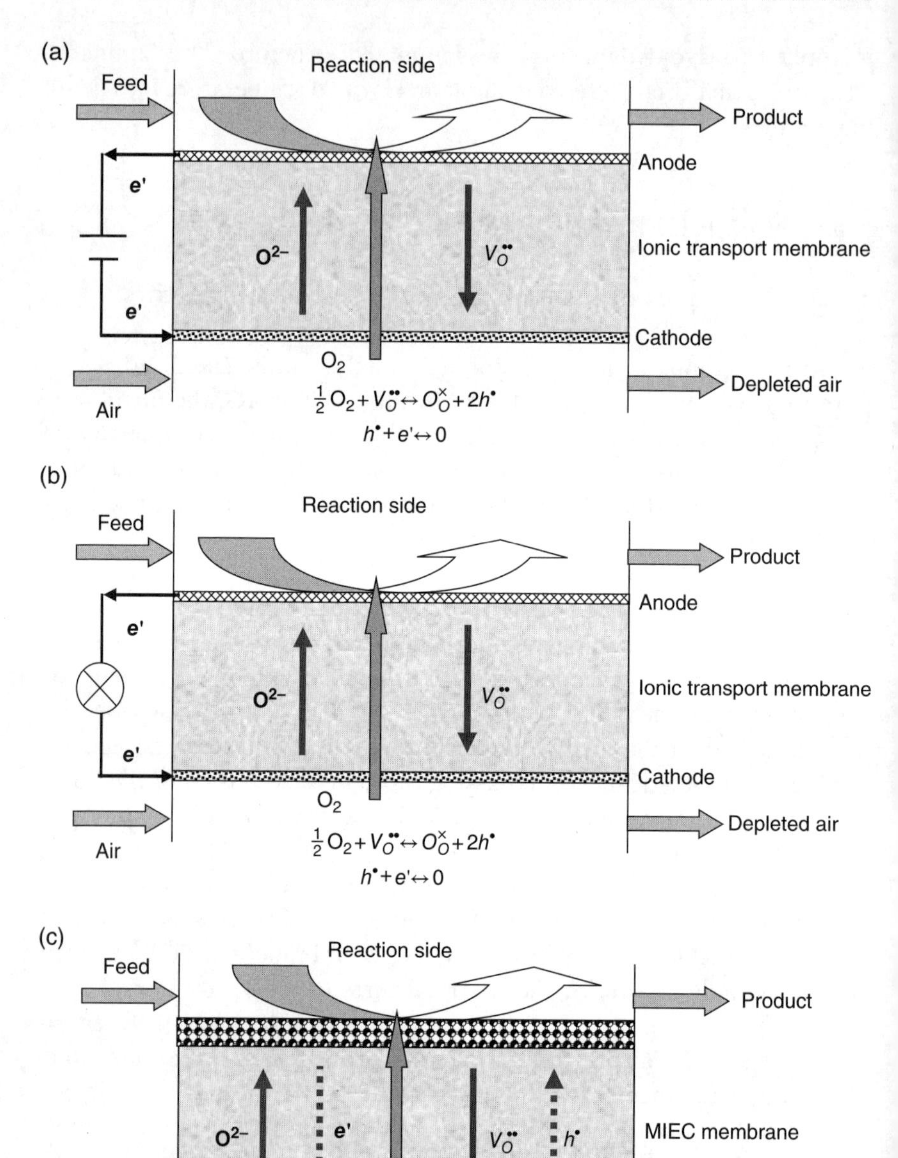

Figure 5.5 Principles of dense ceramic membrane reactors: (a) electrochemical pump membrane reactor (EP-MR); (b) solid oxide fuel cell membrane reactor (SOFC-MR); (c) mixed ionic–electronic conducting membrane reactor (MIEC-MR).

and external circuits. The membrane itself serves as the internal circuit for electron transport, as shown in Figure 5.5(c). The oxygen on the air side is adsorbed and dissociated into oxygen ions, which are then transported by diffusion to the reaction side to react with the reactant feed. Therefore, the structure of MIEC-MRs is much simplified and thus is applied most often to various reactions.

5.4.2 Configurations of Dense Ceramic Membrane Reactors

There are three main types of dense ceramic membranes: disk/flat sheet, tubular, and hollow fibers. The disk/flat sheet membranes are applied mostly in research work because they can be fabricated easily in laboratories with a small amount of membrane material. Comparatively, the hollow fiber membranes can provide the largest membrane area per volume but low mechanical strength, while the tubular membranes possess a satisfactory specific membrane area, high mechanical strength, and are easy to assemble in membrane reactors. Dense ceramic MRs can be constructed and operated in either packed bed MR or catalytic MR configurations.

PBMRs. This configuration is applied mostly in practical use of dense ceramic MRs. The reactions take place in the catalyst bed while the membrane functions mainly as an oxygen distributor or extractor. Since the catalyst is separated physically from the membrane, the separation function of the membrane and the catalytic properties of catalysts can be modulated separately so that the MR performance will be optimized.

CMRs. It is well established that MIECs are inherently catalytic to oxidation reactions. Therefore, MIEC-derived membranes may serve as both catalyst and oxygen separator, and no other catalysts are used in the dense ceramic MRs. Since chemical reactions take place on the membrane surface, it is required to have a much more porous membrane surface so as to contain a sufficient quantity of active sites. This can be achieved in the membrane preparation process, or by coating a porous membrane material after preparation. The main potential problems for this configuration are that the membrane may not have sufficient catalytic activity and the catalytic selectivity cannot be modulated with respect to the reactions considered.

In order to overcome this problem, an additional catalyst can be coated on the membrane surface to form a new catalytic dense membrane [36, 37]. The catalyst layer is generally porous and integrated with the membrane into a single body. The catalyst layer can also be formed together with the membrane in a single step as shown by Wu *et al.* [39].

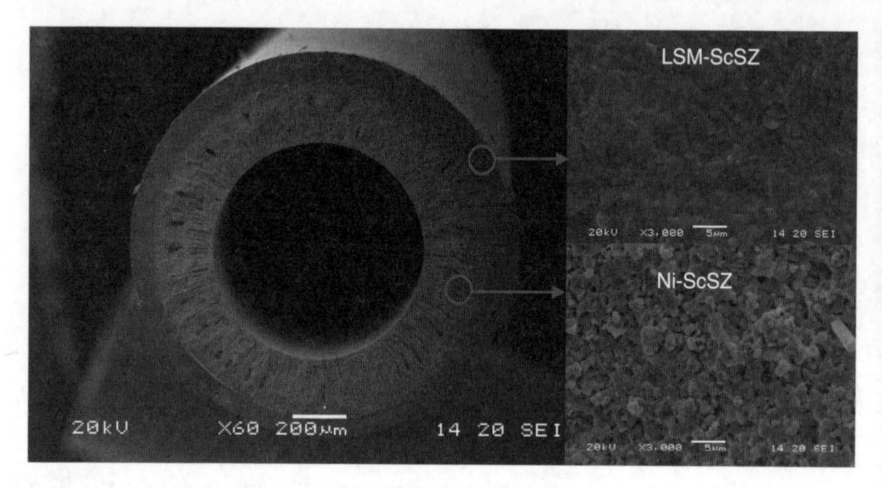

Figure 5.6 SEM images of LSM–ScSZ/Ni–ScSZ hollow fiber membrane. Reproduced from [39]. With permission from Elsevier.

In their study, the catalyst and membrane powders are made into the spinning suspension with different compositions. The two suspensions are co-extruded through a triple-orifice spinneret into a coagulant bath for solvent exchange to form hollow fiber precursors. Thereafter, high-temperature co-sintering is performed to obtain the catalyst/membrane dual-layer hollow fibers. Figure 5.6 shows the morphology of the LSM–ScSz/Ni–ScSZ dual-layer hollow fibers, where LSC–ScSZ serves as oxygen separation membrane and Ni–ScSZ as catalyst for methane conversion. This method requires that the catalyst and the membrane layers should have matching sintering behaviors to avoid cracks and achieve greater adhesion between layers.

5.5 APPLICATIONS OF DENSE CERAMIC OXYGEN PERMEABLE MEMBRANE REACTORS

Dense ceramic oxygen permeable MRs have been studied extensively for potential applications in the partial oxidation of hydrocarbons, where the membrane acts as an oxygen distributor. From a membrane performance perspective, the motivations include: (i) to supply oxygen for the reaction system in a more controllable manner and to maintain the oxygen concentration at a uniformly low value, thus resulting in higher selectivity; (ii) to use air directly as the oxygen source without contaminating the products with nitrogen and nitrogen oxides, leading to remarkably reduced capital investment and operation costs; (iii) to avoid the premixing of hydrocarbon feed with oxygen and thus to reduce the formation of hot spots as encountered in a co-feed reactor, leading to safer operation; and (iv) to strongly reduce the atmosphere created by the reaction products and provide a large oxygen potential gradient to facilitate oxygen transport through the membrane.

5.5.1 Partial Oxidation of Methane to Syngas

The direct catalytic partial oxidation of methane (POM) is a mildly exothermic reaction:

$$CH_4 + \frac{1}{2}O_2 = CO + 2H_2, \Delta H^0_{298} = -36\,\text{kJ}\,\text{mol}^{-1}$$

It yields an ideal feedstock with H_2/CO ratio of 2:1 for methanol synthesis and the Fisher–Tropsch reaction to produce linear hydrocarbons. The use of dense ceramic MRs makes it possible to combine oxygen separation from air, partial oxidation, and reforming of methane in a single step, thus enabling significant reductions of capital investment in the gas-to-liquid industry [40]. Although perovskite membranes may exhibit catalytic activity in the POM reaction, a POM catalyst is usually applied to improve the methane conversion and CO selectivity. Figure 5.7

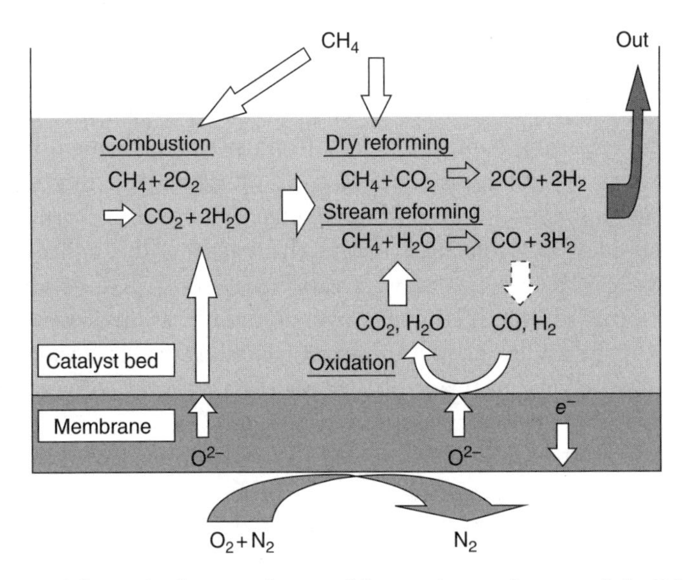

Figure 5.7 Schematic diagram for possible reaction pathways of the POM in a dense ceramic MR.

shows the process of the POM reaction in a dense ceramic MR packed with POM catalysts. Methane and air are fed respectively to opposite sides of the membrane reactor. Under an electrochemical potential gradient, the oxygen in air is permeated from the air side through the membrane at a high temperature to the reaction side to react with methane into syngas.

In the POM process, the oxide membrane provides only molecular oxygen to react at the catalyst because the lattice oxygen of the membrane is not active in breaking the C–H bond but oxidizes preferentially hydrogen to water [41]. Some of the CO and H_2 formed by the reforming reactions is oxidized at the membrane surface into CO_2 and H_2O:

$$CO + O^{2-} = CO_2 + 2e'$$

$$H_2 + O^{2-} = H_2O + 2e'$$

As a result, the oxygen permeation flux under POM reaction conditions is much higher than that when helium is used as sweep gas at the same temperature [42]. Furthermore, the POM catalyst also has a significant influence on the oxygen permeation of the membrane, but the mechanism is not very clear [43].

Of all the potential applications of dense ceramic MRs, the POM to syngas is thought to be the most commercially important one. Considerable efforts have been made in this field in the last two decades, as summarized in Table 5.3.

The overall performance of the dense ceramic MRs in terms of methane conversion and CO selectivity is strongly dependent on the reactor design and operating conditions [44]. In most cases the membrane acts mainly as an oxygen supplier and distributor, while its catalytic properties are less important because of the high activity of the reforming catalyst. The simulation results indicate that the reactor with smaller diameter (D) and greater length-to-diameter ratio (L/D) may give better performance in terms of high methane conversion and high hydrogen concentration [45]. In addition, the amount of catalyst packed and the feed contact time have significant effects on the methane conversion, CO selectivity, and oxygen permeation rate [64, 65]. In order to achieve better production, the amount of catalyst must match well with the available membrane area. Nevertheless, it is generally considered that the methane conversion is controlled mainly by the oxygen permeation rate rather than the reaction rate at the catalyst.

MIEC MRs for the partial oxidation of methane to syngas

Membrane composition	Reactor configuration	Catalyst	Temperature (°C)	Main results	Stability (h)	Ref.
$La_{0.8}Sr_{0.2}MnO_{3-\delta}$–YSZ	Hollow fiber; CMR	NiO–YSZ	950	$S_{CO} = 90\%$; $H_2/CO \approx 2$		[39]
$La_{0.4}Ba_{0.6}Fe_{0.8}Zn_{0.2}O_{3-\delta}$	Disk; PBMR	Ni–Ca	900	$X \rightarrow 100\%$; $S_{CO} > 95\%$	500	[42]
$BaCo_{0.4}Fe_{0.4}Zr_{0.2}O_{3-\delta}$	Disk; PBMR		850	$X = 98\%$; $S_{CO} = 100\%$; $J_{O_2} = \mathbf{5.6\ ml\ cm^{-2}\ min^{-1}}$	2200	[43]
$La_{0.8}Sr_{0.2}Fe_{0.7}Ga_{0.3}O_{3-\delta}$	Tube; CMR	$La_{0.8}Sr_{0.2}Fe_{0.7}Ni_{0.3}O_{3-\delta}$	900	$X = 74\%$; $S_{H2} > 50\%$	142	[44]
$SrCo_{0.4}Fe_{0.5}Zr_{0.1}O_{3-\alpha}$	Disk; PBMR	NiO/Al_2O_3	950	$S_{CO} > 90\%$		[45]
$SrFeCo_{0.5}O_x$; $SrFe_{0.2}Co_{0.8}O_x$	Tube; PBMR	Ru-based	850	$X > 99\%$; $S_{CO} > 98\%$	>1000	[46]
$Ce_{0.8}Sm_{0.2}O_{2-\delta}$–$La_{0.8}Sr_{0.2}CrO_{3-\delta}$	Tube; PBMR	$Ca_{0.8}Sr_{0.2}TiO_3$	950	$X = 17\%$; $S_{H2} > 75\%$		[47]
$Sm_{0.4}Ba_{0.6}Co_{0.2}Fe_{0.8}O_{3-\delta}$	Disk; PBMR	Rh/MgO	900	$X = 90\%$; $S_{CO} = 98\%$		[48]
$La_{0.6}Sr_{0.4}Co_{0.2}Fe_{0.8}O_{3-\delta}$	Tube; PBMR	Ni/γ-Al_2O_3	825–885	$X > 96\%$; $S_{CO} > 97\%$	3–7	[49]
La_2NiO_4	Tube		900	$X = 89\%$; $S_{CO} = 96\%$; $H_2/CO = 1.5$		[50]
$Ba_{0.5}Sr_{0.5}Co_{0.8}Fe_{0.2}O_{3-\delta}$	Tube; PBMR	LiLaNiO/γ-Al_2O_3	875	$X = 94\%$; $S_{CO} > 95\%$; $J_{O_2} = \mathbf{8.0\ ml\ cm^{-2}\ min^{-1}}$	500	[51]
YSZ–$SrCo_{0.4}Fe_{0.6}O_{3-\delta}$	Disk; PBMR	NiO/Al_2O_3	750–850	$X = 64\%$; $S_{CO} \sim 100\%$	220	[52]
$La_{0.5}Sr_{0.5}Fe_{0.8}Ga_{0.2}O_{3-\delta}$/α-$Al_2O_3$	Tube; PBMR	Ru-based	850	$X = 97\%$; $S_{CO} = 100\%$; $H_2/CO = 1.76$		[53]
$Ca_{0.8}Sr_{0.2}Ti_{0.7}Fe_{0.3}O_{3-\alpha}$	Disk; CMR	Ni–$Ca_{0.8}Sr_{0.2}Ti_{0.9}Fe_{0.1}O_{3-\alpha}$	900	$X = 13.7\%$; $S_{CO} = 98\%$		[54]
$YBa_2Cu_3O_{7-x}$	Disk; PBMR and CMR	Ni/ZrO_2	875	$X = 100\%$; $S_{CO} = 95\%$	5	[55]
$SrFe_{0.7}Al_{0.3}O_{3-\delta}$	Disk; PBMR	$SrFe_{0.7}Al_{0.3}O_{3-\delta}$	950	$X = 65\%$; $S_{CO} = 48\%$		[56]
Ba(Co, Fe, Zr)$O_{3-\delta}$	Hollow fiber; PBMR	Ni-catalyst	925	$X = 96\%$; $S_{CO} = 97\%$; $H_2/CO \approx 2$		[57]

(Continued)

Table 5.3 (*Cont'd*)

Membrane composition	Reactor configuration	Catalyst	Temperature (°C)	Main results	Stability (h)	Ref.
$Sm_{0.15}Ce_{0.85}O_{1.925}$/ $Sm_{0.6}Sr_{0.4}Fe_{0.7}Al_{0.3}O_{3-\delta}$	Disk; PBMR	LiLaNiO/γ-Al_2O_3	950	$X > 98\%$; $S_{CO} > 98\%$; $H_2/CO \approx 2$; J_{O_2} = **4.3 ml cm^{-2} min^{-1}**	1100	[58]
$BaCo_{0.7}Fe_{0.2}Nb_{0.1}O_{3-\delta}$	Disk; PBMR	Ni-catalyst	875	$X = 92\%$; $S_{H2} > 90\%$ J_{O_2} = **15 ml cm^{-2} min^{-1}**	550	[59]
$Ce_{0.85}Sm_{0.15}O_{1.925}$/ $Sm_{0.6}Sr_{0.4}FeO_{3-\delta}$	Disk; PBMR	LiLaNiO/γ-Al_2O_3	950	$X > 98\%$; $S_{CO} > 98\%$	500	[60]
$BaCo_{0.7}Fe_{0.2}Ta_{0.1}O_{3-\delta}$	Disk; PBMR	Ni-catalyst	900	$X = 99\%$; $S_{H2} > 94\%$; J_{O_2} = **16.2 ml cm^{-2} min^{-1}**	400	[61]
$BaCe_{0.1}Co_{0.4}Fe_{0.5}O_{3-\delta}$	Disk; PBMR	LiLaNiO/γ-Al_2O_3	950	$X = 99\%$; $S_{CO} > 93\%$; J_{O_2} = **9.5 ml cm^{-2} min^{-1}**	1000	[62]
3% Al_2O_3-doped $SrCo_{0.8}Fe_{0.2}O_3$	Tube; PBMR	Ni-catalyst	900	$X = 99\%$; $S_{CO} > 93\%$		[63]

5.5.2 Oxidative Coupling of Methane

Oxidative coupling of methane (OCM) to C_2 products (C_2H_4 and C_2H_6) represents one of the most effective ways to convert natural gas to more useful products:

$$2CH_4 + \frac{1}{2}O_2 \longrightarrow C_2H_6 + H_2O,\ \Delta H^0_{298} = -177\,\mathrm{kJ\,mol^{-1}}$$

$$2CH_4 + O_2 \longrightarrow C_2H_4 + 2H_2O,\ \Delta H^0_{298} = -282\,\mathrm{kJ\,mol^{-1}}$$

It is considered that the OCM technology may be utilized commercially if a single-pass conversion of 35–37% and selectivity of 88–85%, equivalent to a C_2 yield of 30+%, is achieved. Most of the previous studies have focused on finding suitable catalysts for the selective methane conversion. However, it is hard to obtain C_2 yields higher than 25% in a conventional fixed bed reactor. This may be attributed to the competition between the coupling and the combustion reactions. In general, it is accepted that the initial step in the catalytic oxidative coupling of CH_4 involves the hemolytic cleavage of a C–H bond on the catalyst surface to form $\bullet CH_3$ radicals, which may undergo coupling to form ethane in the gaseous phase. The intermediate radicals and their products may undergo deep oxidation to carbon oxides in the presence of molecular oxygen. In order to improve the OCM selectivity, the oxygen concentration in the gas phase should be as low as possible, while the amount of methane provided should also be enough for high methane conversion.

Applications of porous or dense inorganic membrane reactors to control oxygen concentration along the reactors offer the possibility of achieving much higher C_2 hydrocarbon selectivity and yield for OCM [1]. Porous membranes such as alumina, zirconia, and vycor glass possess high stability but low oxygen selectivity. Comparatively, the use of dense ceramic membranes is more effective because oxygen may be distributed finely along the reactor, and methane loss due to back-permeation is prevented. Moreover, the ionic conduction of the ceramic membranes delivers the oxygen into the reaction compartment in the form of dissociated and ionized oxygen. This ionized oxygen reacts with methane on the membrane surface, following a different reaction mechanism so that the formation of CO_x from by-reactions due to the presence of gas-phase oxygen is suppressed. Figure 5.8 demonstrates the OCM process in the MIEC membrane reactors. Methane is adsorbed and reacts with the lattice oxygen ($O_O^{\times}$) and electron holes ($h^{\bullet}$) to form methyl

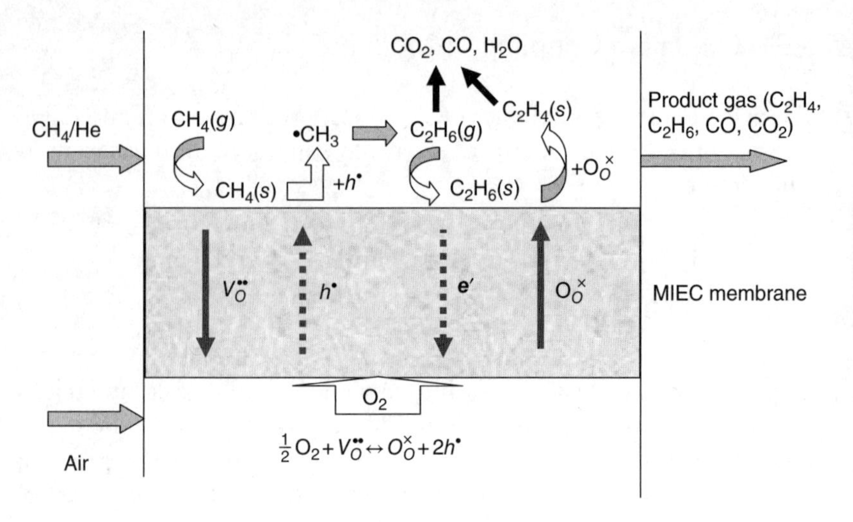

Figure 5.8 The mechanism of OCM in the MIEC-MR.

radicals, $CH_4(s) + h^{\bullet} + \frac{1}{2}O_O^{\times} \rightarrow \bullet CH_3 + \frac{1}{2}H_2O + \frac{1}{2}V_O^{\bullet\bullet}$, which are then coupled in gas phase to form C_2 products, or further react with gaseous oxygen to form carbon oxides.

Table 5.4 summarizes the literature results of OCM in oxygen–MIEC membrane reactors. It shows that the MIEC membrane may exhibit intrinsically catalytic activity toward the OCM reaction, since no OCM catalyst is applied. The C_2 yield is very sensitive to the membrane characteristics, reaction conditions, and reactor dimensions [66]. Because the complete oxidation reactions occurring in the gas phase and partially on the catalyst surface may lower the C_2 selectivity substantially, especially under conditions of high temperature and pressure, in order to achieve high C_2 yields the oxygen permeation flux, methane flow rate, and intrinsic reaction rate must match well with each other. Insufficient oxygen supply leads to poor conversion but a high oxygen flux may result in low selectivity because of the complete oxidation reactions – especially at high temperature and pressure [67]. This implies that the oxygen permeability of the membrane has to match with the catalytic activation of the membrane surface [68]. For a given composite membrane, the oxygen flux can readily be improved by decreasing the membrane thickness or improving the surface exchange kinetics. Therefore, the selection of a membrane material with good intrinsic catalytic properties or the modification of these high-oxygen-permeable ceramic membrane surfaces with an appropriate OCM catalyst (such as lead oxides and alkali

Table 5.4 MIEC membrane reactors for methane oxidative coupling

Membrane	Reactor configuration	Temperature (°C)	Main results	Ref.
$Ba_{0.5}Sr_{0.5}Co_{0.8}Fe_{0.2}O_{3-\delta}$	Disk coated with La–Sr/CaO catalyst	950	$Y_{C2} = 18\%$; $S_{C2} > 65\%$	[68]
$Ba_{0.5}Sr_{0.5}Co_{0.8}Fe_{0.2}O_{3-\delta}$	Tube, no catalyst, or packed with La–Sr/CaO catalyst	800–900	$S_{C2} = 62\%$; or $Y_{C2} = 13–15\%$; $S_{C2} = 54–58\%$	[70]
$BaCe_{0.8}Gd_{0.2}O_{3-\delta}$	Tube	778	$Y_{C2} = 16.5\%$; $S_{C2} = 62.5\%$	[71]
$La_{0.6}Sr_{0.4}Co_{0.8}Fe_{0.2}O_{3-\delta}$	Disk, no catalyst	800–900	$Y_{C2} = 1–3\%$; $S_{C2} \rightarrow 70\%$	[72]
Y-doped Bi_2O_3	Disk	750–950	$Y_{C2} = 16–14\%$; $S_{C2} = 20–90\%$	[73]
$La_{0.8}Sr_{0.2}Co_{0.6}Fe_{0.4}O_{3-\delta}$	Disk	850	$Y_{C2} = 10–18\%$; $S_{C2} = 70–90\%$	[74]
$La_{0.8}Sr_{0.2}CoO_3$	Disk	800–850	$Y_{C2} = 12–14$; $S_{C2} = 40–56\%$	[75]
$Bi_{1.5}Y_{0.3}Sm_{0.2}O_{3-\delta}$	Tube, no catalyst	900	$Y_{C2} = 35\%$; $S_{C2} = 54\%$	[69]
$La_{0.6}S_{r0.4}Co_{0.2}Fe_{0.8}O_3$	Hollow fiber packed with $SrTi_{0.9}Li_{0.1}O_3$ catalyst	780–980	$Y_{C2} \rightarrow 21\%$; $S_{C2} \rightarrow 71.9\%$	[76]

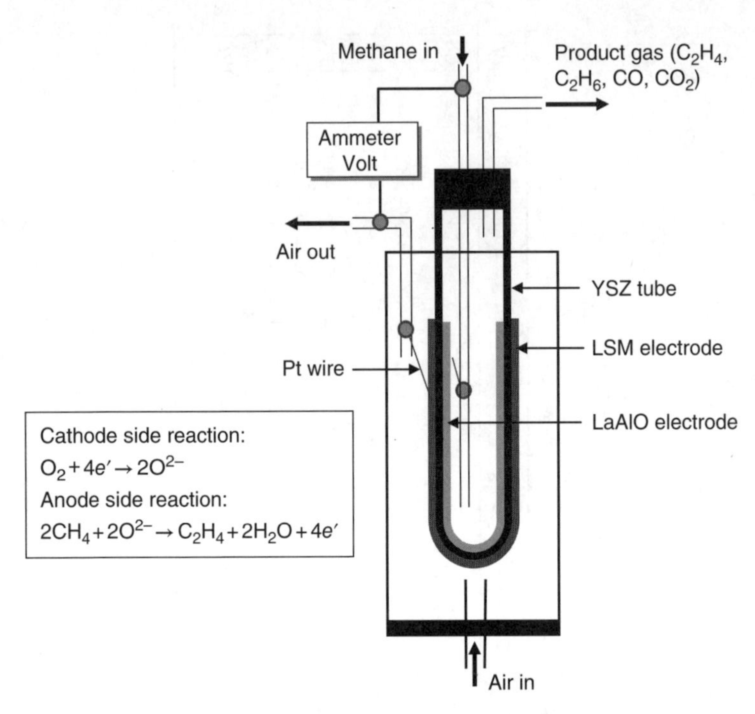

Figure 5.9 Schematic diagram of the YSZ-based SOFC membrane reactor for OCM. Reproduced from [77]. With permission from Elsevier.

compounds) has become the most critical step in the development of dense membrane reactors for OCM. Among the various membranes developed, $Bi_{1.5}Y_{0.3}Sm_{0.2}O_{3-\delta}$ exhibits not only high oxygen permeability and catalytic activity but also high chemical and mechanical stability under OCM conditions. The C_2 yield in the $Bi_{1.5}Y_{0.3}Sm_{0.2}O_{3-\delta}$ membrane reactor reached 35% [69].

In the SOFC-type membrane reactor made of pure ionic conducting membranes such as yttria-stabilized zirconia (YSZ, 8% Y_2O_3–ZrO_2), electrical power can be co-generated along with the OCM reaction. Figure 5.9 illustrates schematically a tubular YSZ-based SOFC-type membrane reactor for OCM reaction [77]. A YSZ tube with one dead-end is used as the electrolyte. $La_{0.85}Sr_{0.15}MnO_3$ powder is pounded and mixed with glycerol, pasted into thin film on the outside of the YSZ tube, and heated at elevated temperature to form the cathode. $La_{1.8}Al_{0.2}O_3$ prepared on the inside of the YSZ tube by a mist pyrolysis method is used as the OCM catalyst as well as the anode. Pt wire is connected to platinum mesh placed on both electrodes to serve as the current collector. Oxygen ions are transferred from the cathode through the membrane to the anode side and react with CH_4 to yield C_2 products.

The theoretical electromotive force is giving by

$$E = -\frac{\Delta G}{nF} \tag{5.14}$$

where ΔG is the Gibbs free energy and n is the number of electrons. The oxygen permeation rate is determined by the electrical current I:

$$F_{O_2} = \frac{I}{4F} \tag{5.15}$$

The anode catalyst plays a key role in the C_2 selectivity. For example, when silver was used as electrode and 1 wt% Sr/La_2O_3–Bi_2O_3 as catalyst, an electric current of 20–40 mA with C_2 selectivity of 90–94% and C_2 yield of 0.2–1% was obtained at 730°C [79]. When using $La_{1.8}Al_{0.2}O_3$ as anode catalyst, the electric current and C_2 yield could reach 180 mA and 4%, respectively [77]. However, all the membrane reactors tested in practice so far have not shown very high C_2 yields. This was attributed to the low oxygen permeation flux which did not match the catalytic activation of methane on the membrane surface. In fact, if an external power source is applied to form an EOPMR, the catalytic activity and C_2 selectivity of the metal and metal oxide catalysts can be altered dramatically and reversibly due to supplying more active oxygen species [80], leading to much higher C_2 yields. In general, the SOFC-type MR requires an operating temperature approximately 200 K higher than the others; the electricity generated simultaneously as a by-product still makes it attractive.

In summary, compared with the SOFC membrane reactor, the MIEC membrane reactor gives better performance in terms of C_2 yield due to the much higher methane conversion when operating at the same temperature. Furthermore, the MIEC-MR is superior to the porous MR at high temperature (>1150 K), but inferior at low temperature (<1150 K) [81]. Although the SOFC-MR requires an operating temperature approximately 200 K higher than the others, the electricity generated simultaneously as a by-product might still make it attractive.

5.5.3 Oxidative Dehydrogenation of Alkanes (Ethane and Propane)

Selective oxidation of alkanes such as ethane and propane to corresponding olefins is an important catalytic process:

$$C_2H_6 + \frac{1}{2}O_2 \Leftrightarrow C_2H_4 + H_2O,\ \Delta H^0_{298} = -105\,\mathrm{kJ\,mol^{-1}}$$

$$C_3H_8 + \frac{1}{2}O_2 \Leftrightarrow C_3H_6 + H_2O$$

The principle of the MIEC membrane reactor for oxidative dehydrogenation of ethane and propane is similar to the OCM process, but without the presence of methane coupling reactions. On the oxygen-rich side, gaseous O_2 is first adsorbed on the membrane surface, reduced to O^{2-}, and then transported through the bulk of the membrane to the reaction side surface. On the reaction side, ethane is oxidized by the surface O^{2-}. As the surface oxygen is depleted, the bulk O_2 diffuses from the oxygen-rich side to fill in the oxygen vacancies. The reaction mechanism on the reaction side is

$$C_2H_6(s) + O^{2-} \Leftrightarrow C_2H_4 + H_2O + 2e'$$

or

$$C_2H_6(s) + O_O^{\times} + 2h^{\bullet} \rightarrow C_2H_4 + H_2O + V_O^{\bullet\bullet}$$

Therefore, this type of operation allows complete control over the contact mode of reactants with each other, and with the catalytically active surface, and the selectivity of the oxidation reaction can be controlled at a very high level [78]. Table 5.5 summarizes the results of oxidative dehydrogenation of ethane/propane in dense ceramic MRs. The performance of the membrane reactor can be changed with application of surface catalyst. For example, by using BSCF membranes with V/MgO micrometer grain or Pd nanocluster-modified surfaces, the ethylene yield could reach 75% at 1040 or 1050 K, respectively. However, Ni cluster deposition leads to a decrease in ethane conversion compared with the bare membrane without changing the ethylene selectivity [27]. In addition, the contact time between the reactant and the membrane plays an important role in selectivity. Therefore, hollow fiber membrane reactors give lower selectivity than disk-shaped ones because of their longer contact time [82].

5.5.4 Decomposition of H_2O, NO_x, and CO_2

Oxygen-permeable membranes can also be used as extractors to selectively remove the oxygen produced in reactions so as to overcome the thermodynamic limitation and/or kinetic limitation and improve the

Table 5.5 Oxidative dehydrogenation of ethane/propane in dense ceramic MRs

Reaction	Membrane	Reactor configuration	Catalyst	Temperature (°C)	Main results	Ref.
$C_2H_6 \rightarrow C_2H_4$	BSCF	Tube	—	650	$S=90\%$	[78]
	BYS	Tube with a dead-end	—	875	$Y=56\%$; $S=80\%$	[83]
	BCFZ	Hollow fiber	—	800	$S=64\%$; $X=63\%$	[27]
	BSCF	Disk; coated catalyst	V/MgO	770	$Y=75\%$; $S>92\%$	[82]
$C_3H_8 \rightarrow C_3H_6$	BSCF	Tube with a dead-end	—	750	$S=23.8$–40.2%; $Y=17.1$–11.7%; $X=71.8$–29.0%	[84]

BSCF = $Ba_{0.5}Sr_{0.5}Co_{0.8}Fe_{0.2}O_{3-\delta}$; BCFZ = $BaCo_xFe_yZr_zO_{3-\delta}$ $(x+y+z=1)$; BYS = $Bi_{1.5}Y_{0.3}Sm_{0.2}O_3$.

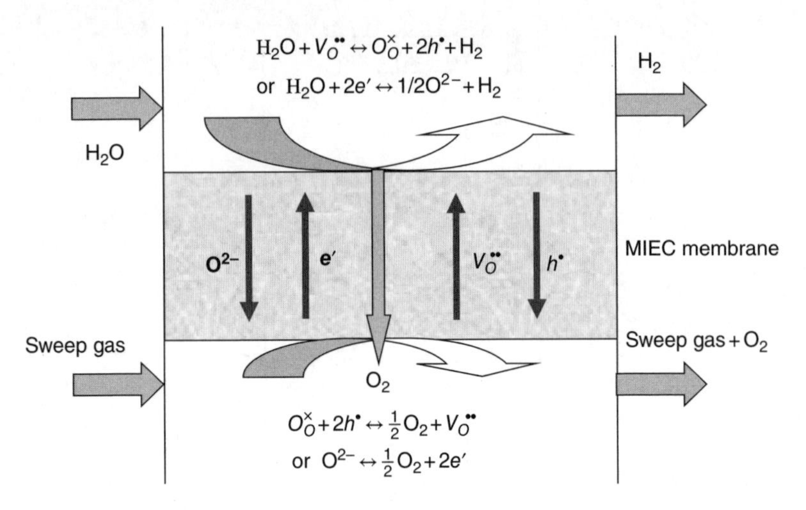

Figure 5.10 Hydrogen production by water dissociation with the help of MIEC ceramic membrane.

product yields. Table 5.6 summarizes the application of dense ceramic MRs as an oxygen extractor.

At high temperatures, water dissociates into oxygen and hydrogen. Generally, very low concentrations of hydrogen and oxygen are generated even at high temperatures (e.g., 0.1 and 0.042% for hydrogen and oxygen, respectively, at 1600°C), because of the small equilibrium constant. If the equilibrium is shifted toward dissociation by removing either oxygen or hydrogen, significant amounts of hydrogen or oxygen can be generated at moderate temperatures. This can be achieved using an MIEC membrane without the need for electrical power or circuitry, as shown in Figure 5.10. The driving force for oxygen permeation may be achieved by using an inert sweep gas or a reducing gas such as methane. The hydrogen production rate depends largely on the rate at which oxygen is removed from the water dissociation zone. In order to obtain a high hydrogen production rate, the membrane should possess high electron and oxygen-ion conductivities and good surface exchange properties. In addition, the hydrogen production rate can also be improved by decreasing the membrane thickness, increasing the active surface area of the membrane (coating a porous layer), or applying a water-dissociation catalyst to the surface of the membrane. Several MIEC membranes – such as Ga-doped CeO_2–Ni, $Gd_{0.2}Ce_{0.8}O_{1.9-\delta}$–$Gd_{0.08}Sr_{0.88}Ti_{0.95}Al_{0.05}O_{3-\delta}$ – were tested for hydrogen production and the results are summarized in Table 5.6. The maximum hydrogen production rate reached 7.44 μmol cm^{-2} s^{-1}.

Table 5.6 Applications of dense ceramic MRs as oxygen extractor

Reaction	Membrane	Configuration	Catalyst	Temperature (°C)	Main results	Ref.
$H_2O \rightarrow H_2 + 1/2O_2$	Gd-doped CeO_2–40% Ni	Disk (0.13 mm)	—	900	$r_{H2} = 4.46\ \mu mol\ cm^{-2}\ s^{-1}$	[85]
	$SrFeCo_{0.5}O_x$	Disk (0.09 mm)	—	900	$r_{H2} > 7.44\ \mu mol\ cm^{-2}\ s^{-1}$	[86]
	GDC–GSTA	Disk (25 μm coated support)	—	900	$r_{H2} = 7\ \mu mol\ cm^{-2}\ s^{-1}$	[87]
	BCFZ	Hollow fiber (0.17 mm)	—	950	$r_{H2} = 2.31\ \mu mol\ cm^{-2}\ s^{-1}$	[88]
$N_2O \rightarrow N_2$ $C_2H_6 \rightarrow C_2H_4$	BCFZ	Hollow fiber (0.17 mm)	Ni/Al_2O_3	875	$X_{N2O} = 100\%$; $X_{N2O} = 91\%$; $S_{C2H4} = 80\%$	[88]
$CO_2 \rightarrow CO$ $CH_4 \rightarrow$ syngas	SCFA	Tube	Ni/Al_2O_3	900	$X_{CO2} = 12.4\%$; $X_{CH4} = 86\%$; $S_{CO} = 93\%$; $H_2/CO = 1.8$	[63]

GDC = $Gd_{0.2}Ce_{0.8}O_{1.9-\delta}$; GSTA–$Gd_{0.08}Sr_{0.88}Ti_{0.95}Al_{0.05}O_{3-\delta}$; BCFZ = $BaCo_xFe_yZr_{1-x-y}O_{3-\delta}$; SCFA = 3% Al_2O_3-doped $SrCo_{0.8}Fe_{0.2}O_{3-\delta}$.

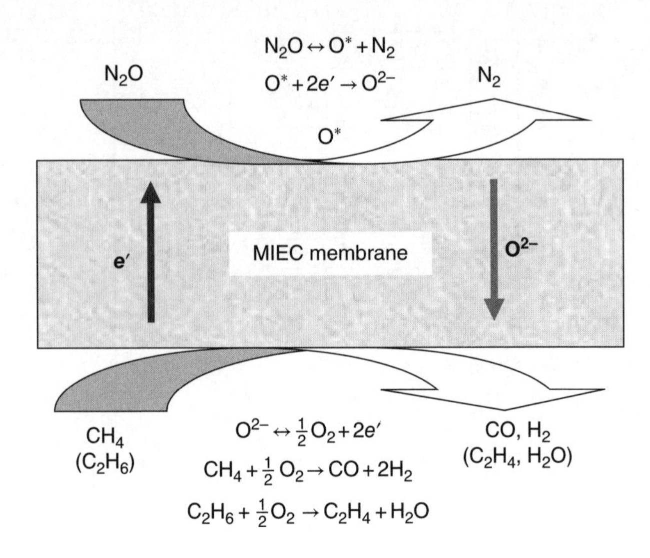

Figure 5.11 N_2O decomposition in the MIEC membrane reactor enhanced by coupling with the partial oxidation of hydrocarbons.

Nitrogen oxides (NO_x, i.e., NO, NO_2, N_2O) are considered as major air pollutants responsible for photochemical smog, acid rain, ozone depletion, as well as climate change. The conventional route to eliminate NO_x pollution is to reduce NO_x catalytically into N_2 using NH_3, urea, H_2, CO, and hydrocarbons as reducing agents. The decomposition of N_2O is a kinetically limited reaction and inhibited by the oxygen molecule produced:

$$N_2O(g) \Leftrightarrow N_2 + \frac{1}{2}O_2$$

Most perovskite catalysts cannot tolerate the co-existence of O_2 because the adsorbed oxygen blocks the catalytically active sites for N_2O decomposition. Using an MIEC membrane reactor, the inhibitor oxygen can be removed as oxygen ion (O^{2-}) through the membrane:

$$N_2O(g) \Leftrightarrow N_2 + O^* \longrightarrow N_2 + \frac{1}{2}O_2$$

Accordingly, the total decomposition of N_2O can be achieved in the membrane reactor [88]. To increase the driving force for oxygen transport through the membrane, methane or ethane can be fed to the permeate side of the membrane to consume the permeated oxygen. Figure 5.11

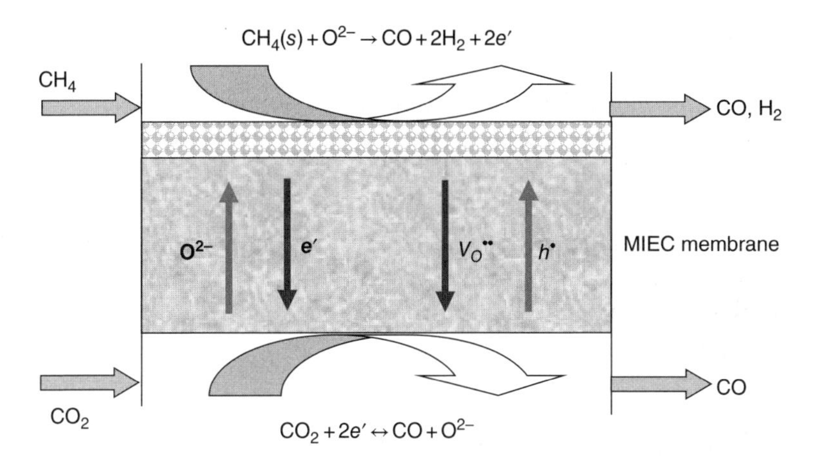

Figure 5.12 Decomposition of CO_2 coupled with POM in the MIEC membrane reactor.

shows the principle of the MIEC membrane reactor for N_2O decomposition combined with the partial oxidation of hydrocarbons. It should be mentioned that the decomposition of water for hydrogen production can also be enhanced by coupling with the partial oxidation of hydrocarbons.

Recently, greater attention has been focused on CO_2 capture and sequestration. One potential route for the consumption of CO_2 is the thermal decomposition of carbon dioxide to CO and O_2, because CO can be utilized as a raw material in the synthesis of important basic chemical products:

$$2CO_2 \Leftrightarrow 2CO + O_2,\ \Delta H^0_{298} = 552\,\mathrm{kJ\,mol^{-1}}$$

However, this reaction is highly endothermic, taking place only at high temperature, and is not easy to realize in conventional fixed bed reactors. In a dense ceramic MR, the CO_2 decomposition reaction can be coupled with POM to syngas, as shown in Figure 5.12. In this process, the decomposition reaction takes place on one side of the membrane. The oxygen released from the decomposition permeates through the oxygen-permeable membrane to the other side of the membrane and reacts with CH_4 to produce syngas over catalyst. Therefore, CO_2 actually acts as the oxygen source for the POM reaction. The oxygen permeation rate in the MIEC membrane plays a key role in the CO_2 conversion [63].

5.6 PROSPECTS AND CHALLENGES

Dense ceramic membranes allow oxygen separation with extremely high selectivity and can be incorporated into membrane reactors for a variety of oxygen-related reactions. The applications of dense ceramic MRs will bring many economic and environmental benefits, with improved selectivity and yields. However, in order to realize the potential benefits of MRs and commercialize them successfully, there are still many challenges that have to be faced not only from membrane materials but also from engineering aspects.

- **Membrane materials**

  For practical applications, the membrane materials used in MRs must meet a number of requirements, including: (i) considerable high-oxygen permeability under operating conditions; (ii) high stability for long-term operation under strongly reducing atmosphere, such as the mixture of carbon monoxide and hydrogen, at elevated temperatures (>700°C); (iii) enough mechanical strength to construct MRs; (iv) low fabrication costs for large-scale industrial applications.

  Permeation rate. From an economic point of view, the oxygen fluxes in the dense ceramic membranes are required to be higher than 5 mL(STP) $cm^{-2} min^{-1}$ (or $1.9 \times 10^{-3} mol\, cm^{-2} s^{-1}$) so as to achieve commercial benefits [21]. However, most currently developed membranes exhibit much lower permeation rate than this.

  Stability. In addition to high permeability, the membrane materials must possess good thermal and chemical stability whilst also maintaining high ionic and electronic conductivities, related directly to the material nature. Most currently available oxygen separation membranes are not sufficiently stable under the wide range of oxygen partial pressures and especially at highly reducing atmosphere. The membrane degradation during the reaction process can be caused by: (i) reaction of the membrane with catalyst or support, or with gas contaminants (e.g., H_2S); (ii) decomposition by reduction in contact with reducing gases (H_2, CO_x, H_2O); (iii) kinetic demixing and decomposition due to the difference in mobility of the cations at high temperatures; (iv) structural failure occurring due to the chemical expansion.

  Mechanical strength. Membrane strength is of critical importance when considering the integration of the membrane in the reactor and especially when using high-temperature sealing. High-temperature sealing usually induces additional stresses on the membrane due to the assembly of dissimilar materials. Creep also occurs due to the long-term operation of the membrane under a pressure gradient at high temperatures. Such mechanical properties data is currently very limited.

Permeation temperature. Many important catalytic reactions, such as the selective oxidation of alkanes, are operated at relatively low temperatures (<500°C), but the operation temperatures to obtain acceptable oxygen fluxes for the ITMs are usually high (>600°C). The development of such membranes operating at moderate temperatures poses a key challenge to material scientists in this area.

Membrane costs. Commercially affordable membrane costs are determined by raw material costs, the fabrication process, and the resulting membrane performance. Economic analysis indicates that, for a membrane with oxygen flux of ca. 10 $mL\,cm^{-2}\,min^{-1}$, the membrane material costs to compete with the traditional route must be lower than 1600 $\$\,m^{-2}$ [89]. This material cost target is out of reach currently.

- **Engineering aspects**

  The scaling up of membrane technologies is the most challenging yet important task. Design and engineering problems have to be overcome before dense ceramic MRs will find extensive commercial applications.

  High-temperature sealing and connection. For the fabrication of industrial dense ceramic MRs, a key problem is the development of seals for membrane integrity with other reactor components that are able to operate at high temperatures. One solution to seal ceramic membranes with metal is the use of ceramic glass. These materials usually require high temperatures to melt, and creep of the membrane can be detrimental to the sealing [5].

  Large-scale membrane fabrication. Preparation of dense ceramic membranes is a complex multi-step process. Many factors may influence the microstructure and performance of the resulting membranes. This leads to low reproducibility and low yields to fabricate high-quality membranes on a large scale.

To overcome these challenges, skills are required from different disciplines, including materials and ceramic science, chemistry, and chemical engineering. Considerable future work should be focused on:

1. Further identifying new membrane materials that combine high oxygen flux with sufficient reliability and durability. We need an improved understanding of the role of extended defect interactions, and the onset of locally ordered arrangements of oxygen vacancies,

in determining the oxygen fluxes and how these can be manipulated when optimized performance is to be achieved [21]. The membrane stability must also be assessed under relevant industrial constraints. It is believed that Co- and alkaline earth metal-free membranes, double-perovskites like doped La_2NiO_4 oxides, and dual-phase membranes have higher CO_2 resistance.

2. Developing more efficient routes to prepare high-quality supported membranes on a large scale with improved yields. Composite membranes with a dense, thin MIEC film on a porous substrate can be fabricated by various methods such as slurry coating and chemical vapor deposition. The use of thin-film membranes will allow a substantial reduction in operating temperature [4]. With such composite membranes, a multiple planar stack can be employed to enlarge the membrane area to that required on a plant scale [90]. However, many problems – such as sealing, connecting, and pressure resistance – have to be faced. In the last decade, the combined phase-inversion and sintering technique has been applied extensively to fabricate dense ceramic hollow fiber membranes. The fabrication process is fast and simple without the need to use any expensive apparatus, leading to considerably low fabrication costs. More importantly, the resulting hollow fiber membranes have an asymmetric structure (i.e., a thin separating dense layer integrated with porous layers on either side or both sides), and hence the permeability can be improved significantly. However, the hollow fiber ceramic membranes also demonstrate low mechanical strength. Tubular composite membranes are desired for high mechanical strength but compromised by the lower packing density.
3. Deepen our theoretical and experimental investigation into the design and control of membrane permeability to match the catalytic activity of catalysts in membrane reactors. This is critical for the design of industrial membrane reactors. Experimental procedures in combination with mathematical modeling will provide a better understanding.

NOTATION

a_i activity of charged defect i
C_i concentration of charged defect i (mol m^{-3})
D_i diffusion coefficient of charged defect i (m^2 s^{-1})

E	activation energy ($J\ mol^{-1}$)
F	Faraday constant
I	electrical current (A)
J	permeation flux ($mol\ m^{-2} s^{-1}$)
k_f	rate constant for the forward surface exchange reaction ($cm\ Pa^{-0.5} s^{-1}$)
k_r	rate constant for the reverse surface exchange reaction ($mol\ cm^{-2} s^{-1}$)
L	membrane thickness (m)
p'_{O_2}, p''_{O_2}	oxygen partial pressure on the upstream and downstream side (Pa)
R	ideal gas constant, $8.314\,J\,mol^{-1}\,K^{-1}$
R_{in}, R_o	inner and outer radius of membrane tube (m)
R_m	logarithmic mean radius of dense membrane, $R_m = \delta/\ln(1+\delta/R_o)$ (m)
r	radius of oxygen, A or B ions in perovskite (m)
T	temperature (K)
t_i	transport number of defect i
z_i	charge number of defect i

Greeks

μ_i	electrochemical potential of defect i
μ_i^0	standard chemical potential
σ_i	conductivity of defect i
ϕ	local electrostatic potential or Galvanic (internal) potential
η	tolerance factor of perovskite
υ	local velocity of inert marker

Subscripts and superscripts

e	electron
i, j	i, j component
h	electron hole
V	oxygen vacancy
0	pre-exponential factor

REFERENCES

[1] Liu, S., Tan, X., Li, K. and Hughes, R. (2001) Methane coupling using catalytic membrane reactors. *Catalysis Reviews*, 43, 147–198.

[2] Jiang, Q., Faraji, S., Slade, D.A., Stagg-Williams, S.M. (2011) A review of mixed ionic and electronic conducting ceramic membranes as oxygen sources for high-temperature

reactors. In *Membrane Science and Technology*, Elsevier B.V., Amsterdam, pp. 235–273.

[3] Wei, Y., Yang, W., Caro, J. and Wang, H. (2013) Dense ceramic oxygen permeable membranes and catalytic membrane reactors. *Chemical Engineering Journal*, 220, 185–203.

[4] Yang, W., Wang, H., Zhu, X. and Lin, L. (2005) Development and application of oxygen permeable membrane in selective oxidation of light alkanes. *Topics in Catalysis*, 35, 155–167.

[5] Geffroy, P.M., Fouletier, J., Richet, N. and Chartier, T. (2013) Rational selection of MIEC materials in energy production processes. *Chemical Engineering Science*, 87, 408–433.

[6] Liu, Y., Tan, X. and Li, K. (2006) Mixed conducting ceramics for catalytic membrane processing. *Catalysis Reviews*, 48, 145–198.

[7] Mundschau, M.V., Xie, X., Evenson, C.R. and Sammells, A.F. (2006) Dense inorganic membranes for production of hydrogen from methane and coal with carbon dioxide sequestration. *Catalysis Today*, 118, 12–23.

[8] Hendriksen, P.V., Larsen, P.H., Mogensen, M., Poulsen, F.W. and Wiik, K. (2000) Prospects and problems of dense oxygen permeable membranes. *Catalysis Today*, 56, 283–295.

[9] Chen, T., Zhao, H., Xu, N. *et al.* (2011) Synthesis and oxygen permeation properties of a $Ce_{0.8}Sm_{0.2}O_{2-\delta}$–$LaBaCo_2O_{5+\delta}$ dual-phase composite membrane. *Journal of Membrane Science*, 370, 158–165.

[10] Kim, J. and Lin, Y.S. (2000) Synthesis and oxygen permeation properties of ceramic–metal dual-phase membranes. *Journal of Membrane Science*, 167, 123–133.

[11] Xue, J., Liao, Q., Wei, Y., Li, Z. and Wang, H. (2013) A CO_2-tolerance oxygen permeable $60Ce_{0.9}Gd_{0.1}O_{2-\delta}$–$40Ba_{0.5}Sr_{0.5}Co_{0.8}Fe_{0.2}O_{3-\delta}$ dual phase membrane. *Journal of Membrane Science*, 443, 124–130.

[12] Liu, J.-j., Liu, T., Wang, W.-d., Gao, J.-f. and Chen, C.-s. (2012) $Zr_{0.84}Y_{0.16}O_{1.92}$–$La_{0.8}Sr_{0.2}Cr_{0.5}Fe_{0.5}O_{3-\delta}$ dual-phase composite hollow fiber membrane targeting chemical reactor applications. *Journal of Membrane Science*, 389, 435–440.

[13] Yang, C., Xu, Q., Liu, C., Liu, J., Chen, C. and Liu, W. (2011) $Bi_{1.5}Y_{0.3}Sm_{0.2}O_3$–$La_{0.8}Sr_{0.2}MnO_{3-\delta}$ dual-phase composite hollow fiber membrane for oxygen separation. *Materials Letters*, 65, 3365–3367.

[14] Ishihara, T., Kilner, J.A., Honda, M., Sakai, N., Yokokawa, H. and Takita, Y. (1998) Oxygen surface exchange and diffusion in $LaGaO_3$ based perovskite type oxides. *Solid State Ionics*, 113–115, 593–600.

[15] Tan, X., Liu, S., Li, K. and Hughes, R. (2000) Theoretical analysis of ion permeation through mixed conducting membranes and its application to dehydrogenation reactions. *Solid State Ionics*, 138, 149–159.

[16] Kharton, V.V., Naumovich, E.N. and Nikolaev, A.V. (1996) Materials of high-temperature electrochemical oxygen membranes. *Journal of Membrane Science*, 111, 149–157.

[17] Qiu, L., Lee, T.H., Liu, L.M., Yang, Y.L. and Jacobson, A.J. (1995) Oxygen permeation studies of $SrCo_{0.8}Fe_{0.2}O_{3-\delta}$. *Solid State Ionics*, 76, 321–329.

[18] Xu, S.J. and Thomson, W.J. (1999) Oxygen permeation rates through ion-conducting perovskite membranes. *Chemical Engineering Science*, 54, 3839–3850.

[19] Tan, X. and Li, K. (2002) Modeling of air separation in a $La_{0.6}Sr_{0.4}Co_{0.2}Fe_{0.8}O_{3-\delta}$ hollow fiber membrane module. *AIChE Journal*, 48, 1469–1477.

[20] Kharton, V.V., Naumovich, E.N., Kovalevsky, A.V. *et al.* (2000) Mixed electronic and ionic conductivity of LaCo(M)O_3 (M = Ga, Cr, Fe or Ni): IV. Effect of preparation method on oxygen transport in $LaCoO_{3-\delta}$. *Solid State Ionics*, 138, 135–148.

[21] Bouwmeester, H.J.M. (2003) Dense ceramic membranes for methane conversion. *Catalysis Today*, 82, 141–150.

[22] Li, S., Jin, W., Xu, N. and Shi, J. (1999) Synthesis and oxygen permeation properties of $La_{0.2}Sr_{0.8}Co_{0.2}Fe_{0.8}O_{3-\delta}$ membranes. *Solid State Ionics*, 124, 161–170.

[23] Aasland, S., Tangen, I.L., Wiik, K. and Ødegård, R. (2000) Oxygen permeation of $SrFe_{0.67}Co_{0.33}O_{3-\delta}$. *Solid State Ionics*, 135, 713–717.

[24] Kim, S., Yang, Y.L., Jacobson, A.J. and Abeles, B. (1998) Diffusion and surface exchange coefficients in mixed ionic electronic conducting oxides from the pressure dependence of oxygen permeation. *Solid State Ionics*, 106, 189–195.

[25] Lee, T.H., Yang, Y.L., Jacobson, A.J., Abeles, B. and Milner, S. (1997) Oxygen permeation in $SrCo_{0.8}Fe_{0.2}O_{3-\delta}$ membranes with porous electrodes. *Solid State Ionics*, 100, 87–94.

[26] Tan, L., Gu, X., Yang, L., Jin, W., Zhang, L. and Xu, N. (2003) Influence of powder synthesis methods on microstructure and oxygen permeation performance of $Ba_{0.5}Sr_{0.5}Co_{0.8}Fe_{0.2}O_{3-\delta}$ perovskite-type membranes. *Journal of Membrane Science*, 212, 157–165.

[27] Akin, F.T. and Lin, Y.S. (2002) Selective oxidation of ethane to ethylene in a dense tubular membrane reactor. *Journal of Membrane Science*, 209, 457–467.

[28] Lu, Y., Dixon, A.G., Moser, W.R., Ma, Y.H. and Balachandran, U. (2000) Oxygen-permeable dense membrane reactor for the oxidative coupling of methane. *Journal of Membrane Science*, 170, 27–34.

[29] Li, S.G., Qi, H., Xu, N.P. and Shi, J. (1999) Tubular dense perovskite type membrane. Preparing, sealing, and oxygen permeation properties. *Industrial Engineering Chemistry Research*, 38, 5029.

[30] Tan, X., Liu, Y. and Li, K. (2005) Preparation of $La_{0.6}Sr_{0.4}Co_{0.2}Fe_{0.8}O_{3-\delta}$ hollow fiber membranes for oxygen production by a phase-inversion/sintering technique. *Industrial Engineering Chemistry Research*, 44, 61–66.

[31] Kingsbury, B.F.K. and Li, K. (2009) A morphological study of ceramic hollow fibre membranes. *Journal of Membrane Science*, 328, 134–140.

[32] Tan, X., Liu, N., Meng, B. and Liu, S. (2011) Morphology control of the perovskite hollow fibre membranes for oxygen separation using different bore fluids. *Journal of Membrane Science*, 378, 308–318.

[33] Itoh, H., Asano, H., Fukuroi, K., Nagata, M. and Iwahara, H. (1997) Spin coating of a Ca(Ti,Fe)O_3 dense film on a porous substrate for electrochemical permeation of oxygen. *Journal of the American Ceramic Society*, 80, 1359–1365.

[34] Jin, W., Li, S., Huang, P., Xu, N. and Shi, J. (2001) Preparation of an asymmetric perovskite-type membrane and its oxygen permeability. *Journal of Membrane Science*, 185, 237–243.

[35] Watanabe, K., Yuasa, M., Kida, T., Shimanoe, K., Teraoka, Y. and Yamazoe, N. (2008) Preparation of oxygen evolution layer/$La_{0.6}Ca_{0.4}CoO_3$ dense membrane/porous support asymmetric structure for high-performance oxygen permeation. *Solid State Ionics*, 179, 1377–1381.

[36] Juste, E., Julian, A., Geffroy, P.M. *et al.* (2010) Influence of microstructure and architecture on oxygen permeation of $La_{(1-x)}Sr_xFe_{(1-y)}(Ga, Ni)_yO_{3-\delta}$ perovskite catalytic membrane reactor. *Journal of the European Ceramic Society*, 30, 1409–1417.

[37] Kaiser, A., Foghmoes, S., Chatzichristodoulou, C. *et al.* (2011) Evaluation of thin film ceria membranes for syngas membrane reactors–preparation, characterization and testing. *Journal of Membrane Science*, 378, 51–60.
[38] Sobyanin, V.A., Belyaev, V.D. and Gal'vita, V.V. (1998) Syngas production from methane in an electrochemical membrane reactor. *Catalysis Today*, 42, 337–340.
[39] Wu, Z., Wang, B. and Li, K. (2010) A novel dual-layer ceramic hollow fibre membrane reactor for methane conversion. *Journal of Membrane Science*, 352, 63–70.
[40] Ishihara, T. and Takita, Y. (2000) Partial oxidation of methane into syngas with oxygen permeating ceramic membrane reactors. *Catalysis Surveys from Japan*, 4, 125–133.
[41] Czuprat, O., Caro, J., Kondratenko, V.A. and Kondratenko, E.V. (2010) Dehydrogenation of propane with selective hydrogen combustion: A mechanistic study by transient analysis of products. *Catalysis Communications*, 11, 1211–1214.
[42] Gong, Z. and Hong, L. (2011) Integration of air separation and partial oxidation of methane in the $La_{0.4}Ba_{0.6}Fe_{0.8}Zn_{0.2}O_{3-\delta}$ membrane reactor. *Journal of Membrane Science*, 380, 81–86.
[43] Ishihara, T., Tsuruta, Y., Todaka, T., Nishiguchi, H. and Takita, Y. (2002) Fe doped $LaGaO_3$ perovskite oxide as an oxygen separating membrane for CH_4 partial oxidation. *Solid State Ionics*, 152, 709–714.
[44] Delbos, C., Lebain, G., Richet, N. and Bertail, C. (2010) Performances of tubular $La_{0.8}Sr_{0.2}Fe_{0.7}Ga_{0.3}O_{3-\delta}$ mixed conducting membrane reactor for under pressure methane conversion to syngas. *Catalysis Today*, 156, 146–152.
[45] Hoang, D. and Chan, S. (2006) Effect of reactor dimensions on the performance of an O_2 pump integrated partial oxidation reformer–a modelling approach. *International Journal of Hydrogen Energy*, 31, 1–12.
[46] Balachandran, U., Dusek, J.T., Maiya, P.S. *et al.* (1997) Ceramic membrane reactor for converting methane to syngas. *Catalysis Today*, 36, 265–272.
[47] Tian, T., Wang, W., Zhan, M. and Chen, C. (2010) Catalytic partial oxidation of methane over $SrTiO_3$ with oxygen-permeable membrane reactor. *Catalysis Communications*, 11, 624–628.
[48] Ikeguchi, M., Mimura, T., Sekine, Y., Kikuchi, E. and Matsukata, M. (2005) Reaction and oxygen permeation studies in $Sm_{0.4}Ba_{0.6}Fe_{0.8}Co_{0.2}O_{3-\delta}$ membrane reactor for partial oxidation of methane to syngas. *Applied Catalysis A: General*, 290, 212–220.
[49] Jin, W.Q., Li, S.G., Huang, P., Xu, N.P., Shi, J. and Lin, Y.S. (2000) Tubular lanthanum cobaltite perovskite-type membrane reactors for partial oxidation of methane to syngas. *Journal of Membrane Science*, 166, 13–22.
[50] Zhu, D.C., Xu, X.Y., Feng, S.J., Liu, W. and Chen, C.S. (2003) La_2NiO_4 tubular membrane reactor for conversion of methane to syngas. *Catalysis Today*, 82, 151–156.
[51] Wang, H., Cong, Y. and Yang, W. (2003) Investigation on the partial oxidation of methane to syngas in a tubular $Ba_{0.5}Sr_{0.5}Co_{0.8}Fe_{0.2}O_{3-\delta}$ membrane reactor. *Catalysis Today*, 82, 157–166.
[52] Gu, X.H., Jin, W.Q., Chen, C.L., Xu, N.P., Shi, J. and Ma, Y.H. (2002) YSZ–$SrCo_{0.4}Fe_{0.6}O_{3-\delta}$ membranes for the partial oxidation of methane to syngas. *AIChE Journal*, 48, 2051–2060.
[53] Ritchie, J.T., Richardson, J.T. and Luss, D. (2001) Ceramic membrane reactor for synthesis gas production. *AIChE Journal*, 47, 2092–2101.

[54] Hamakawa, S., Hayakawa, T., Suzuki, K. *et al.* (2000) Methane conversion into synthesis gas using an electrochemical membrane reactor. *Solid State Ionics*, 136, 761–766.
[55] Hu, J., Xing, T., Jia, Q. *et al.* (2006) Methane partial oxidation to syngas in $YBa_2Cu_3O_{7-x}$ membrane reactor. *Applied Catalysis A: General*, 306, 29–33.
[56] Kharton, V.V., Yaremchenko, A.A., Valente, A.A. *et al.* (2005) Methane oxidation over Fe-, Co-, Ni- and V-containing mixed conductors. *Solid State Ionics*, 176, 781–791.
[57] Wang, H., Tablet, C., Schiestel, T., Werth, S. and Caro, J. (2006) Partial oxidation of methane to syngas in a perovskite hollow fiber membrane reactor. *Catalysis Communications*, 7, 907–912.
[58] Zhu, X., Li, Q., Cong, Y. and Yang, W. (2008) Syngas generation in a membrane reactor with a highly stable ceramic composite membrane. *Catalysis Communications*, 10, 309–312.
[59] Zhang, Y., Liu, J., Ding, W. and Lu, X. (2011) Performance of an oxygen-permeable membrane reactor for partial oxidation of methane in coke oven gas to syngas. *Fuel*, 90, 324–330.
[60] Zhu, X., Li, Q., He, Y., Cong, Y. and Yang, W. (2010) Oxygen permeation and partial oxidation of methane in dual-phase membrane reactors. *Journal of Membrane Science*, 360, 454–460.
[61] Luo, H., Wei, Y., Jiang, H. *et al.* (2010) Performance of a ceramic membrane reactor with high oxygen flux Ta-containing perovskite for the partial oxidation of methane to syngas. *Journal of Membrane Science*, 350, 154–160.
[62] Li, Q., Zhu, X., He, Y. and Yang, W. (2010) Partial oxidation of methane in $BaCe_{0.1}Co_{0.4}Fe_{0.5}O_{3-\delta}$ membrane reactor. *Catalysis Today*, 149, 185–190.
[63] Zhang, C., Jin, W., Yang, C. and Xu, N. (2009) Decomposition of CO_2 coupled with POM in a thin tubular oxygen-permeable membrane reactor. *Catalysis Today*, 148, 298–302.
[64] Tan, X. and Li, K. (2009) Design of mixed conducting ceramic membranes/reactors for the partial oxidation of methane (POM) to syngas. *AIChE Journal*, 55, 2675–2685.
[65] Zhang, P., Chang, X.F., Wu, Z.T., Jin, W.Q. and Xu, N.P. (2005) Effect of the packing amount of catalysts on the partial oxidation of methane reaction in a dense oxygen-permeable membrane reactor. *Industrial and Engineering Chemistry Research*, 44, 1954–1959.
[66] Wang, W. and Lin, Y.S. (1995) Analysis of oxidative coupling of methane in dense oxide membrane reactors. *Journal of Membrane Science*, 103, 219–233.
[67] Haag, S., van Veen, A.C. and Mirodatos, C. (2007) Influence of oxygen supply rates on performances of catalytic membrane reactors. Application to the oxidative coupling of methane. *Catalysis Today*, 127, 157–164.
[68] Olivier, L., Haag, S., Mirodatos, C. and van Veen, A.C. (2009) Oxidative coupling of methane using catalyst modified dense perovskite membrane reactors. *Catalysis Today*, 142, 34–41.
[69] Akin, F.T. and Lin, Y.S. (2002) Oxidative coupling of methane in dense ceramic membrane reactor with high yields. *AIChE Journal*, 48, 2298–2306.
[70] Wang, H., Cong, Y. and Yang, W. (2005) Oxidative coupling of methane in $Ba_{0.5}Sr_{0.5}Co_{0.8}Fe_{0.2}O_{3-\delta}$ tubular membrane reactors. *Catalysis Today*, 104, 160–167.

[71] Lu, Y.P., Dixon, A.G., Moser, W.R., Ma, Y.H. and Balachandran, U. (2000) Oxidative coupling of methane using oxygen-permeable dense membrane reactors. *Catalysis Today*, 56, 297–305.
[72] ten Elshof, J.E., Bouwmeester, H.J.M. and Verweij, H. (1995) Oxidative coupling of methane in a mixed-conducting perovskite membrane reactor. *Applied Catalysis A: General*, 130, 195–212.
[73] Zeng, Y. (2000) Oxygen permeation and oxidative coupling of methane in yttria doped bismuth oxide membrane reactor. *Journal of Catalysis*, 193, 58–64.
[74] Zeng, Y., Lin, Y.S. and Swartz, S.L. (1998) Perovskite-type ceramic membrane: Synthesis, oxygen permeation and membrane reactor performance for oxidative coupling of methane. *Journal of Membrane Science*, 150, 87–98.
[75] Lin, Y.S. and Zeng, Y. (1996) Catalytic properties of oxygen semipermeable perovskite-type ceramic membrane materials for oxidative coupling of methane. *Journal of Catalysis*, 164, 220–231.
[76] Tan, X., Pang, Z., Gu, Z. and Liu, S. (2007) Catalytic perovskite hollow fibre membrane reactors for methane oxidative coupling. *Journal of Membrane Science*, 302, 109–114.
[77] Tagawa, T., Moe, K.K., Ito, M. and Goto, S. (1999) Fuel cell type reactor for chemicals–energy co-generation. *Chemical Engineering Science*, 54, 1553–1557.
[78] Bhatia, S., Thien, C.Y. and Mohamed, A.R. (2009) Oxidative coupling of methane (OCM) in a catalytic membrane reactor and comparison of its performance with other catalytic reactors. *Chemical Engineering Journal*, 148, 525–532.
[79] Xui-Mei, G., Hidajat, K. and Ching, C.-B. (1999) Simulation of a solid oxide fuel cell for oxidative coupling of methane. *Catalysis Today*, 50, 109–116.
[80] Eng, D. and Stoukides, M. (1991) Catalytic and electrochemical oxidation of methane on platinum. *Journal of Catalysis*, 130, 306–309.
[81] Kiatkittipong, W., Tagawa, T., Goto, S., Assabumrungrat, S., Silpasup, K. and Praserthdam, P. (2005) Comparative study of oxidative coupling of methane modeling in various types of reactor. *Chemical Engineering Journal*, 115, 63–71.
[82] Rebeilleau-Dassonneville, M., Rosini, S., Veen, A.C.v., Farrusseng, D. and Mirodatos, C. (2005) Oxidative activation of ethane on catalytic modified dense ionic oxygen conducting membranes. *Catalysis Today*, 104, 131–137.
[83] Wang, H., Cong, Y. and Yang, W. (2002) High selectivity of oxidative dehydrogenation of ethane to ethylene in an oxygen permeable membrane reactor. *Chemical Communications*, 14, 1468–1469.
[84] Wang, H., Tablet, C., Schiestel, T. and Caro, J. (2006) Hollow fiber membrane reactors for the oxidative activation of ethane. *Catalysis Today*, 118, 98–103.
[85] Balachandran, U. (2004) Use of mixed conducting membranes to produce hydrogen by water dissociation. *International Journal of Hydrogen Energy*, 29, 291–296.
[86] Balachandran, U., Lee, T. and Dorris, S. (2007) Hydrogen production by water dissociation using mixed conducting dense ceramic membranes. *International Journal of Hydrogen Energy*, 32, 451–456.
[87] Wang, H., Gopalan, S. and Pal, U.B. (2011) Hydrogen generation and separation using $Gd_{0.2}Ce_{0.8}O_{1.9-\delta}$–$Gd_{0.08}Sr_{0.88}Ti_{0.95}Al_{0.05}O_{3\pm\delta}$ mixed ionic and electronic conducting membranes. *Electrochimica Acta*, 56, 6989–6996.
[88] Jiang, H., Wang, H., Liang, F. *et al.* (2010) Improved water dissociation and nitrous oxide decomposition by in situ oxygen removal in perovskite catalytic membrane reactor. *Catalysis Today*, 156, 187–190.

[89] Hendriksen, P.V., Larsen, P.H., Mogensen, M., Poulsen, F.W. and Wiik, K. (2000) Prospects and problems of dense oxygen permeable membranes. *Catalysis Today*, 56, 283–295.

[90] Lu, G.Q., Diniz da Costa, J.C., Duke, M. *et al.* (2007) Inorganic membranes for hydrogen production and purification: A critical review and perspective. *Journal of Colloid and Interface Science*, 314, 589–603.

6

Proton-Conducting Ceramic Membrane Reactors

6.1 INTRODUCTION

Proton-conducting dense ceramic membranes (PCMs) are made from ceramics exhibiting predominant protonic conduction under hydrogen-containing atmosphere at high temperatures. Under an electrochemical potential gradient, protons can be transported through the membrane from one side to the other of the membrane. Accordingly, various kinds of applications can be devised using proton-conducting ceramic membranes – such as hydrogen or hydrogen-containing compound sensors, hydrogen analyzers, fuel cells, steam electrolysis for hydrogen production, hydrogen or steam pumps, and hydrogenation and dehydrogenation reactions [1–4].

This chapter will give an extensive description of the proton-conducting dense ceramic membrane reactors (PCMRs), including their principles, fabrication, and design as well as their applications. The prospects and challenges of PCMRs will also be discussed at the end of the chapter.

6.2 PROTON/HYDROGEN PERMEATION IN PROTON-CONDUCTING CERAMIC MEMBRANES

6.2.1 Proton-Conducting Ceramics

The first proton-conducting ceramic, $SrCe_{0.95}Yb_{0.05}O_{3-\delta}$, was discovered by Iwahara *et al.* in the early 1980s [5]. Since then, many other proton-conducting ceramics have been synthesized, mostly based on $SrCeO_3$,

Inorganic Membrane Reactors: Fundamentals and Applications, First Edition. Xiaoyao Tan and Kang Li.

$BaCeO_3$, $CaZrO_3$, or $SrZrO_3$ perovskite-type oxide solid solutions. A trivalent cation is partially substituted for Ce or Zr to increase the oxygen vacancy concentration in the composite. The general formula of these doped perovskites can be written as $AB_{1-x}M_xO_{3-\delta}$, where A is Ca, Sr, or Ba; B is Ce or Zr; M is taken from the group consisting of Tm, Nd, Gd, Y, Yb, In, Tm, Tb, and so on; x is less than the upper limit of its solid solution formation range (usually less than 0.2); and δ denotes the number of oxygen deficiencies per unit formula of the perovskite. These oxides exhibit p-type (electron hole) conduction in an atmosphere free from hydrogen or water vapor at elevated temperatures. However, when water vapor or hydrogen is introduced into the atmosphere surrounding the specimen at high temperature, electronic conductivity decreases and protonic conduction appears [3]. The protonic conductivities of doped Ce-based perovskites in hydrogen atmosphere are on the order of 10^{-2} to $10^{-3}\,S\,cm^{-1}$ at 1000–600°C [4], but of the doped zirconates based on $CaZrO_3$, $SrZrO_3$, or $BaZrO_3$, perovskites are about one order of magnitude lower than those of the cerates. A series of complex perovskites based on $Ba_3(CaNb_2)O_9$ or $Sr_2(ScNb)O_6$ were reported to exhibit protonic conductivity as high as that of $BaCeO_3$-based ceramics [6, 7].

Among the above-described perovskite oxides, $BaCeO_3$-based ceramics show the highest conductivity. However, the contribution of oxide ions to the conduction grows markedly as the temperature is raised. The conductivity of $SrCeO_3$-based ceramic is rather low, but the transport number of protons is higher than that of $BaCeO_3$-based ones. The conductivity of zirconate-based ceramics is much lower than that of the cerates, but they are superior with respect to their chemical and mechanical strength. For example, the cerates dissolve easily in strong acids, but the zirconates hardly react with acid solution and they are stable against CO_2, which reacts easily with cerate ceramics below 800°C to form carbonates [4].

The doping cation has an important effect on the electrical (ionic) conductivity and chemical stability. It was found that the two properties are both governed by the ionic radii of the dopants and are in a trade-off relation. The electrical conductivity seems also to be affected by the electronegativity of the dopant element; that is, the conductivity increases with increasing ionic radius and decreasing electronegativity (increasing basicity) [8]. Table 6.1 lists the electronic conductivity of $SrCeO_3$ (SC)-based perovskites doped with different cations. It shows that the electronic conductivity increases with decreasing ionization potential. For example, SCTb has much lower electronic conductivity than that of SCYb, and thus can be used as the electrolyte for a fuel cell, while SCTm

Table 6.1 Electronic conductivity of $SrCeO_3$(SC)-based perovskite with different doping cations (900°C) [9]

Composition	Charge transfer equation	Ionization potential of the doping ions (eV)	Electronic conductivity ($S\,cm^{-1}$)
$SrCe_{0.95}Tb_{0.05}O_3$ (SCTb)	$Tb^{3+} + h^{\bullet} \rightarrow Tb^{4+}$	39.8	1.21×10^{-5}
$SrCeO_3$ (SC)	$Ce^{3+} + h^{\bullet} \rightarrow Ce^{4+}$	36.8	2.82×10^{-4}
$SrCe_{0.95}Yb_{0.05}O_3$ (SCYb)	$Yb^{2+} + h^{\bullet} \rightarrow Yb^{3+}$	25.0	8.50×10^{-3}
$SrCe_{0.95}Tm_{0.05}O_3$ (SCTm)	$Tm^{2+} + h^{\bullet} \rightarrow Tm^{3+}$	23.7	1.21×10^{-2}

possesses a higher electronic conductivity and can be a good membrane material for hydrogen permeation [9]. Several types of non-perovskite oxides – such as $Ce_{0.8}M_{0.2}O_{2-\delta}$ (M = La, Y, Gd, Sm) [10], $La_{1.9}Ca_{0.1}Zr_2O_{6.95}$ [11] – are also reported to exhibit protonic conduction at elevated temperatures, but their protonic conductivities are rather low compared with those of the perovskite-type oxides.

6.2.2 Hydrogen/Proton Permeation in Mixed Conducting Membranes

For hydrogen permeation applications, the proton-conducting ceramics should have high protonic and electronic conductivity (i.e., MIECs). In the proton-conducting perovskite oxides, protons are formed in the water vapor or hydrogen-containing environment at high temperature according to the following reactions:

$$H_2O + V_O^{\bullet\bullet} \leftrightarrow O_O^{\times} + 2H^{\bullet} \tag{6.1}$$

$$\frac{1}{2}H_2 + h^{\bullet} \leftrightarrow H^{\bullet} \tag{6.2a}$$

or

$$\frac{1}{2}H_2 \leftrightarrow H^{\bullet} + e' \tag{6.2b}$$

where $H^{\bullet}$ stands for a proton.

The protons are also considered to bond to oxygen ions, forming substitutional hydroxyl, $OH_O^{\bullet}$, namely

$$H_2O + V_O^{\bullet\bullet} + O_O^{\times} \leftrightarrow 2OH_O^{\bullet} \tag{6.3a}$$

$$H_2 + 2O_O^{\times} + 2h^{\bullet} \leftrightarrow 2OH_O^{\bullet} \tag{6.3b}$$

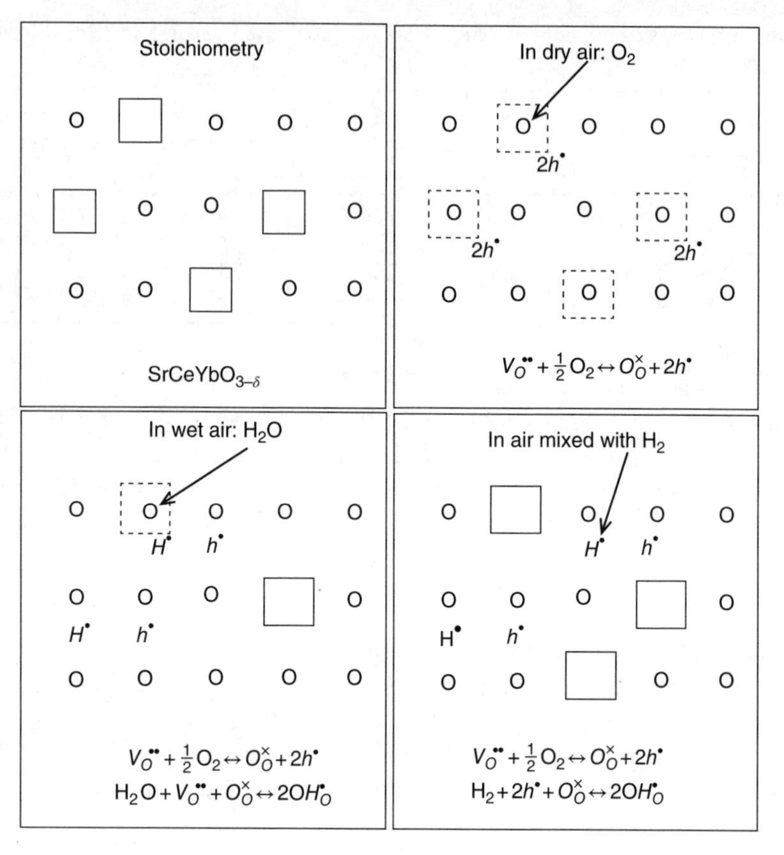

Figure 6.1 Formation of charged defects in proton-conducting ceramic membranes. Reproduced with permission from [13]. Taylor & Francis Ltd, http://www.tandf.co.uk/journals.

This bonding effect is supported by infrared studies [12].

Figure 6.1 shows the formation of charged defects in the proton-conducting membranes exemplified by $SrCe_{0.95}Yb_{0.05}O_{3-\delta}$ [13]. As can be seen, four types of charged defects – including oxygen vacancies, protons, electrons, and holes – may co-exist in the membrane at high temperature. Two mechanisms are usually applied to describe the proton transport in the proton-conducting membranes: (i) free proton transport and (ii) vehicle proton transport [14]. In free proton transport, the protons jump between stationary oxygen ions. Each jump is followed by a rotation around the oxygen ion to get into position for the next jump. The jump is normally considered the rate-limiting step, as the rotation is easier. In vehicle proton transport, the proton moves as a passenger on

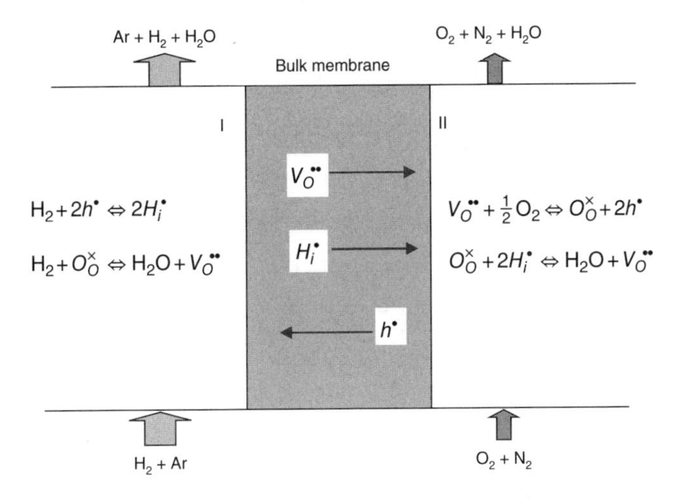

Figure 6.2 Transport diagram of ion permeation in PCMs.

a larger species like oxygen ions. This mechanism can only account for dominant long-range transport processes in oxides when proceeding on interstitial lattice sites. Therefore, hydroxyl ion transport cannot provide a dominant conductivity but accomplishes minority oxygen transport in a proton conductor. The mobility of a proton in the oxide is limited by the proton transfer between fixed oxygen sites, but is facilitated by the thermal fluctuation of the oxygen ion separation [15]. In fact, free proton transport is related strongly to oxygen sublattice vibrations and thus has activation energies close to (and generally somewhat lower than) those for oxygen vacancy mobility.

Like oxygen-permeable ceramic membranes, hydrogen permeation through proton-conducting membranes can be achieved by applying a hydrogen partial pressure gradient or by using an external power source across the membrane. When the mixed proton-conducting membrane is exposed to gaseous hydrogen on one side and to oxygen on the other side at high temperatures (i.e., asymmetrical gas environment), the charged defects transport in certain directions under the electrochemical gradient, as shown in Figure 6.2.

The transport fluxes of charged defects at steady state in the MIEC membrane are given by Eq. (5.7), namely

$$J_i = -D_i C_i \left[\frac{1-t_i}{C_i} \cdot \frac{dC_i}{dx} - \sum_{j \neq i} \frac{z_i}{z_j} \cdot \frac{t_j}{C_j} \cdot \frac{dC_j}{dx} \right] \quad (i = V, p, h) \tag{6.4}$$

Along with the local electric neutrality condition, where SCYb is taken as an example and oxygen vacancy, proton, and hole are included:

$$[Yb'_{Ce}] = [H_i^{\bullet}] + [h^{\bullet}] + 2[V_O^{\bullet\bullet}] \tag{6.5}$$

The transport equations can be given in the form of a matrix as

$$\bar{A} \cdot \frac{d\bar{C}}{dx} = \bar{B} \tag{6.6}$$

where

$$\bar{A} = \begin{bmatrix} 1-t_V & -2t_p\dfrac{C_V}{C_p} & -2t_h\dfrac{C_V}{C_h} \\ -\dfrac{1}{2}t_V\dfrac{C_p}{C_V} & 1-t_p & -t_h\dfrac{C_p}{C_h} \\ 2 & 1 & 1 \end{bmatrix}; \ \bar{B} = \begin{bmatrix} -\dfrac{J_V}{D_V} \\ -\dfrac{J_p}{D_p} \\ 0 \end{bmatrix}; \ \bar{C} = \begin{bmatrix} C_V \\ C_p \\ C_h \end{bmatrix} \tag{6.6a}$$

The boundary conditions for Eq. (6.6) are given by

$$x = 0, \bar{C}_i = C_i(I); \ x = L, \bar{C}_i = C_i(II) \tag{6.7}$$

At steady state, the defect permeation fluxes are constant and can be obtained by integration of Eq. (6.6) over the thickness of the membrane:

$$J_V = -\frac{D_V}{L}\left[\int_{C_V(I)}^{C_V(II)}(1-t_V)dC_V - \int_{C_p(I)}^{C_p(II)}\frac{C_V}{C_p}t_p dC_p - \int_{C_h(I)}^{C_h(II)}\frac{C_V}{C_h}t_h dC_h\right] \tag{6.8a}$$

$$J_p = -\frac{D_p}{L}\left[\int_{C_p(I)}^{C_p(II)}(1-t_p)dC_p - \frac{1}{2}\int_{C_V(I)}^{C_V(II)}\frac{C_p}{C_V}t_V dC_V - \int_{C_h(I)}^{C_h(II)}\frac{C_p}{C_h}t_h dC_h\right] \tag{6.8b}$$

When only the proton and electron hole are considered to be the primary mobile charge carriers, the transport equation Eq. (6.6) may be reduced to

$$J_p = -\frac{D_p D_h (C_p + C_h)}{D_p C_p + D_h C_h} \cdot \frac{dC_p}{dx} \tag{6.9}$$

Integration of the above equation gives the hydrogen permeation flux as

$$J_{H_2} = -\frac{D_p D_h \left(C_p(II) + C_h(II)\right)}{2L\left(D_p - D_h\right)} \cdot \ln\left(\frac{D_p C_p(II) + D_h C_h(II)}{D_p C_p(I) + D_h C_h(I)}\right) \quad (6.10)$$

For simplification, the charged defect concentrations on membrane surfaces can be obtained under the assumption of quasi-equilibrium achieved for both surface exchange reactions [16]. Table 6.2 lists the hydrogen permeation fluxes in some PCMs.

6.3 PREPARATION OF PROTON-CONDUCTING CERAMIC MEMBRANES

Dense PCMs can be fabricated using the methods described in Chapter 5 for dense ceramic oxygen-permeable membranes; that is, cold isostatic pressing for disk/flat sheet membranes, plastic extrusion for tubular membranes, phase invasion for hollow fiber membranes, and suspension coating for composite membranes. It is noteworthy that disk composite membranes can easily be prepared by the co-pressing/co-sintering method [20]. Here, the suspension coating method is highlighted again because the composite PCMs are particularly important to achieve high hydrogen fluxes [17, 28].

6.3.1 Suspension Coating

The preparation of composite PCMs starts with the synthesis of porous substrates, followed by the formation of thin PCM films. Detailed procedures are illustrated in Figure 6.3. An exact procedure should be conducted to obtain thin, dense ceramic films [17]. The carbon black content for the synthesis of porous substrates can be varied so as to determine conditions that match accurately the shrinkage profile of the porous substrates with that of the deposited perovskite films during the final sintering of the composite membrane. Partial sintering of the green substrates is conducted to match their subsequent shrinkage with that of coated powders so that fissures and cracks can be minimized.

The synthesis of thin films involves grinding of oxide powders in a mortar, ultrasonically dispersing the powders in isopropanol to make slurry, and spinning the colloidal dispersion onto the pre-sintered porous substrate. After each coating step, the sample is dried in ambient air at

Table 6.2 Hydrogen permeation flux in PCMs or metal–PCM dual-phase membrane

Membrane composition	Configuration	Operating conditions	H_2 flux ($mol\,cm^{-2}\,s^{-1}$)	Ref.
$SrCe_{0.95}Yb_{0.05}O_{3-\delta}$	Disk; 2 μm	H_2–He/N_2–O_2; 900 K	1×10^{-5}	[17]
$SrCe_{0.95}Tm_{0.05}O_{3-\delta}$	Disk; 1–3 mm	10% H_2–He/air; 900°C	$(2.0–3.3) \times 10^{-8}$	[18]
$SrCe_{0.95-x}Zr_xTm_{0.05}O_{3-\delta}$	Disk; 1.2 mm	80% H_2–He/Ar; 900°C	3.12×10^{-8}	[19]
$SrCe_{0.95}Y_{0.05}O_{3-\delta}$	Disk; 50 μm	80% H_2–He/Ar; 950°C	7.6×10^{-8}	[20]
$BaCe_{0.95}Nd_{0.05}O_{3-\delta}$	Disk; 1.0 mm	80% H_2–He/Ar; 925°C	1.26×10^{-8}	[21]
$BaCe_{0.95}Tb_{0.05}O_{3-\delta}$	Hollow fiber	50% H_2–He/N_2; 1000°C	4.22×10^{-7}	[22]
$BaCe_{0.85}Tb_{0.05}Co_{0.1}O_{3-\delta}$	Hollow fiber	50% H_2–He/N_2; 1000°C	2.66×10^{-7}	[23]
Pd–$CaZr_{0.9}Y_{0.1}O_{3-\delta}$	Disk; 0.5 mm	20–80% H_2–N_2; 600–900°C	$(0.97–1.71) \times 10^{-7}$	[24]
Ni-Ba$(Zr_{0.1}Ce_{0.7}Y_{0.2})O_{3-\delta}$	Disk; 30 μm	80% H_2/N_2(3% H_2O); 900°C	2.4×10^{-7}	[25]
Ni–$BaCe_{0.8}Y_{0.2}O_{3-\delta}$	Disk; 0.38 mm	3.8% H_2–N_2/N_2; 900°C	5.36×10^{-8}	[26]
Ni–$BaCe_{0.95}Tb_{0.05}O_{3-\delta}$	Disk; 0.65 mm	50% H_2–He/N_2; 850°C	6.80×10^{-7}	[27]

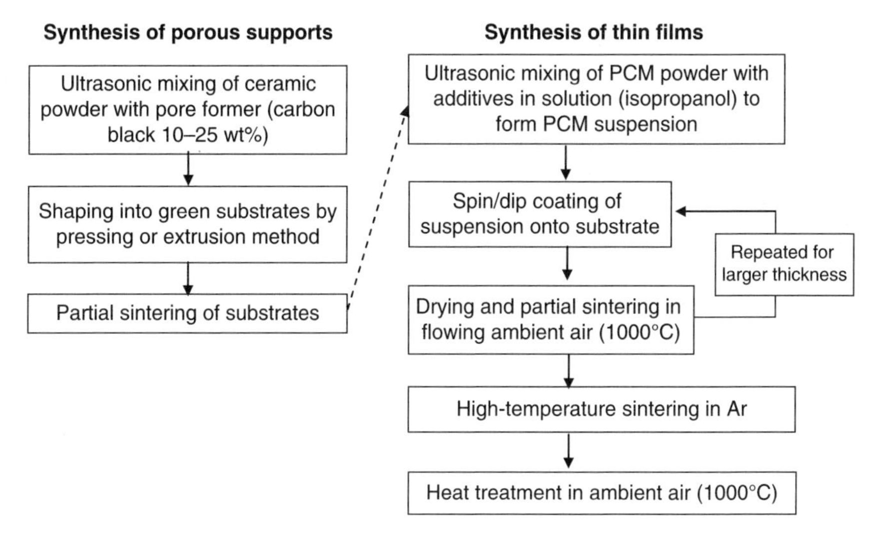

Figure 6.3 Synthesis procedures of composite PCMs by suspension coating. Reproduced from [17]. With permission from Elsevier.

room temperature. It is further partially sintered in ambient air, at the same heating rate and final pre-sintering temperature and time used for the synthesis of the substrate. This is done in order to prevent the development of cumulative stresses during sequential coating steps. The coating/drying/sintering operation is repeated several times in order to vary the film thickness. After coating, the film–substrate composite is sintered in Ar at high temperature (i.e., 1400–1600°C depending on the membrane material) and then cooled to ambient temperature at a slow rate in order to prevent thermal stresses. The membranes are finally treated at 1000°C in ambient air so as to replace any lattice oxygen atoms removed during the Ar treatment and to restore the stoichiometric perovskite composition.

6.4 CONFIGURATION OF PROTON-CONDUCTING MEMBRANE REACTORS

PCMRs are based on proton transport in the membrane under an electrochemical potential gradient, and thus are applied mainly to hydrogen-related reactions. In PCMRs, hydrogen can be supplied to the hydrogenation reaction or extracted from the dehydrogenation reaction through the membrane in the form of protons, driving the reaction to the

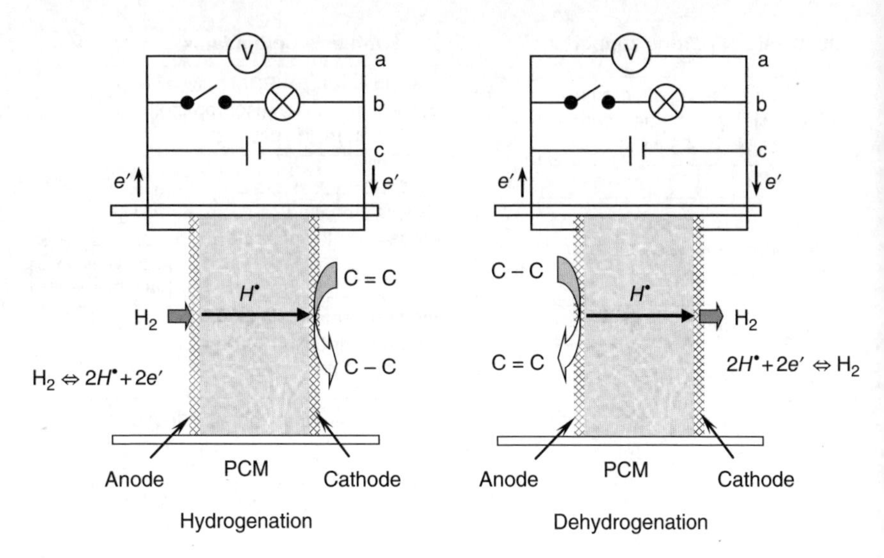

Figure 6.4 Principles of PCMRs for hydrogen/dehydrogenation reaction: (a) open circuit operation; (b) fuel cell mode; (c) pumping mode. Reproduced from [29]. With permission from Elsevier.

product side. Figure 6.4 shows the principle of PCMRs, with two chambers separated by the ceramic membrane.

The two electrodes are connected to a voltmeter (a), or to an external resistive load (b), or to an external power source (c). As long as the chemical potential of hydrogen is different at the two sides (electrodes) of the cell, a driving force for hydrogen transport across the PCM exists and the cell may operate in one of the following modes [29]:

(a) *Open-circuit operation.* There is no net current through the electrolyte. The difference in chemical potential is converted into the open-circuit voltage (OCV) of the cell. Reaction kinetics can be combined with OCV data in order to elucidate the reaction mechanism.
(b) *Fuel cell mode.* In this operation, the electrons released by the anode reaction are transferred to the cathode through the external circuit to form electrical current. Therefore, part of the chemical energy is converted directly into electrical energy.
(c) *Pumping mode.* In this operation, an external power source is used to impose a current (and equivalently a hydrogen flux) through the cell in the desired direction regardless of the hydrogen concentration gradient. The chemical potential of hydrogen in the reaction sites at the electrode can be controlled by electrode

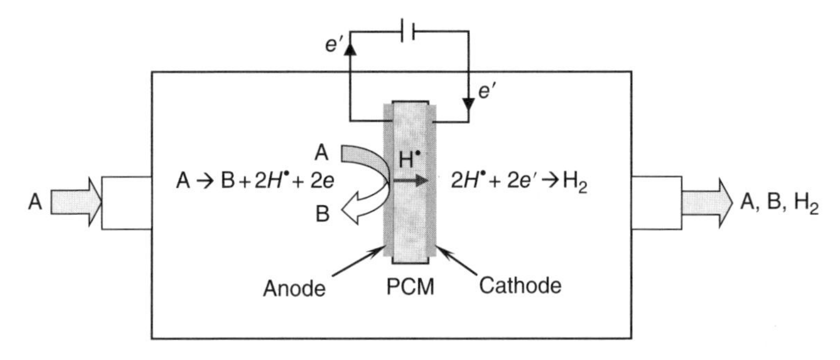

Figure 6.5 Schematic of a single-chamber PCMR. The proton-conducting disk is suspended in a flow of the reacting mixture.

potential and the reaction rate can be controlled by electric current. Furthermore, the catalytic activity and selectivity may also be controlled by the electrode potential [3].

In addition to the normal double-chamber reactor, the PCMRs can be constructed in a single-chamber configuration, as shown in Figure 6.5. The proton-conducting electrolyte membrane is suspended in a flow of the reacting mixture. A big advantage of this design is that it is easy to apply to existing catalytic processes as it does not require reactants to be separated. The proton-conducting ceramic simply replaces the conventional catalyst support [29].

PCMRs can be used to promote catalytic reaction rates electrochemically. If the catalyst to be promoted is one of the electrodes, protons can be "pumped" to or away from the catalyst during reaction. This can alter the catalytic activity and/or the selectivity of the reaction under study. The effect of electrochemical pumping can be expressed quantitatively by

$$\Lambda = \frac{r - r_{OCV}}{I/2F} \tag{6.11}$$

where I is the current imposed, F is Faraday's constant, r is the catalytic reaction rate obtained at closed circuit, and r_{OVC} is the open circuit catalytic rate. If $\Lambda = 1$, the effect is Faradaic; that is, the increase in reaction rate equals the rate of proton transport through the electrolyte. In many cases, a strong non-Faradaic effect has been observed with Λ values on the order of 10^5. This is the NEMCA (non-Faradaic electrochemical modification of catalytic activity) phenomenon. NEMCA can have a significant impact on catalytic research because: (i) unlike traditional catalytic promoters, the surface modification can be monitored electrochemically; and (ii) the product yield may be altered by imposing an electric current orders of magnitude lower than that stoichiometrically required. It is noteworthy that Eq. (6.11) should be modified for mixed proton–electron conductors as

$$\Lambda = \frac{r - r_{OCV}}{I/2F} \cdot t_p \tag{6.12}$$

where t_p is the proton transport number, which is the fraction of the total current that is carried through the solid electrolyte in the form of H^+:

$$t_p = \frac{n_{H_2}}{I/2F} \tag{6.13}$$

where n_{H_2} is the moles of hydrogen transferred through the electrolyte membrane per second. A more useful design of PCMRs for dehydrogenation reactions uses air as oxidant to promote hydrogen permeation. As shown in Figure 6.6(a), the hydrocarbon reactant is fed in on one side while the other side of the membrane is exposed to the air stream. Protons formed by dehydrogenation permeate though the membrane to react with oxygen on the cathode, and electrons go through the external circuit to form electrical current. Two electrodes are applied, as catalyst and electron distributor/collector. Therefore, electrical power can be co-generated with the hydrocarbon products. If the PCM is an MIEC, the reactor will be much simplified because no external circuitry is needed. The membrane itself serves as the internal circuit for electron transport, as shown in Figure 6.6(b).

6.5 APPLICATIONS OF PROTON-CONDUCTING CERAMIC MEMBRANE REACTORS

PCMRs can be applied to various hydrogen-related reactions due to their unique properties [29]. In addition to the high perm-selectivity toward hydrogen, the oxide membrane may serve as an electrolyte to supply or remove ions to or from the catalyst surface, leading to noticeably enhanced catalytic activity and selectivity. Furthermore, compared with dense metal membranes, dense ceramic PCMRs have higher thermal resistance and better chemical stability, and thus are more attractive for use in high-temperature and harsh environments. The principal applications of PCMRs, with recent important findings for each reaction system, are summarized below.

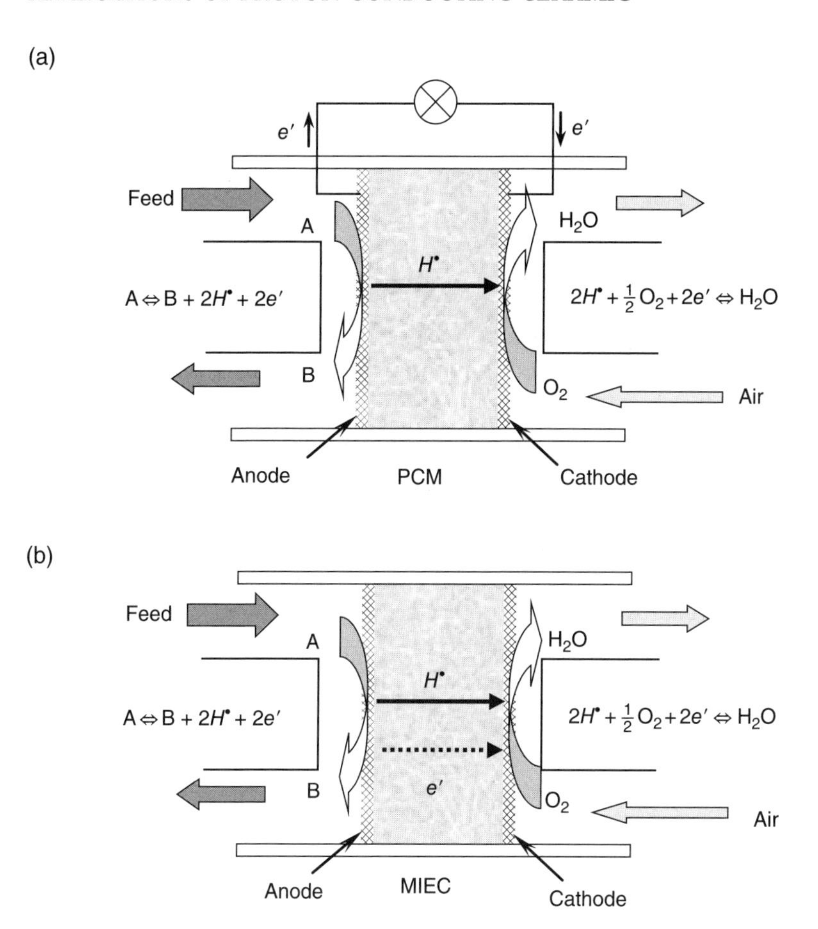

Figure 6.6 PCMRs for dehydrogenation reaction promoted by air feed: (a) fuel cell mode; (b) permeation mode.

6.5.1 Dehydrogenation Coupling of Methane

One of the most important applications of PCMRs is for various dehydrogenation reactions, typically methane coupling to C_2 hydrocarbons. The oxidative coupling of methane in dense ceramic oxygen-permeable MRs faces two severe problems: (i) deep oxidation of C_2 products in the presence of permeated oxygen gas, leading to low C_2 selectivity and yield; and (ii) formation of water by-product on the product side, leading to more complexity of the reactor system and loss of catalytic activities in the membrane and catalyst. As an alternative to the oxidative process, methane coupling can also be achieved through a dehydrogenation process where gaseous oxygen can be avoided:

$$2CH_4 \Leftrightarrow C_2H_6 + H_2, \Delta G^0_{1000} = 71.1\,\text{kJ}\,\text{mol}^{-1}$$

$$2CH_4 \Leftrightarrow C_2H_4 + 2H_2, \Delta G^0_{1000} = 57.2\,\text{kJ}\,\text{mol}^{-1}$$

However, the thermodynamics of these reactions are unfavorable and display equilibrium-limited conversion. By in-situ removing the hydrogen from the reaction site through a proton-conducting membrane, the equilibrium of the methane coupling may be driven toward the C_2 product side, leading to enhanced conversion [30, 31]. Figure 6.7 shows the mechanism of methane coupling in the PCMRs. On the anode side, methane is adsorbed and converted catalytically into methyl radicals, which enter into the gas phase for a coupling reaction. The protons released from the methane coupling reaction are driven electrochemically to the cathode side of the membrane, where they combine with electrons into hydrogen or react with oxygen to form water.

In the pumping operation, the hydrogen separation rate can be controlled by modulation of the applied current. When the methane feed rate is constant, a high yield requires a high proton permeation rate, or equivalently a high current density. However, the imposed current density usually cannot exceed an upper limit due to the thick membranes used [31]. Therefore, the C_2 yields are usually very low, which limits the practicality of the PCMRs severely. The proton transference number of the membrane oxides may be varied depending primarily on temperature and the partial pressures of hydrogen and oxygen. When the membrane is a mixed proton–electron conductor, the hydrogen can be extracted in the reaction system by the self-discharge phenomenon. In this case, the external electric source and the electrode materials – as well as current collectors – are unnecessary for transporting protons across the solid electrolyte, and the construction of the reactor is much simplified (permeation mode, Figure 6.7(b)). However, the methane activation and dimerization rate may be many times lower than that in the pump operating mode [30, 32].

Catalysts are usually applied to activate methane dissociation and hydrogen combination. Such catalysts may be of Pt, Ag, or other specially designed oxides. It is worth noting that several mixed conducting perovskites (e.g., strontium cerates) are very efficient methane-coupling catalysts and have been used as such in regular catalytic reactors. Studies on methane coupling in PCMRs are summarized in Table 6.3. As can be seen, most of the reported C_2 yields are very low (less than 2%). The

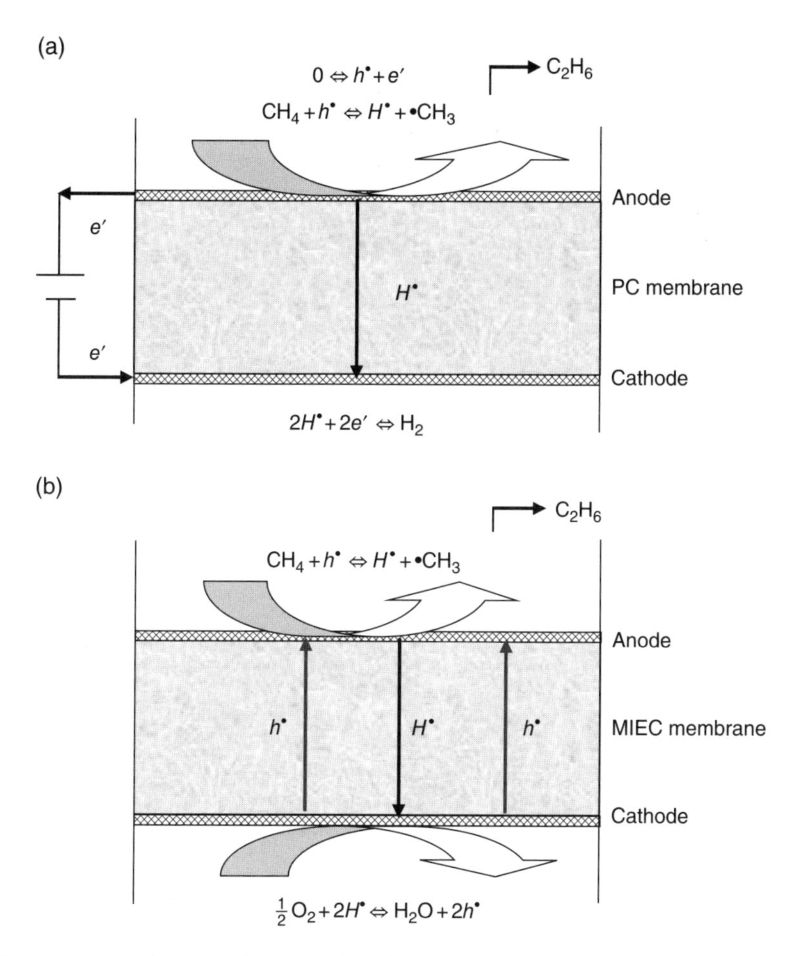

Figure 6.7 Mechanism of methane dehydrogenation coupling in PCMRs: (a) pumping mode; (b) permeation mode.

reasons for this include the low catalytic activity of the membrane surface and catalyst, the large membrane thickness, and the methane pyrolysis in the absence of oxygen.

6.5.2 Dehydrogenation of Alkanes into Alkenes

The catalytic dehydrogenation of propane is a potential method for the production of propylene, a key chemical in the polymerization and organic synthesis industries:

$$C_3H_8 \Leftrightarrow C_3H_6 + H_2,\ \ \Delta H^0_{298} = 124\,\mathrm{kJ\,mol^{-1}}$$

Table 6.3 Dehydrogenation reactions in PCMRs

Membrane	Configuration	Catalyst/electrode	Operating mode	Temperature (°C)	Main results	Ref.
Methane coupling to C_2 hydrocarbons: $CH_4 \leftrightarrow \cdot CH_3 + H^{\bullet} + e'$; $2H^{\bullet} + 1/2O_2 + 2e' \leftrightarrow H_2O$						
SCYb	Disk	Ag	Pumping	900	0.23–0.52 μmol min^{-1} cm^{-2}	[30, 32]
SCYb	Disk	Ag	Pumping	750	—	[32]
SCYb	Disk (1 mm)	Ag	Permeation	900	0.23–1.33 μmol min^{-1} cm^{-2}; $S_{C2} \rightarrow 100\%$	[33]
BCM	Disk (1 mm)	—	Permeation	950	8.9 μmol min^{-1} cm^{-2}	[34]
SCYb	Disk (1.5 mm)	Ag	Pumping	750	$Y_{C2} = 0.55\%$; $S_{C2} = 64\%$	[35]
SCYb	Hollow fiber	—	Permeation	950	$Y_{C2} = 13.4\%$; $S_{C2} = 21\%$	[36]
Propane to propylene: $C_3H_8 \leftrightarrow C_3H_6 + 2H^{\bullet} + 2e'$						
SCYb	Disk (1.5 mm)	Pt, Pd/Ag	Pumping	650–750		[37, 38]
Ethane to ethylene: $C_2H_6 \leftrightarrow C_2H_4 + 2H^{\bullet} + 2e'$; $2H^{\bullet} + 1/2O_2 + 2e' \leftrightarrow H_2O$						
BCY	Disk (0.5 mm)	Pt	Fuel cell	700	$X = 34\%$; $S = 96\%$; $P = 174$ mW cm^{-2}	[39]
BCY	Disk (30 μm on porous support)	Pt	Fuel cell	700	$X = 36.7\%$; $S = 90.5\%$; $P = 216$ mW cm^{-2}	[40]

SCYb = $SrCe_{0.95}Yb_{0.05}O_{3-\delta}$; BCM = $BaCe_{0.95}Mn_{0.05}O_{3-\delta}$; BCY = $BaCe_{0.85}Y_{0.15}O_{3-\delta}$.

This reaction requires high temperatures (500–700°C) and low pressures (0.3–1 atm). At such high temperatures, however, the side-reaction of C_3H_8 thermal decomposition takes place:

$$C_3H_8 \Rightarrow C_2H_4 + CH_4$$

With a PCMR to withdraw the hydrogen from the reactor, the limited conversion of the catalytic dehydrogenation of propane dictated by the thermodynamic equilibrium can be shifted, leading to promotion of the propylene selectivity and yield. The driving force is the difference in electrochemical potential of protons at the two sides of the membrane, which is generated using an external power source (e.g., galvanostat or potentiostat). In addition, high-purity hydrogen can be obtained as a product at the cathode. Polycrystalline Pt is superior to Pd as the anode/catalyst to yield higher rates of propane decomposition [35]. When a propane/steam mixture is introduced instead of pure propane in the reactor, up to 90% of the hydrogen produced could be separated electrochemically [36].

Similarly, ethane can also be converted into ethylene in PCMR with enhanced yields [37, 38]. Electrical power can be co-generated with ethylene product in the fuel cell operating mode. The performance of the fuel cell MR can be improved by reducing the membrane thickness [40]. The results of propane/ethane dehydrogenation in PCMRs are also summarized in Table 6.3.

6.5.3 WGS Reaction and Water Electrolysis for Hydrogen Production

The WGS reaction is one of the key technologies in the hydrogen purification processes of syngas obtained by steam reforming or partial oxidation of hydrocarbons:

$$CO + H_2O \Leftrightarrow CO_2 + H_2 \quad \Delta H^0_{298} = -41.1\,\text{kJ}\,\text{mol}^{-1}$$

PCMRs have been tried in the WGS reaction. $SrCe_{0.95}Yb_{0.05}O_{3-\delta}$ (SCYb) can be operated as the electrolyte membrane, but it is not adequate because of its reaction with CO_2 [41]. $SrZr_{0.95}Y_{0.05}O_{3-\delta}$ electrolyte demonstrates high tolerance to CO_2 exposure (500 h) but low hydrogen permeability [42]. In order to separate the hydrogen with higher efficiency, the applied voltage must be as low as possible, using thinner

Table 6.4 PCMRs for hydrogen production

Membrane	Configuration	Electrode/catalyst	Operating mode	Temperature (°C)	Main results	Ref.
WGS process: $CO + H_2O \leftrightarrow CO_2 + 2H^{\bullet} + 2e'$						
SCYb	Disk (0.5 mm)	Pt	Pumping	800		[41]
SZY	Disk	Pd	Pumping	600–750	$r_{H2} = 0.8–3.2\ mol\ s^{-1}$	[42]
SCEu	Tube (23 μm on porous support)	Ni–$SrCeO_3$	Permeation	900	$X_{CO} = 46\%$; $Y_{H2} = 32\%$	[43]
Steam electrolysis: $H_2O \leftrightarrow 2H^{\bullet} + 2e' + 1/2O_2$; $2H^{\bullet} + 2e' \leftrightarrow H_2$						
SCYb	Disk (0.5 mm)	Pt	Pumping	800		[41]
SZYb	Tube with one closed end	Pt cermet	Pumping	460		[44]

SCYb = $SrCe_{0.95}Yb_{0.05}O_{3-\delta}$; SZY = $SrZr_{0.95}Y_{0.05}O_{3-\delta}$; SCEu = $SrCe_{0.9}Eu_{0.1}O_{3-\delta}$; SZYb = $SrZr_{0.9}Yb_{0.1}O_{3-\delta}$.

electrolytes and an improved anode. In addition, $SrCe_{0.9}Eu_{0.1}O_{3-\delta}$ oxide exhibits mixed proton–electron conduction, and thus can be conducted for hydrogen production under permeation mode [43].

Another effective way to produce hydrogen is by the electrolysis of water if electric power is available, preferably from renewable or sustainable energies, such as solar and wind powers, bio-energy, and so on [41]. The hydrogen can be pumped out electrochemically through a proton-conducting membrane using an external current. The current efficiency increases with increasing partial pressure of water vapor and temperature and decreases with increasing current density [44]. Studies on hydrogen production in PCMRs are summarized in Table 6.4.

6.5.4 Decomposition of NO_x

Proton-conducting membranes can be applied to reduce NO_x emission by combining heterogeneous catalysis and solid-state electrochemistry. The solid electrolytes in membrane reactors serve to control chemisorptive bonds electrochemically and enhance catalytic activity. Figure 6.8 shows a schematic diagram of a steam electrolysis cell constructed with a proton conductor for reducing NO. Steam is electrolyzed at the anode. The electrode reactions are

$$\text{Anode reaction: } H_2O \leftrightarrow 2H^{\bullet} + \tfrac{1}{2}O_2 + 2e'$$

$$\text{Cathode reaction: } NO + 2H^{\bullet} + 2e' \rightarrow \tfrac{1}{2}N_2 + H_2O$$

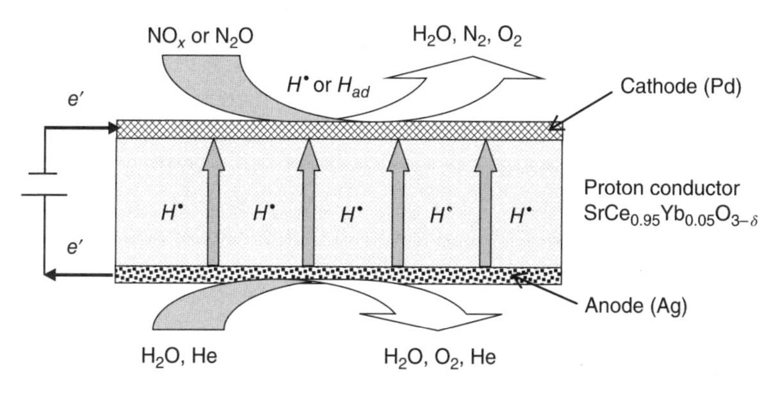

Figure 6.8 Schematic principle of the reduction of nitrogen oxides in a PCMR.

Table 6.5 PCMRs for the reduction of nitrogen oxides

Membrane	Configuration	Electrode/ catalyst	Temperature (°C)	Main results	Ref.
SZYb	Tube with one closed end	$Pt/Sr–Al_2O_3$	350	50% removal, 20% efficiency	[45]
SCYb, SZYb	Tube with one closed end	$Pt/Ba/Al_2O_3$	450	Removal → 100%	[46]
SCYb	Disk	Ag/Pd	500–700	$S_{N2} = 100\%$	[47]

SZYb = $SrZr_{0.9}Yb_{0.1}O_{3-\delta}$; SCYb = $SrCe_{0.95}Yb_{0.05}O_{3-\delta}$.

Strontium cerate or strontium zirconate proton conductors have been tried as electrolyte membrane, with the results of nitrogen oxide reduction as summarized in Table 6.5. When $Pt/Ba/Al_2O_3$ or $Pt/Sr/Al_2O_3$ is used as working electrode, it is possible to reduce the NO_x even in the presence of excess O_2. The reduction of NO proceeds through the electrochemical reduction of NO absorbed in Sr/Al_2O_3 but not through the chemical reduction of NO by H_2 gas.

6.5.5 Synthesis of Ammonia

The synthesis of ammonia involves the reaction of gaseous nitrogen and hydrogen on an Fe-based catalyst at high pressures (150–300 bar):

$$N_2 + 3H_2 \xleftrightarrow{\text{Fe-cat}} 2NH_3$$

The conversion to ammonia is limited by thermodynamics. Since the gas volume decreases with the reaction, very high pressures must be used in order to push the equilibrium to the ammonia side. Although the reaction is exothermic and therefore conversion increases with decreasing temperature, the reaction temperature must be high so as to achieve industrially acceptable reaction rates. Owing to the application of more active catalysts, most modern synthesis loops nowadays are operated at 100–300 bar and 450–500°C, with equilibrium conversion on the order of 10–15%.

In a PCMR, ammonia synthesis can be achieved at atmospheric pressure and a higher conversion may be attained than the equilibrium value of the reaction. Figure 6.9 illustrates schematically the principle of ammonia synthesis in a typical PCMR. Pd–Ag alloy usually serves as

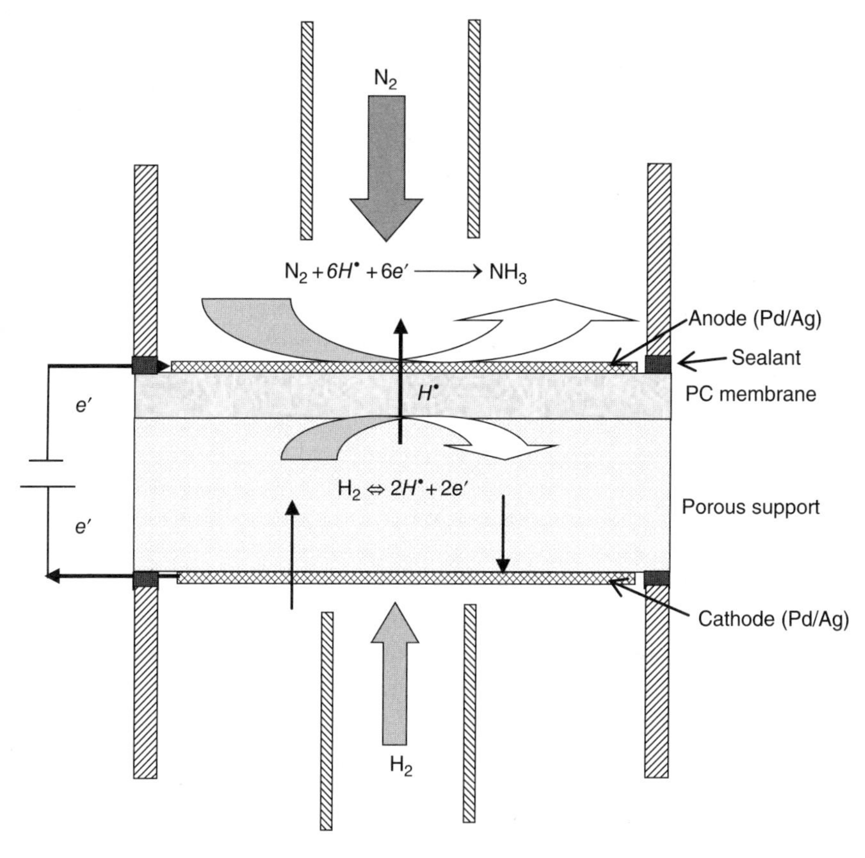

Figure 6.9 Principle of ammonia synthesis in a PCMR.

the working electrode. Gaseous H_2 passing over the anode of the cell reactor is converted to protons:

$$3H_2 \rightarrow 6H^{\bullet} + 6e'$$

The protons ($H^{\bullet}$) are then transported through the solid electrolyte to the cathode, where the following half-cell reaction takes place:

$$N_2 + 6H^{\bullet} + 6e' \rightarrow 2NH_3$$

Since hydrogen is supplied electrochemically to the catalyst surface in the form of protons, the reaction may follow a different mechanism, or at least the rate-determining step is different from that in a typical catalytic system (where it is dissociative adsorption of gaseous molecular nitrogen onto the catalyst's active sites). As a result, the NH_3 production rates in the membrane reactor may exceed those in normal reactors by at least three orders of magnitude, and the conversion of H^+ into NH_3 may be as high as 78% [48]. Upon "pumping" H^+ to the catalyst surface, the reaction rate could increase by as much as 1300%, while the catalytic activity would decrease and eventually be lost completely upon "pumping" H^+ away from the catalyst. Some studies on ammonia synthesis in PCMRs are summarized in Table 6.6. Generally, the reaction rate is limited by the rate of proton transport through the membrane. Therefore, methods related to improving the proton transport rate – such as decreasing the membrane thickness and promoting the surface exchange kinetics – are able to increase the ammonia yields.

6.5.6 Challenges and Future Work

It has been well established that PCMRs possess many distinctive advantages over other reactors, including: (i) running dehydrogenation reactions while simultaneously producing valuable chemicals and pure hydrogen; (ii) operating in fuel cell mode with co-generation of electricity and useful chemicals; (iii) modifying the intrinsic catalytic activity in the desired direction; (iv) replacing high-pressure (NH_3 synthesis) operation with the supply of electrical energy [29]. However, PCMRs have not been used in industrial processes. The hurdles that have to be overcome on route to their commercialization include:

1. The construction of PCMRs with electrodes and current connectors is very complicated.
2. The protonic conductivity of the currently developed ceramic conductors is not sufficiently high and thus considerably high temperatures are required to raise it.
3. In order to improve the conductivity and reduce the operating temperature it is necessary to make thin ceramic films, but to prepare thin and dense films on a porous support is technologically challenging.
4. The operating temperature is too high for most organic reactions.

In fact, all the challenges faced by dense ceramic oxygen-permeable membrane reactors as described in Chapter 5 have to be faced by PCMRs. Searching for better membrane materials, developing effective membrane synthesis methods, and improving chemical and structural stability of the current membrane materials will be the focus of active

Table 6.6 Ammonia synthesis in proton-conducting membrane reactors

Membrane	Configuration	Electrode/ catalyst	Temperature (°C)	Main results	Ref.
SCYb	Tube with one closed end	Pd	570–750	r_{NH3} increased at least 3 orders of magnitude	[48]
LCZ	Disk (0.8 mm)	Pd–Ag	520	$r_{NH3} = 2.0 \times 10^{-9}$ mol s^{-1} cm^{-2}	[11]
BCNb	Disk (0.8 mm)	Pd–Ag	620	$r_{NH3} = 2.16 \times 10^{-9}$ mol s^{-1} cm^{-2}	[7]
CM	Disk (0.8 mm)	Pd–Ag	400–800	$r_{NH3} = (7.2–8.2) \times 10^{-9}$ mol s^{-1} cm^{-2}	[10]
LSGM	Disk	Pd–Ag	550	$r_{NH3} = 2.37 \times 10^{-9}$ mol s^{-1} cm^{-2}	[49]
BCY	Disk on NiO–BCY support	Pd–Ag/BSCF	530	$r = 4.1 \times 10^{-9}$ mol s^{-1} cm^{-2}	[50]

SCYb = $SrCe_{0.95}Yb_{0.05}O_{3-\delta}$; LCZ = $La_{1.9}Ca_{0.1}Zr_2O_{6.95}$; BCNb = $Ba_3(Ca_{1.18}Nb_{1.82})O_{9-\delta}$; CM = $Ce_{0.8}M_{0.2}O_{2-\delta}$ (M = La, Y, Gd, Sm); LSGM = $La_{0.9}Sr_{0.1}Ga_{0.8}Mg_{0.2}O_{3-\alpha}$; BCY = $BaCe_{0.85}Y_{0.15}O_{3-\delta}$; BSCF = $Ba_{0.5}Sr_{0.5}Co_{0.8}Fe_{0.2}O_{3-\alpha}$.

research in these areas. In addition, more future work should be focused on: (i) the development of mixed proton–electron conductors for methane coupling; (ii) the conversion of alkanes into alkenes (e.g., propane into propene) with simultaneous separation and production of pure hydrogen; (iii) the application of PCMR in the elucidation of reaction mechanisms.

NOTATION

C_i concentration of charged defect i (mol m^{-3})
D_i diffusion coefficient of charged defect i (m^2 s^{-1})
F Faraday constant
J permeation flux (mol m^{-2} s^{-1})
L membrane thickness (m)
r reaction rate (mol s^{-1})
t_i transport number of defect i
z_i charge number of defect i
Λ NEMCA constant

Subscripts
e electron
h electron hole
V oxygen vacancy
p proton
OCV open circuit voltage

REFERENCES

[1] Iwahara, H. (1995) Technological challenges in the application of proton conducting ceramics. *Solid State Ionics*, 77, 289–298.
[2] Iwahara, H. (1996) Proton conducting ceramics and their applications. *Solid State Ionics*, 86–88, 9–15.
[3] Iwahara, H. (1999) Hydrogen pumps using proton-conducting ceramics and their applications. *Solid State Ionics*, 125, 271–278.
[4] Iwahara, H. (2004) Prospect of hydrogen technology using proton-conducting ceramics. *Solid State Ionics*, 168, 299–310.
[5] Iwahara, H., Esaka, T., Uchida, H. and Maeda, N. (1981) Proton conduction in sintered oxides and its application to steam electrolysis for hydrogen production. *Solid State Ionics*, 3/4, 359–363.
[6] Nowick, A.S. and Yang, D. (1995) High-temperature protonic conductors with perovskite-related structures. *Solid State Ionics*, 77, 137–146.

[7] Li, Z., Liu, R., Xie, Y., Feng, S. and Wang, J. (2005) A novel method for preparation of doped $Ba_3(Ca_{1.18}Nb_{1.82})O_{9-\delta}$: Application to ammonia synthesis at atmospheric pressure. *Solid State Ionics*, 176, 1063–1066.
[8] Matsumoto, H., Kawasaki, Y., Ito, N., Enoki, M. and Ishihara, T. (2007) Relation between electrical conductivity and chemical stability of $BaCeO_3$-based proton conductors with different trivalent dopants. *Electrochemical and Solid State Letters*, 10, B77–B80.
[9] Lin, Y.S. (2001) Microporous and dense inorganic membranes: Current status and prospective. *Separation and Purification Technology*, 25, 39–55.
[10] Liu, R., Xie, Y., Wang, J., Li, Z. and Wang, B. (2006) Synthesis of ammonia at atmospheric pressure with $Ce_{0.8}M_{0.2}O_{2-\delta}$ (M = La, Y, Gd, Sm) and their proton conduction at intermediate temperature. *Solid State Ionics*, 177, 73–76.
[11] Xie, Y. (2004) Preparation of $La_{1.9}Ca_{0.1}Zr_2O_{6.95}$ with pyrochlore structure and its application in synthesis of ammonia at atmospheric pressure. *Solid State Ionics*, 168, 117–121.
[12] Hibino, T., Mizutani, K., Yajima, T. and Iwahara, H. (1992) Characterization of proton in Y-doped $SrZrO_3$ polycrystal by IR spectroscopy. *Solid State Ionics*, 58, 85–88.
[13] Uchida, H., Maeda, N. and Iwahara, H. (1983) Relation between proton and hole conduction in $SrCeO_3$ based solid electrolytes under water-containing atmospheres at high temperatures. *Solid State Ionics*, 11, 117–124.
[14] Liu, Y., Tan, X. and Li, K. (2006) Mixed conducting ceramics for catalytic membrane processing. *Catalysis Reviews*, 48, 145–198.
[15] Kreuer, K.D., Fuchs, A. and Maier, J. (1995) H/D isotope effect of proton conductivity and proton conduction mechanism in oxides. *Solid State Ionics*, 77, 157–162.
[16] Tan, X., Liu, S., Li, K. and Hughes, R. (2000) Theoretical analysis of ion permeation through mixed conducting membranes and its application to dehydrogenation reactions. *Solid State Ionics*, 138, 149–159.
[17] Hamakawa, S., Li, L., Li, A. and Iglesia, E. (2002) Synthesis and hydrogen permeation properties of membranes based on dense $SrCe_{0.95}Yb_{0.05}O_{3-\alpha}$ thin films. *Solid State Ionics*, 148, 71–81.
[18] Qi, X. and Lin, Y.S. (2000) Electrical conduction and hydrogen permeation through mixed proton–electron conducting strontium cerate membranes. *Solid State Ionics*, 130, 149–156.
[19] Liang, J., Mao, L., Li, L. and Yuan, W. (2010) Protonic and electronic conductivities and hydrogen permeation of $SrCe_{00.95-x}Zr_xTm_{0.05}O_{3-\delta}$ ($0 \leq x \leq 0.40$) membrane. *Chinese Journal of Chemical Engineering*, 18, 506–510.
[20] Zhan, S., Zhu, X., Ji, B. *et al.* (2009) Preparation and hydrogen permeation of $SrCe_{0.95}Y_{0.05}O_{3-\delta}$ asymmetrical membranes. *Journal of Membrane Science*, 340, 241–248.
[21] Cai, M., Liu, S., Efimov, K., Caro, J., Feldhoff, A. and Wang, H. (2009) Preparation and hydrogen permeation of $BaCe_{0.95}Nd_{0.05}O_{3-\delta}$ membranes. *Journal of Membrane Science*, 343, 90–96.
[22] Tan, X., Song, J., Meng, X. and Meng, B. (2012) Preparation and characterization of $BaCe_{00.95}Tb_{0.05}O_{3-\alpha}$ hollow fibre membranes for hydrogen permeation. *Journal of the European Ceramic Society*, 32, 2351–2357.
[23] Song, J., Li, L., Tan, X. and Li, K. (2013) $BaCe_{00.85}Tb_{0.05}Co_{0.1}O_{3-\delta}$ perovskite hollow fibre membranes for hydrogen/oxygen permeation. *International Journal of Hydrogen Energy*, 38, 7904–7912.

[24] Jeon, S.Y., Lim, D.K., Choi, M.B., Wachsman, E.D. and Song, S.J. (2011) Hydrogen separation by Pd–$CaZr_{0.9}Y_{0.1}O_{3-\delta}$ cermet composite membranes. *Separation and Purification Technology*, 79, 337–341.

[25] Zhu, Z., Sun, W., Yan, L., Liu, W. and Liu, W. (2011) Synthesis and hydrogen permeation of Ni–$Ba(Zr_{0.1}Ce_{0.7}Y_{0.2})O_{3-\delta}$ metal–ceramic asymmetric membranes. *International Journal of Hydrogen Energy*, 36, 6337–6342.

[26] Song, S.J., Moon, J.H., Lee, T.H., Dorris, S.E. and Balachandran, U. (2008) Thickness dependence of hydrogen permeability for Ni–$BaCe_{0.8}Y_{0.2}O_{3-\delta}$. *Solid State Ionics*, 179, 1854–1857.

[27] Meng, X., Song, J., Yang, N. *et al.* (2012) Ni–$BaCe_{00.95}Tb_{0.05}O_{3-\delta}$ cermet membranes for hydrogen permeation. *Journal of Membrane Science*, 401/402, 300–305.

[28] Kosacki, I. and Anderson, H.U. (1997) The structure and electrical properties of $SrCe_{0.95}Yb_{0.05}O_3$ thin film protonic conductors. *Solid State Ionics*, 97, 429–436.

[29] Kokkofitis, C., Ouzounidou, M., Skodra, A. and Stoukides, M. (2007) High temperature proton conductors: Applications in catalytic processes. *Solid State Ionics*, 178, 507–513.

[30] Chiang, P.-H., Eng, D. and Stoukides, M. (1993) Electrocatalytic nonoxidative dimerization of methane over Ag electrodes. *Solid State Ionics*, 61, 99–103.

[31] Chiang, P.-H., Eng, D., Tsiakaras, P. and Stoukides, M. (1995) Ion transport and polarization studies in a proton conducting solid electrolyte cell. *Solid State Ionics*, 77, 305–310.

[32] Hamakawa, S., Hibino, T. and Iwahara, H. (1993) Electrochemical methane coupling using protonic conductors. *Journal of the Electrochemical Society*, 140, 459–462.

[33] Hamakawa, S., Hibino, T. and Iwahara, H. (1994) Electrochemical methane coupling in a proton–hole mixed conductor and its application to a membrane reactor. *Journal of the Electrochemical Society*, 141, 1720–1726.

[34] J.H. White, M. Schwartz, A.F. Sammells, Solid state proton and electron mediating membrane and use in catalytic membrane reactors. US Patent 5821185 A, Oct. 13, 1998.

[35] Langguth, J., Dittmeyer, R., Hofmann, H. and Tomandl, G. (1997) Studies on oxidative coupling of methane using high-temperature proton-conducting membranes. *Applied Catalysis A: General*, 158, 287–305.

[36] Liu, Y., Tan, X. and Li, K. (2006) Non-oxidative methane coupling in the $SrCe_{0.95}Yb_{0.05}O_{3-\alpha}$ (SCYb) hollow fibre membrane reactor. *Industrial and Engineering Chemistry Research*, 45, 3782–3790.

[37] Karagiannakis, G., Kokkofitis, C., Zisekas, S. and Stoukides, M. (2005) Catalytic and electrocatalytic production of H_2 from propane decomposition over Pt and Pd in a proton-conducting membrane-reactor. *Catalysis Today*, 104, 219–224.

[38] Karagiannakis, G., Zisekas, S., Kokkofitis, C. and Stoukides, M. (2006) Effect of H_2O presence on the propane decomposition reaction over Pd in a proton conducting membrane reactor. *Applied Catalysis A: General*, 301, 265–271.

[39] Shi, Z., Luo, J.-L., Wang, S., Sanger, A.R. and Chuang, K.T. (2008) Protonic membrane for fuel cell for co-generation of power and ethylene. *Journal of Power Sources*, 176, 122–127.

[40] Fu, X.-Z., Luo, J.-L., Sanger, A.R., Xu, Z.-R. and Chuang, K.T. (2010) Fabrication of bi-layered proton conducting membrane for hydrocarbon solid oxide fuel cell reactors. *Electrochimica Acta*, 55, 1145–1149.

[41] Matsumoto, H., Okubo, M., Hamajima, S., Katahira, K. and Iwahara, H. (2002) Extraction and production of hydrogen using high-temperature proton conductor. *Solid State Ionics*, 152, 715–720.

[42] Kokkofitis, C., Ouzounidou, M., Skodra, A. and Stoukides, M. (2007) Catalytic and electrocatalytic production of H_2 from the water gas shift reaction over Pd in a high temperature proton-conducting cell-reactor. *Solid State Ionics*, 178, 475–480.

[43] Li, J., Yoon, H., Oh, T.-K. and Wachsman, E.D. (2009) High temperature $SrCe_{0.9}Eu_{0.1}O_{3-\delta}$ proton conducting membrane reactor for H_2 production using the water–gas shift reaction. *Applied Catalysis B: Environmental*, 92, 234–239.

[44] Kobayashi, T., Abe, K., Ukyo, Y. and Matsumoto, H. (2001) Study on current efficiency of steam electrolysis using a partial protonic conductor $SrZr_{0.9}Yb_{0.1}O_{3-\alpha}$. *Solid State Ionics*, 138, 243–251.

[45] Kobayashi, T., Abe, K., Ukyo, Y. and Iwahara, H. (2000) Reduction of nitrogen oxide by steam electrolysis cell using a protonic conductor $SrZr_{0.9}Yb_{0.1}O_{3-\alpha}$ and the catalyst Sr/Al_2O_3. *Solid State Ionics*, 134, 241–247.

[46] Kobayashi, T., Abe, K., Ukyo, Y. and Iwahara, H. (2002) Performance of electrolysis cells with proton and oxide-ion conducting electrolyte for reducing nitrogen oxide. *Solid State Ionics*, 154, 699–705.

[47] Kalimeri, K.K., Athanasiou, C.I. and Marnellos, G.E. (2010) Electro-reduction of nitrogen oxides using steam electrolysis in a proton conducting solid electrolyte membrane reactor (H^+-SEMR). *Solid State Ionics*, 181, 223–229.

[48] Marnellos, G., Zisekas, S. and Stoukides, M. (2000) Synthesis of ammonia at atmospheric pressure with the use of solid state proton conductors. *Journal of Catalysis*, 193, 80–87.

[49] Zhang, F., Yang, Q., Pan, B., Xu, R., Wang, H. and Ma, G. (2007) Proton conduction in $La_{0.9}Sr_{0.1}Ga_{0.8}Mg_{0.2}O_{3-\alpha}$ ceramic prepared via microemulsion method and its application in ammonia synthesis at atmospheric pressure. *Materials Letters*, 61, 4144–4148.

[50] Wang, W.B., Cao, X.B., Gao, W.J., Zhang, F., Wang, H.T. and Ma, G.L. (2010) Ammonia synthesis at atmospheric pressure using a reactor with thin solid electrolyte $BaCe_{0.85}Y_{0.15}O_{3-\alpha}$ membrane. *Journal of Membrane Science*, 360, 397–403.

7

Fluidized Bed Membrane Reactors

7.1 INTRODUCTION

Fluidized bed membrane reactors, also called membrane assisted fluidized bed reactors, are a special type of reactor in which the permeable membranes are integrated inside a fluidized reaction bed to intensify the reaction process. Such integration not only provides the advantages of both the fluidized bed and the membrane reactor, but also accomplishes a synergistic effect. The main advantages of FBMRs include [1]:

- Isothermal operation. The fluidized bed reactor has excellent tube-to-bed heat transfer properties, allowing a safe and efficient reactor operation even for highly exothermic reactions. Also, for highly endothermic reactions, where the hot catalyst is circulated between the reactor and the regenerator, the excellent gas-to-solid heat transfer characteristics of the fluidized beds can be exploited effectively. The intense macroscale solids mixing induced by the rising bubbles results in a remarkable temperature uniformity.
- Negligible pressure drop, and no internal mass and heat transfer limitations because of the small particle sizes of the catalysts employed.
- Flexibility in membrane and heat transfer surface area and arrangement of the membranes. Optimal concentration profiles can be created via distributive feeding of one of the reactants or selective withdrawal of one of the products.

Inorganic Membrane Reactors: Fundamentals and Applications, First Edition. Xiaoyao Tan and Kang Li.

- Improved fluidization behavior due to the presence of the inserts and gas permeation through the membranes, leading to reduced axial gas back-mixing and enhanced bubble breakage [2], while the bubble-to-emulsion mass transfer and membrane permeation are also improved due to the reduced bubble size.

Therefore, large improvements in conversion and selectivity may be achieved in FBMRs. Moreover, the membrane area can be reduced remarkably compared with the packed bed membrane reactors [3]. However, there are some disadvantages of FBMRs, including:

- Difficulties in reactor construction and membrane sealing at the wall.
- Erosion of reactor internals and catalyst attrition.

Nevertheless, the FBMR has become the most recent trend to overcome the limitations often prevailing in packed bed membrane reactors. Despite the advances in recent years, many challenges and difficulties still face the commercialization of FBMRs.

This chapter will present a description of the FBMRs with special emphasis on their construction, operation, and applications. The prospects and challenges for FBMRs will also be addressed at the end of the chapter.

7.2 CONFIGURATIONS AND CONSTRUCTION OF FBMRs

In FBMRs, the membranes are inserted inside the fluidized catalyst bed, serving as a product extractor or a reactant distributor. Figure 7.1 shows a typical FBMR structure for selective removal of a product (hydrogen) [4, 5]. Pd-membrane tubes are placed vertically in the FBMR. The reactant gas is fed through the gas distribution plate at the bottom of the reactor to fluidize the fine particulate catalysts. Entrained solids are separated from the reaction product gas stream by internal cyclone separator and then returned to the reactor catalyst bed.

In order to achieve rapid heat transfer and temperature uniformity, the reactant gas flow velocity should be high enough at least to form gas bubbles in the catalyst bed. However, the need for large-scale operation and relatively small catalyst particles makes operation in the slug

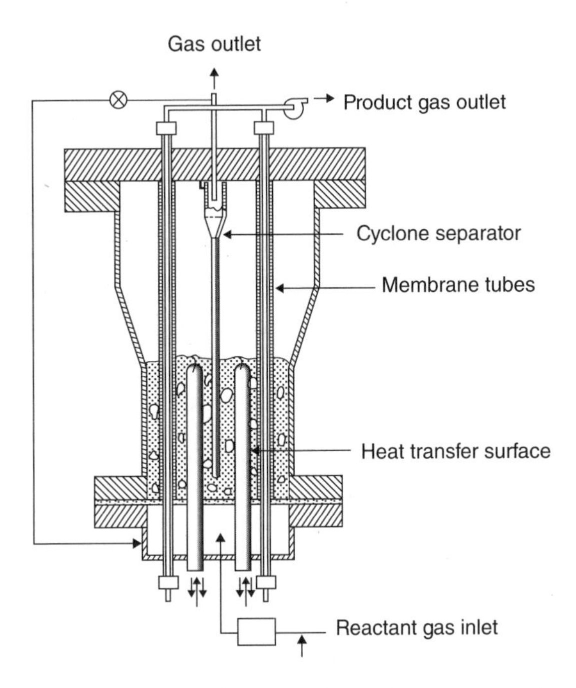

Figure 7.1 Typical structure of FBMR for selective removal of a product. Reproduced from [5]. With permission from Elsevier.

flow regime unlikely. Therefore, the operating regime of FBMRs should be bubbling, turbulent, or fast fluidization – as shown in Figure 7.2. The choice is dictated by competing factors such as compactness, temperature uniformity, bed-to-surface heat transfer, influence of bed hydrodynamics on catalytic properties such as coking, and limitations imposed by the erosion of the membrane surfaces and catalyst attrition, especially in the turbulent or fast fluidization regime [1]. So far, most of the experience in FBMRs has been in the bubbling bed regime. However, the turbulent and fast fluidization regimes in particular may also prove to be advantageous in certain cases.

The vertical arrangement of the membrane and heat-transfer tubes is employed mostly in FBMRs for hydrogen separation because of the ease of construction. In fact, the vertical tube bundle arrangement is selected in most research. However, the horizontal arrangement of the membranes for selective removal of the product might be more effective in certain cases because it gives an extra degree of freedom to add fresh sweep gas along the axis of the reactor, by which higher hydrogen concentration gradients across the membrane tube can be accomplished. Moreover, the horizontal tube configuration might be best suited for

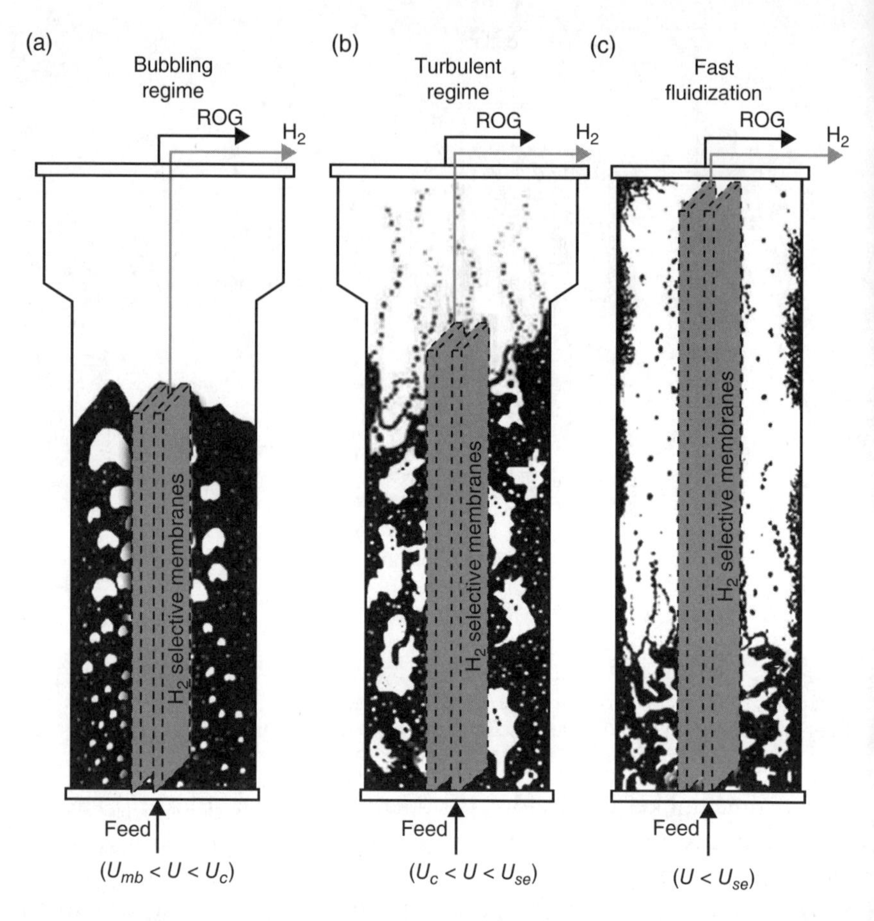

Figure 7.2 Schematic representation of FBMRs for hydrogen production: (a) bubbling fluidization regime; (b) turbulent fluidization regime; (c) fast fluidization regime. U = superficial gas velocity; U_{mb} = minimum bubbling velocity; U_c = velocity of transition from bubbling to turbulent fluidization regime; U_{se} = velocity of transition from turbulent to fast fluidization regime/significant entrainment; ROG = reactor off-gas; V = reactor volume. Reproduced from [6]. With permission from Elsevier.

partial oxidation reactions to achieve large improvements in conversion and selectivity, because the oxygen concentration profile can be optimized by variation of axial flow of oxygen via the membranes, and the effective axial dispersion can be decreased via compartmentalization of the fluidized bed. Figure 7.3 shows a typical representation of an FBMR for distributive addition of oxygen via microporous hollow fiber ceramic membranes with horizontal arrangements. An experimental study indicated that the horizontal membrane arrangement reduced the axial gas back-mixing much more than the vertical arrangement, although

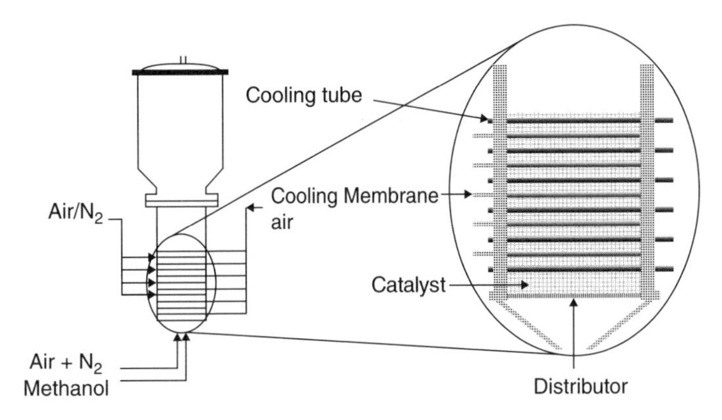

Figure 7.3 Schematic representation of FBMR with horizontal arrangement of the membrane and heat transfer tubes for partial oxidation of methanol. Reproduced from [7]. With permission from Elsevier.

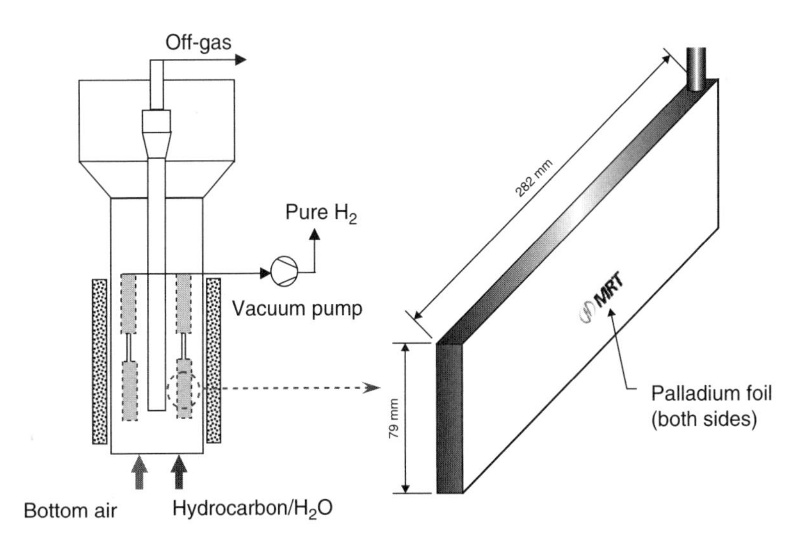

Figure 7.4 Schematic structure of FBMR with Pd-membrane panels for autothermal production of hydrogen. Reproduced from [8]. With permission from Elsevier.

both may result in reduced gas back-mixing compared with a bed having no internals [2].

Figure 7.4 shows the structure of an FBMR with plate-type Pd–Ag dense metal membranes for hydrogen production [8, 9]. Two-sided planar membrane panels are suspended vertically in the reactor. Each side of the panels consists of 25 μm thick Pd–Ag foil mounted on a porous stainless steel base with a barrier layer to prevent interdiffusion

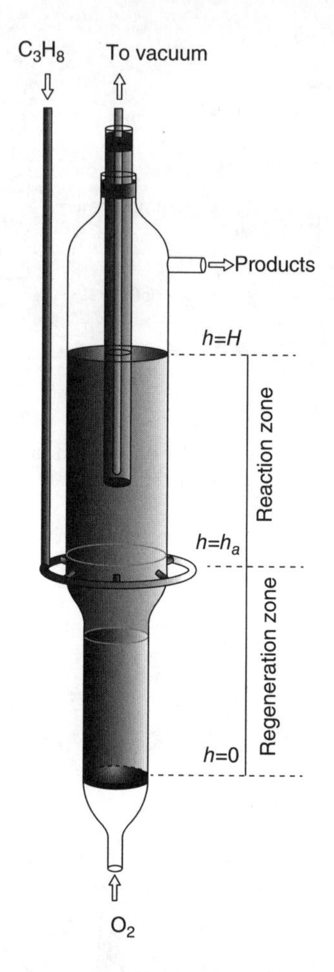

Figure 7.5 Scheme of the two-zone FBMR for propane catalytic dehydrogenation. Reproduced from [11]. With permission from Elsevier.

using proprietary Membrane Reactor Technologies Limited (MRT) sealing and protection techniques. For autothermal steam reforming, the mixture of hydrogen carbon and water vapor together with preheated air is fed from the reactor to fluidize the fine particulate catalysts. A vacuum pump is used to enhance the permeation rate. The preheated air can also be fed through two air distributors, one near the bottom of the bed and the other above the membrane module, into the fluidized bed [10].

Figure 7.5 shows schematically a two-zone FBMR for alkane catalytic dehydrogenation [11]. This configuration aims to combine in-situ catalyst regeneration with hydrogen separation using the Pd membrane. The hydrocarbon reactant is fed at an intermediate point, while a second

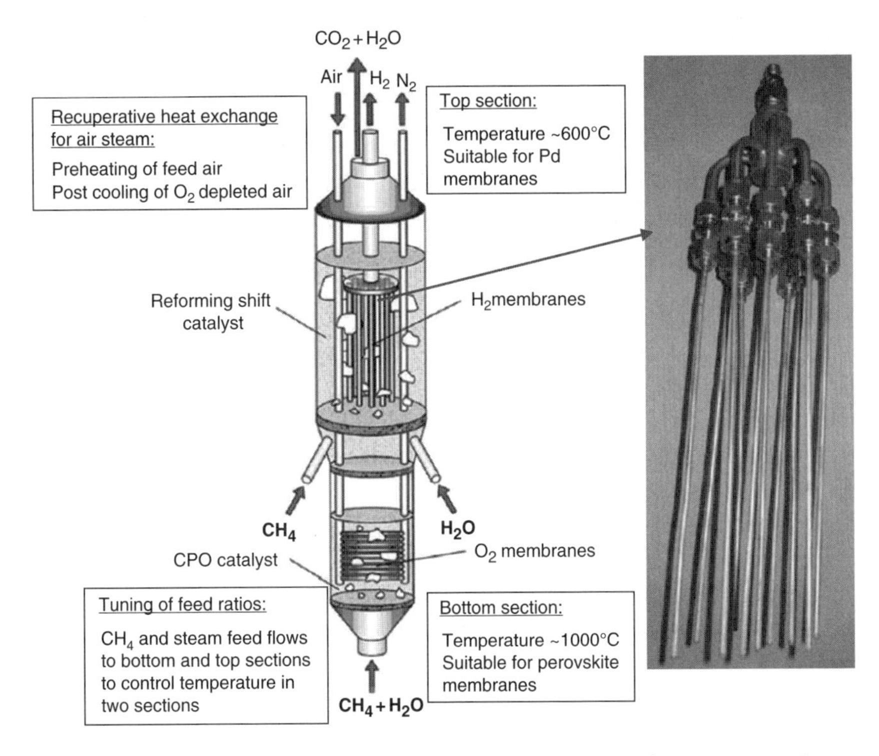

Figure 7.6 Schematic of bi-membrane fluidized bed reactor for pure H_2 and pure CO_2 production. Reproduced from [12]. With permission from Elsevier.

stream containing an oxidizing reactant is fed at the bottom of the bed. In this way, two zones with different atmosphere are created in the bed: a reducing zone, in the upper part, where the desired reaction takes place and coke is formed on the catalyst surface and an oxidizing environment, in the lower part, where the catalyst is regenerated by coke combustion. A continuous circulation of solids between both zones is caused by the bubbles. Under suitable operating conditions, a steady state can be achieved, with most of the oxygen being consumed in the lower part of the bed. Since the reactants are not premixed and usually the oxygen concentration at the hydrocarbon feed inlet is very low, operational safety can be improved.

A bi-functional FBMR integrating the perm-selective Pd metallic membranes with perovskite membranes is shown in Figure 7.6 [12, 13]. The reactor consists of a partial oxidation bottom section and a steam reforming/water gas shift top section. Pd membranes are used to remove hydrogen and perovskite membranes are used to supply oxygen along the length of the reformer for oxidative reforming of the hydrocarbons, providing the necessary heat for the highly endothermic steam reforming reaction. Incorporation of both types of membranes within a single reactor has the clear advantage of producing ultra-pure H_2 and pure CO_2, circumventing expensive CO_2 sequestration. Moreover, by tuning the feed ratios to the bottom and top sections, the temperatures in both sections can effectively be controlled for optimal membrane performance.

7.3 APPLICATIONS

Based on their advantages, FBMRs are applied mainly to reaction systems with highly exothermic or highly endothermic reactions, where isothermal operation is required for high selectivity or severe coking may be present. The applications aim at a major reduction in intra-particle resistance, improved heat transfer characteristics, a beneficial shift from conventional thermodynamic equilibrium constraints, and efficient energy supply. However, due to its complexity in construction, the application studies of FBMRs are quite limited. Table 7.1 summarizes the typical reactions conducted in FBMRs in the literature. The important findings for these reaction systems are given below.

7.3.1 Methane Steam Reforming and Dehydrogenation Reactions

The most extensive application of FBMRs is focused on the production of hydrogen via methane steam reforming. By selective removal of hydrogen through the Pd-based membranes, the thermodynamic limitation can be overcome, leading to increased synthesis gas yields. Furthermore, the problems of thermal control encountered in fixed bed reactors can also be solved. Addition of oxygen into the steam reforming reactors for in-situ generation of heat by partial oxidation of methane to supply the energy for the endothermic reforming reaction (autothermal reforming) may give more hydrogen productivity and higher energy efficiency [9, 10].

For dehydrogenation reactions, the hydrogen depletion in the reaction zone due to permeation through the Pd-based membranes generally enhances the coke formation. The use of a fluidized bed alleviates the coking problem. This can be achieved in FBMRs by: (i) dosing oxygen to the catalyst bed using O_2 perm-selective membranes to remove the deposited carbonaceous materials; or (ii) continuous regeneration of the catalysts in the circulating fluidized bed reactors.

Table 7.1 Typical reactions conducted in FBMRs

Reaction type	Membrane	Catalyst	Operation	Results	Ref.
MSR for hydrogen production	Pd tube (0.2–0.28 mm thickness)	Ni/Al_2O_3; $d = 90–355\ \mu m$	$T = 450–650°C$; $P = 0.7–1.0\ MPa$; $H_2O/CH_4 = 4.1–2.3$	Improved H_2 yield	[5]
	Pd–Ag 25%/PSS panel	Ni-based; $d = 101\ \mu m$	$T = 550°C$; $P = 0.4\ MPa$; $H_2O/CH_4 = 3$	Produced $H_2 = 0.602\ Nm^3\ h^{-1}$	[8]
	Pd–Ag 25%/PSS panel	Ni or precious metal catalyst; $d = 90\ \mu m$	$T = 550°C$; $P = 0.65/0.9\ MPa$; $H_2O/CH_4 = 3$; $O_2/CH_4 = 0$ or 0.35	Produced H_2/CH_4 MSR = 2.07; ATR = 3.03	[9]
	Pd–Ag 25%/PSS panel	Ni-based; $d = 85–90\ \mu m$	$T = 500–600°C$; $P = 1.5–2.6\ MPa$	H_2 production rate ($N\ m^3\ m^{-2} h^{-1}$): MSR = 4.8–12; ATR = 9–18	[10]
Steam reforming of n-heptane	Pd–Ag 23%/PSS panel		$T = 475–550°C$; $P = 0.4–0.8\ MPa$; H_2O/heptane = 5	Produced H_2/heptane: 14.7	[14]
Propane dehydrogenation	Pd- or Pd-alloys/Al_2O_3 hollow fibers	Pt/Al_2O_3; $d = 100–200\ \mu m$	$T = 500°C$; vacuum on permeate side	Maximum C_3H_6 yield = 18%	[11]
Partial oxidation of methane to syngas	silicalite/PSS tube or Pd/PSS tube	5% Ni/α-Al_2O_3; $d = 160–255\ \mu m$	$T = 700–750°C$	No improvement in yield	[15]

PSS = porous stainless steel; MSR = methane steam reforming; ATR = autothermal reforming.

Gas withdrawal through the membranes decreases the superficial gas velocities in the top section of the bed, resulting in smaller gas bubbles, which may increase the inter-phase gas exchange, favoring higher conversions [16]. On the contrary, the gas withdrawal may also lead to local defluidization in the vicinity of the surfaces, depending on the membrane tube arrangement and its spacing.

7.3.2 Partial Oxidation Reactions

The partial oxidation reactions are highly exothermic and require carefully designed reactors because of the large amount of reaction heat liberated and the high selectivity requirement for the intermediate product of interest. Owing to mass transfer limitations, a large part of the catalyst in a PBMR might not be used efficiently. The reduced effectiveness of the catalyst packing influences the distribution of the activity over the length of the packed bed, and thus the distribution of the released reaction heat, which might ultimately affect the product selectivity and even the reactor safety.

In FBMRs, isothermal reactor operation can be achieved more easily, leading to a higher selectivity compared with PBMRs [17, 18]. This is also attributed to the mass-transfer limitations between the bubble and emulsion phases. However, the improvement in selectivity of separation due to the inter-phase gas exchange decreases with increasing pressure. Furthermore, through the controlled dosing of air via porous membranes, hot spots can be minimized and the reactor is inherently safer because the oxygen–hydrocarbon feed separation helps to overcome flammability and explosion limits.

7.4 PROSPECTS AND CHALLENGES

FBMRs have shown the promise of industrial commercialization of MR technology, especially in the production of pure hydrogen for fuel cell applications. However, there are still many issues to be addressed before successful commercialization – such as the complexity of construction, the costs of membranes, and their long-term durability in harsh fluidization conditions. The challenges faced by the commercial viability of FBMRs include:

- Justifying the cost of membrane development and manufacture. Pd membranes may not be economical, except for producing hydrogen for fuel cells. Oxygen-permeable membranes are considered high-cost and high-risk.

- Developing cheap, high-temperature sealing systems for membrane reactors.
- Developing new technologies for heat supply/removal and temperature control for large-scale FBMR modules.
- Further investigating the effects of gas addition, as well as withdrawal through permeable surfaces, and of gas generation due to reactions on the fluidization behavior (heat transfer and bubble size distribution).
- Optimizing the reactor design for maintenance of bed mobility and providing sufficient membrane capacity.

REFERENCES

[1] Deshmukh, S.A.R.K., Heinrich, S., Mörl, L., van Sint Annaland, M. and Kuipers, J.A.M. (2007) Membrane assisted fluidized bed reactors: Potentials and hurdles. *Chemical Engineering Science*, 62, 416–436.

[2] Deshmukh, S.A.R.K., van Sint Annaland, M. and Kuipers, J.A.M. (2007) Gas back-mixing studies in membrane assisted bubbling fluidized beds. *Chemical Engineering Science*, 62, 4095–4111.

[3] Gallucci, F., Van Sintannaland, M. and Kuipers, J.A.M. (2010) Theoretical comparison of packed bed and fluidized bed membrane reactors for methane reforming. *International Journal of Hydrogen Energy*, 35, 7142–7150.

[4] Adris A.M., Grace J.R., Lim C.J., Elnashaie S.S. (1994) Fluidized bed reaction system for steam/hydrocarbon gas reforming to produce hydrogen. US Patent 5326550, July 7, 1994.

[5] Adris, A.M., Lim, C.J. and Grace, J.R. (1994) The fluidized bed membrane reactor system: A pilot scale experimental study. *Chemical Engineering Science*, 49, 5833–5843.

[6] Mahecha-Botero, A., Chen, Z., Grace, J.R. *et al.* (2009) Comparison of fluidized bed flow regimes for steam methane reforming in membrane reactors: A simulation study. *Chemical Engineering Science*, 64, 3598–3613.

[7] Deshmukh, S.A.R.K., Laverman, J.A., Cents, A.H.G., van Sint Annaland, M. and Kuipers, J.A.M. (2005) Development of a membrane assisted fluidized bed reactor. 1: Gas phase back-mixing and bubble to emulsion mass transfer using tracer injection and ultrasound. *Industrial and Engineering Chemistry Research*, 44, 5955–5965.

[8] Andrés, M.-B., Boyd, T., Grace, J.R. *et al.* (2011) In-situ CO2 capture in a pilot-scale fluidized-bed membrane reformer for ultra-pure hydrogen production. *International Journal of Hydrogen Energy*, 36, 4038–4055.

[9] Mahecha-Botero, A., Boyd, T., Gulamhusein, A. *et al.* (2008) Pure hydrogen generation in a fluidized-bed membrane reactor: Experimental findings. *Chemical Engineering Science*, 63, 2752–2762.

[10] Chen, Z., Grace, J., Jimlim, C. and Li, A. (2007) Experimental studies of pure hydrogen production in a commercialized fluidized-bed membrane reactor with SMR and ATR catalysts. *International Journal of Hydrogen Energy*, 32, 2359–2366.

[11] Gimeno, M.P., Wu, Z.T., Soler, J., Herguido, J., Li, K. and Menéndez, M. (2009) Combination of a two-zone fluidized bed reactor with a Pd hollow fibre membrane for catalytic alkane dehydrogenation. *Chemical Engineering Journal*, 155, 298–303.

[12] Patil, C.S., van Sint Annaland, M. and Kuipers, J.A.M. (2007) Fluidised bed membrane reactor for ultrapure hydrogen production via methane steam reforming: Experimental demonstration and model validation. *Chemical Engineering Science*, 62, 2989–3007.

[13] Patil, C.S., Van Sint Annaland, M. and Kuipers, J.A.M. (2006) Experimental study of a membrane assisted fluidized bed reactor for H_2 production by steam reforming of CH_4. *Chemical Engineering Research and Design*, 84, 399–404.

[14] Rakib, M.A., Grace, J.R., Lim, C.J. and Elnashaie, S.S.E.H. (2010) Steam reforming of heptane in a fluidized bed membrane reactor. *Journal of Power Sources*, 195, 5749–5760.

[15] Ostrowski, T., Giroir-Fendler, A., Mirodatos, C. and Mleczko, L. (1998) Comparative study of the partial oxidation of methane to synthesis gas in fixed-bed and fluidized-bed membrane reactors. Part II–Development of membranes and catalytic measurements. *Catalysis Today*, 40, 191–200.

[16] Mleczko, L., Ostrowski, T. and Wurzel, T. (1996) A fluidised-bed membrane reactor for the catalytic partial oxidation of methane to synthesis gas. *Chemical Engineering Science*, 51, 3187–3192.

[17] Ostrowski, T., Giroir-Fendler, A., Mirodatos, C. and Mleczko, L. (1998) Comparative study of the catalytic partial oxidation of methane to synthesis gas in fixed-bed and fluidized-bed membrane reactors. Part I–A modeling approach. *Catalysis Today*, 40, 181–190.

[18] Ostrowski, T., Giroir-Fendler, A., Mirodatos, C. and Mleczko, L. (1998) Comparative study of the partial oxidation of methane to synthesis gas in fixed-bed and fluidized-bed membrane reactors. Part II–Development of membranes and catalytic measurements. *Catalysis Today*, 40, 191–200.

8

Membrane Microreactors

8.1 INTRODUCTION

The term "microreactor" refers to the minimal chemical reactor with a characteristic length of 1–1000 μm [1]. The miniaturization of a reactor offers numerous advantages, such as high surface area per unit volume, improved heat and mass transfer properties due to the short transport distances, enhanced reaction performances (i.e., higher conversion, selectivity, and yield), and better energy and material utilization with the result of more efficient production and less pollution. Other advantages include simpler process optimization, rapid design implementation, easier scale-up through replication, and better safety. The rapid mass and heat transfer rates obtained, even in a laminar flow regime, make microreactors an excellent choice – especially for fast or highly exothermic reaction systems, where the external heat and mass transfer processes are important.

When a functional membrane is incorporated into the microreactor, it becomes a membrane microreactor. The combination integrates the advantages of both the membrane and the microreactor, leading to greatly intensified production processes. It allows the reactor to work not only under optimum reaction conditions due to rapid mass and heat transfer rates, but also at significantly lower operating temperature and/or using less amounts of catalyst than in conventional reactors. The miniaturization makes it possible to obtain a large membrane area in a small compact unit by simply assembling small membrane pieces. As a result, the problems that plague large membrane units – such as deformation and cracks due to accumulated stresses during processing and

Inorganic Membrane Reactors: Fundamentals and Applications, First Edition. Xiaoyao Tan and Kang Li.

operation – can be avoided. Moreover, the selectivity and permeance of the miniature membranes can be enhanced noticeably compared with conventional membrane units [2]. Based on these unique properties, MMRs have been attracting considerable interest in recent years [3–6], and have even been tried for commercial applications [7, 8].

Since the membranes incorporated in MMRs are zeolite or dense metal ones, which have been described extensively in previous chapters, this chapter focuses mainly on the general structure, design, and fabrication of MMRs and their applications. The recent advances and future trends of MMR technology are also given.

8.2 CONFIGURATIONS AND FABRICATION OF MEMBRANE MICROREACTORS

A distinctive feature of microreactors is the microchannels for fluid flow. MMRs are mainly characteristic of such microchannels with anchored catalysts for reactions and miniature membranes to perform separation, which are formed on the porous ceramic or metal supports. Based on the configuration and architecture of the reactor, MMRs can be classified into two categories: plate type and tubular type.

8.2.1 Plate-Type Membrane Microreactors

In plate-type MMRs, the microchannels for catalyst deposition are fabricated on a porous silicon or stainless steel (i.e., SS 316L) plate, while the membrane layer is deposited on the back of the plate to perform separation (membrane/catalyst plate) as shown in Figure 8.1(a). In some cases the membrane layer, which also shows catalytic activity, is deposited on the wall of the microchannels for simultaneous reaction and separation

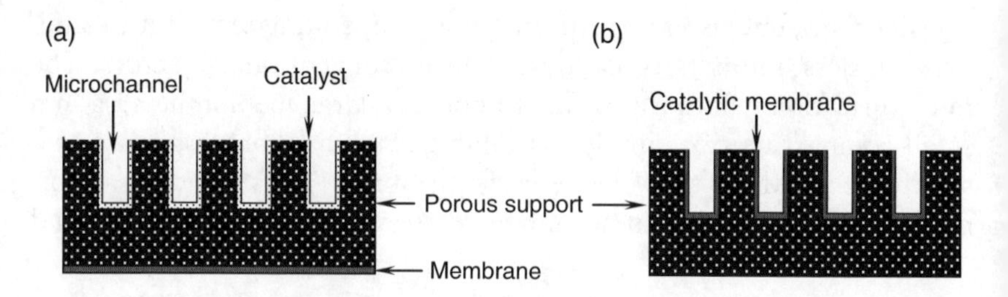

Figure 8.1 Plate-type MMRs showing the microchannels: (a) membrane–catalyst plate; (b) catalytic membrane plate.

(the so-called catalytic membrane plate), as shown in Figure 8.1(b). Both designs promise great numbers of potential applications as MMRs, electrochemical cells, and sensors. Fabrication of MMRs starts with the preparation of microchannel platelets, followed by the incorporation of catalysts and membranes.

- **Fabrication of microreactors**
  Several micro-electro-mechanical system (MEMS) technologies are currently available for the manufacture of plate-type microreactors or microseparators [9]. The most popular method uses a traditional semiconductor fabrication procedure [10]. It is a mature technology, with existing production infrastructure in the form of semiconductor foundries. This fabrication method also allows other devices such as pressure transducers, flow controllers, and temperature control apparatus to be incorporated easily into the system to form micro-systems capable of analyzing or synthesizing chemicals. However, there is a material restriction to silicon and a design limitation to two dimensions. Another process called LIGA, which includes the essential steps of X-ray lithography using synchrotron radiation, electroplating of metal to form the mold, injection molding of plastics, and electroforming with the desired metal, has been used for high-aspect-ratio micromachining [11]. This process is relatively complex and requires sophisticated equipment. Other methods, such as multi-lamination, laser micromachining, and direct-write "dip pen" nanolithography (DPN) techniques, are also becoming more popular because of the increasing maturity of the technologies. DPN employs scanning tunneling and atomic force microprobes to write patterns up to 30 nm in line width [12].
- **Incorporation of catalysts and membranes**
  Currently, the membranes incorporated in MMRs are mainly zeolite and Pd-based dense metal ones. Incorporation of these membranes in microreactors can be achieved using one of the preparation methods described in previous chapters.

  An important issue in catalytic MMRs is the proper incorporation of active catalysts within microchannels. Packing a powdered catalyst in the microchannels presents obvious difficulties in terms of pressure drop and flow distribution. Therefore, the catalyst is usually deposited as a thin catalytic coating on the surface of the microchannels in MMRs. These coatings should not only provide enough catalytic loads, but also have high mechanical stability to withstand extremely severe conditions in terms of temperature and mechanical shock and good accessibility of the reactants to the catalyst (high external surface-to-volume ratio in the coating).

Ideally, the catalyst should be an integral part of the MMR. For metal catalysts, the most direct approach is to deposit a thin layer of active metal (e.g., Pd, Pt, Ag) using thermal deposition, chemical vapor deposition, or sputter-coating methods [13]. However, since the geometric surface of the microchannels is unable to provide sufficient specific surface area for impregnating catalysts, which are necessary for most of the catalytic reactions, chemical treatments, and deposition of thin catalytic coatings in most cases must be applied on the surface of the microchannel walls. In this case, a porous oxide layer is formed by anodization or wash-coating technique (suspension coating method), followed by impregnation with solution of the metal precursor [14–17]. The wash-coating process generally includes the following four steps:

1. *Synthesis of the oxide powder slurry.* The oxide power is dispersed under vigorous stirring in deionized water with acetic acid and acrylic acid as binders. The suspension is kept under stirring for 24 h prior to deposition at room temperature.
2. *Pre-treatment of the microchannel platelets.* The platelets are cleaned with acetone, then with $H_2O_2 + NH_3OH$ aqueous solution, and then with $H_2O_2 + H_3PO_4 + CH_3COOH$ aqueous solution in sequences to eliminate impurities as well as organic compounds. Subsequently, high-temperature (1200°C) annealing is conducted to trigger the segregation of an alumina layer on the metallic surface. Prior to starting the coating procedure, all the supports were washed with ethanol, removing any superficial impurities caused by manipulation.
3. *Coating deposition.* The slurry is applied to the microchannels of the platelets using a syringe. Even though the use of a syringe is not a commercially viable process in the production of bulk quantity of coated platelets, it is a convenient way to achieve catalyst deposition on a laboratory scale with good layer quality. Any excess of suspension should be wiped off.
4. *Drying and calcination at elevated temperature.* The platelets are dried at ambient temperature in air. The coated microchannel platelets are finally calcined at 800°C for 4 h.

The binders in the slurry play important roles in the properties of coatings like coating adhesion and catalytic activity. By optimizing the slurry preparation parameters such as particle size, viscosity, solid loading and/or binder content, porous oxide films at a desired thickness of 25 μm with a good adhesion and reasonable uniformity can be obtained [17]. Cui *et al.* [18] used a different way to incorporate catalysts in the silicon microchannels. Twenty nanometer Ti was first sputtered and oxidized as the adhesion layer and catalyst carrier before a Pt layer of 20 nm thick was sputtered as the catalyst. Ti oxidation using NaOH and H_2O_2 was proved to be a simple and very effective way to get the catalyst carrier compared to anodic oxidation. It is noteworthy that in some cases the membrane itself may function as the catalyst for the reactions, that is, catalytic membrane, thus additional catalysts are not required any more [3, 19].

The incorporation of zeolites in MMRs as catalyst or membrane with controlled particle size, crystal morphology, layer thickness (3–16 μm) and film orientation can be achieved by various strategies: zeolite powder coating, localized zeolite growth, and etching of zeolite/silicon composite film [9]. Among these methods, the powder coating technique is relatively simple and straightforward, and can easily be used to incorporate other types of catalyst in addition to zeolites. The zeolite loading can be controlled by changing the concentration of zeolite in the slurry and through repeated coating of the microchannels. However, the coating adhesion is poor and the zeolites may easily be removed. Using chemical grafting and polymer adhesives can improve the powder adhesion significantly, but can also interfere with the function of the catalyst and the operation of the reactor. Comparatively, the zeolite film growth method shows more advantages since it yields a highly adherent catalytic layer, a binder is not needed (meaning that the layer is 100% catalyst), and the considerable experience gathered in the synthesis of zeolite films allows us to regulate film thickness and crystal orientation [20]. When the microchannels are coated with a 1–2 μm layer of zeolite, almost all of the catalyst is available for reaction at a fraction of the pressure drop. In addition, the large surface-to-volume ratio provides an excellent contact between reactants and catalyst, thus minimizing bypass, and a better approach to matching permeation and reaction rates. Figure 8.2 demonstrates the process for the preparation of a microscale zeolite membrane and catalyst on PSS plates through the local hydrothermal synthesis method. The film thickness, crystal morphology, and intergrowth can be controlled by modulating the seed population, synthesis chemistry, and hydrothermal treatment conditions. A clear advantage of microscale zeolite membranes is the higher probability of obtaining a defect-free interface, since this probability increases for smaller membrane areas. Furthermore, microscale membranes are amenable to preparation methods that use high-throughput synthesis procedures [21].

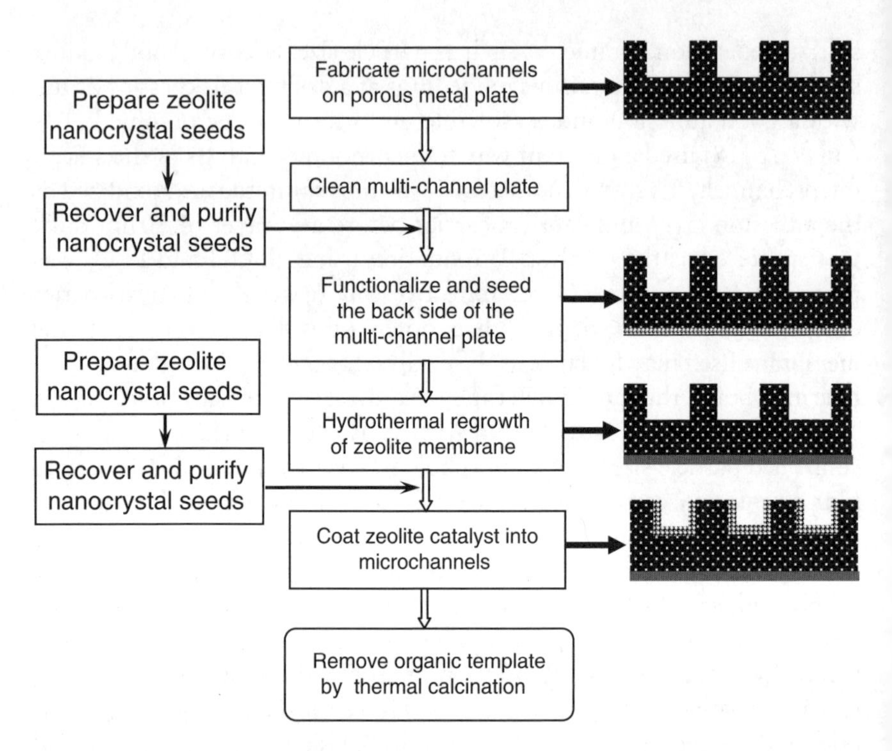

Figure 8.2 Process flow diagram for the preparation of zeolite membrane–catalyst platelets.

Iglesia *et al.* [22] studied the preparation of MFI layers on microchannels by three different synthesis methods: hydrothermal synthesis, seeded hydrothermal synthesis, and steam-assisted crystallization. They found that the seeded hydrothermal method, with a careful control of gel dilution and temperature, was adequate for controlling the thickness of the zeolite layer, resulting in a well inter-grown zeolite coating on the channel surface. On the contrary, the steam-assisted crystallization method allowed better control of the location of the zeolite film and provided better accessibility to the catalyst by maintaining the individuality of the zeolite crystals, but the synthesis time to obtain a zeolite layer was as long as 8–16 days. In addition to MFI, other zeolites including TS-1, ZSM-5, MOR, ETS-10, BEA, LTA, and FAU have been grown onto different supports such as ceramic monoliths, brass, molybdenum, and stainless steel microchannel reactors [23–27]. Figure 8.3 shows the CsX zeolite catalyst coating on the microchannels and the NaA zeolite membrane grown on the back of the porous SS-316L multi-channel plate [28].

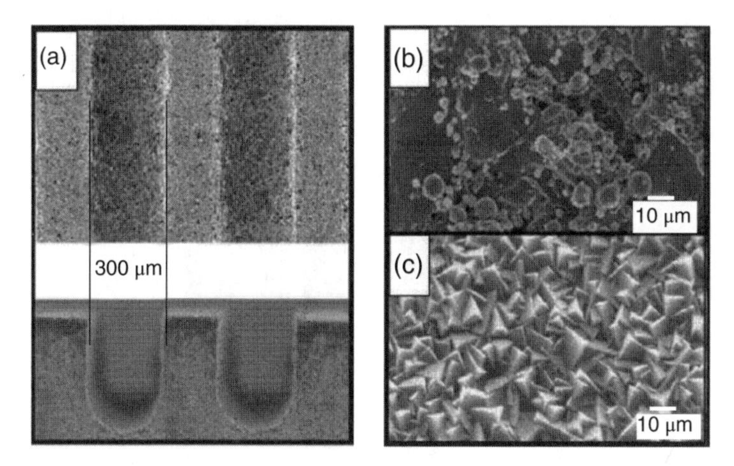

Figure 8.3 SEM images of (a) top and cross-section views of a membrane–catalyst plate with (b) CsX catalysts coated on the microchannels and (c) 6 μm thick NaA separation membrane. Reproduced from [28]. With permission from Elsevier.

In the method of etching zeolite composite film, a highly oriented zeolite film is first grown onto a silicon wafer, which can be achieved by seeding the wafer with colloidal zeolites followed by hydrothermal synthesis. Etching is then performed to form microchannels [9]. Using this fabrication methodology, the shape, morphology, quantity, and individual locations of the zeolite catalysts can be engineered precisely to optimize the microreactor performance.

8.2.2 Tubular Membrane Microreactors

In tubular MMRs, the porous support for loading microchannels and separation membranes is of tubular configuration. Microchannels for fluid flow may be formed due to the small size of the tube or distributed within the tube wall, while the separation membrane is usually coated on the outer surface of the tube. Yamamoto *et al.* [29] constructed a simple MMR just by inserting a stainless steel rod into a tubular membrane reactor to form microchannels for dehydrogenation of cyclohexane to benzene, as shown in Figure 8.4. They investigated the effects of microchannel size by varying the diameter of the stainless steel rod and found that the yield with the biggest rod was approximately twice as high as that without any rods, which could be attributed to the increase in surface area per volume of Pd membrane.

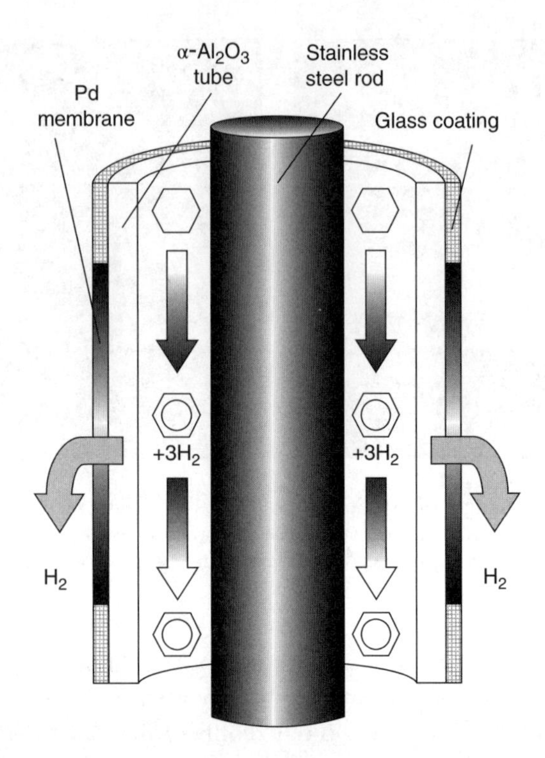

Figure 8.4 Microreactor with a Pd membrane for dehydrogenation of cyclohexane to benzene. Reproduced from [29]. With permission from Elsevier.

When the membrane tube is reduced in diameter to a certain level, that is, ID < 1 mm, it becomes a hollow fiber and the fiber lumen may take on the effect of a microchannel on the fluid flow. The catalyst can be coated on the inner surface of the hollow fiber or impregnated inside the porous wall, while the separation is achieved by the porous hollow fiber itself or by the membrane formed on the outer surface of the hollow fiber, as shown in Figure 8.5. Such catalytic hollow fiber membranes can easily be fabricated into MMRs, called hollow fiber membrane microreactors (HFMMRs).

In the fabrication of HFMMRs, the porous hollow fiber membranes with asymmetric structures are first prepared through a phase-inversion/sintering technique, which has been described in Chapter 2 of this book and in many recent publications [31]. The microstructure of the hollow fibers can be tailored as expected for different applications by modulation of the suspension composition and the spinning parameters [32, 33]. Aran *et al.* [5, 21] developed porous Al_2O_3 HFMMRs with various geometrical parameters for gas–liquid–solid (G–L–S) reactions. The Pd–Al_2O_3 catalyst

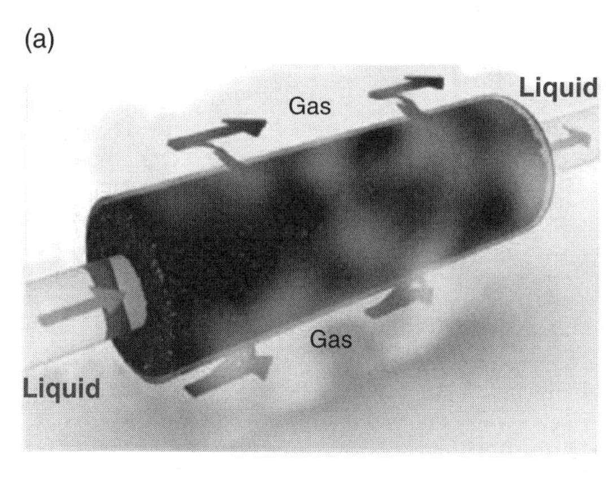

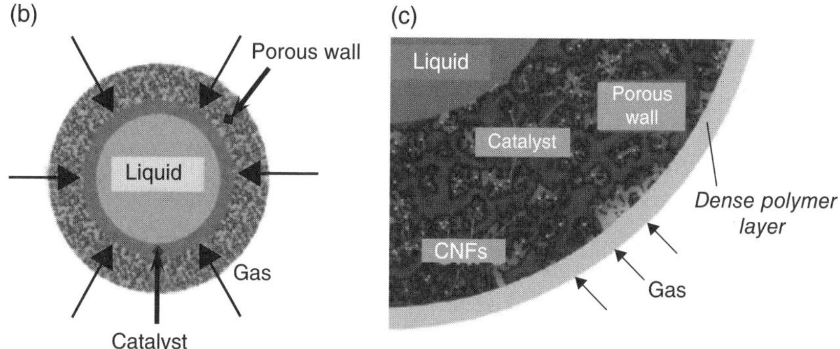

Figure 8.5 Schematic diagram of the catalytic hollow fiber membrane: (a) flow pattern; (b) catalyst coated on inner surface [5]; (c) catalyst impregnated inside the wall. Reproduced from [30]. With permission from Elsevier.

was coated on the inner surface of the α-Al_2O_3 hollow fibers (Figure 8.5(a)), with the following steps: (i) aqueous γ-Al_2O_3 suspension prepared using PVA and acetic acid as additives; (ii) suspension introduced into the hollow fiber lumen and then emptied with a controlled filling/emptying speed; (iii) hollow fiber dried and calcined to form γ-Al_2O_3 coatings; (iv) γ-Al_2O_3 coated hollow fiber immersed into a Pd precursor solution followed by calcination in O_2 and subsequent reduction in H_2 atmosphere. Perfluorinated octyltrichlorosilane (FOTS) was gas-deposited onto the α-Al_2O_3 layer to make it hydrophobic, so as to fit G–L–S reactions. In another work [30], Aran *et al.* fabricated a stainless steel HFMMR to improve the mechanical stability. Carbon nanofibers (CNFs) were grown by a chemical vapor deposition method inside the porous structure of the stainless steel hollow fibers as catalyst support. Pd catalyst was then immobilized in the CNFs by the immersion/drying/calcination process.

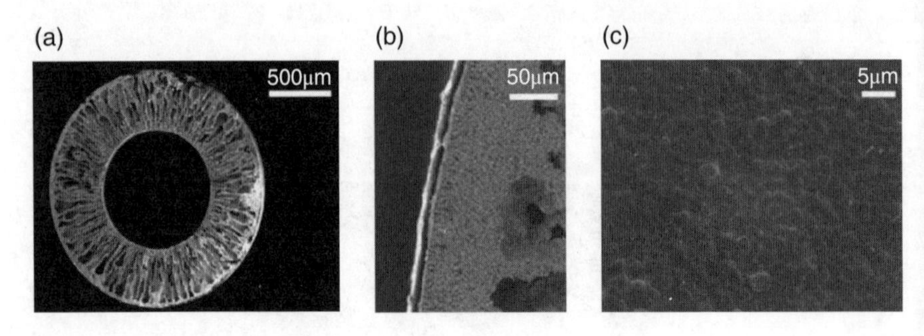

Figure 8.6 SEM image of Pd–Al_2O_3 hollow fiber membrane: (a) cross-sectional area; (b) outside edge of the coated Pd membrane; (c) top surface of the Pd membrane. Reproduced from [35]. With permission from Elsevier.

A dense polymer layer was coated on the outer surface serving as the membrane for gas separation, as shown in Figure 8.5(b).

Recently, a novel Pd-MMR design based on asymmetric ceramic hollow fiber membranes was developed by Li's group for hydrogen production [34–36]. As shown in Figure 8.6(a), YSZ or Al_2O_3 ceramic hollow fiber membranes with open finger-like pores on the inner surface were fabricated by a viscous fingering-induced phase inversion and sintering technique. A Pd/Ag membrane was deposited on the outer surface of the hollow fibers by the electroless plating method for hydrogen separation. Catalysts such as 30% CuO/CeO_2 for WGS or 10 wt% NiO/MgO–CeO_2 for ESR reactions were impregnated in the finger pores by a sol-gel coating technique to form catalytic hollow fiber membranes. As can be seen, the finger-like pores can be viewed as microchannels with diameters around 10 µm, although they are not regular as MEMS-manufactured ones are. Each catalyst-impregnated hollow fiber is actually a catalytic microreactor made up of hundreds of microchannels ($d_p \sim 10\,\mu m$) distributed perpendicularly around the lumen of the fiber. This catalytic HFMMR possesses advantages as an on-board hydrogen generator for vehicular applications due to its compact feature and its ability to produce pure hydrogen for fuel cell systems. Another advantage is that minimal heat losses can be realized due to the low thermal conductivity. However, the mechanical stability of single ceramic hollow fiber membranes is relatively low, limiting maximum pressure differentials for driving mass transfer across metal films and providing significant challenges to fabricating mechanically stable membrane reactors.

Utilization of porous ceramic honeycomb monoliths could realize large networks of microchannel membranes for gas separation with superior mechanical stability. By increasing the thickness of the membrane

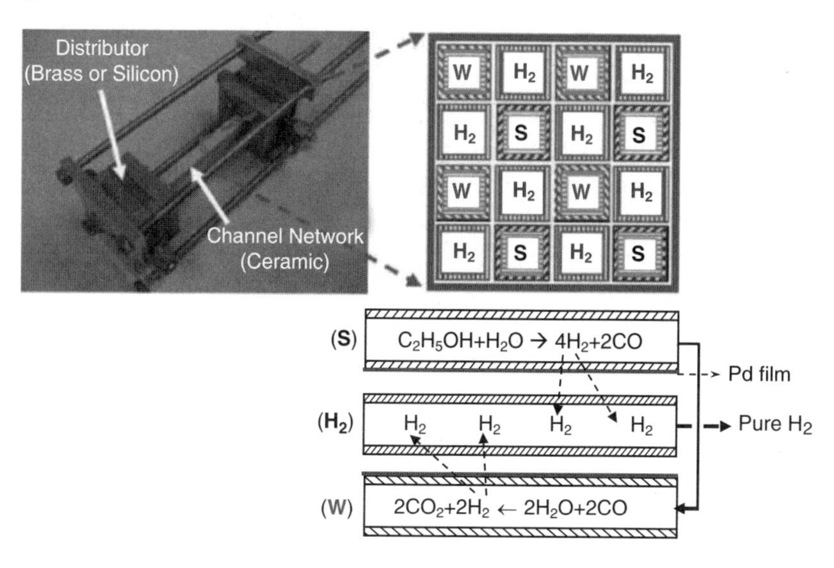

Figure 8.7 Picture of an integrated portable membrane reformer system with inset showing the integration schemes. Reproduced from [37]. With permission from Elsevier.

support to 500–1000 μm, limitations associated with mechanical stability can effectively be removed, while the use of high-porosity ceramic supports minimizes the introduction of substantial mass transfer resistances (concentration polarization). Kim *et al.* [37] fabricated Pd membranes within the extruded cordierite ($2MgO–2Al_2O_3–5SiO_2$) honeycomb monolith for hydrogen purification. Washcoating with micropowder γ-Al_2O_3 was performed first to realize cylindrical surfaces for subsequent deposition of Pd films. A second nanopowder γ-Al_2O_3 layer was washcoated to provide a uniform surface for deposition of defect-free Pd films. Electroless plating under kinetic-limited conditions was conducted finally to deposit thin (8 μm) defect-free Pd films with crystallite sizes ~2 μm. The resulting two-channel Pd-membrane system exhibited hydrogen flux of $(1.0–5.5) \times 10^{-3}\,mol\,m^{-2}\,s^{-1}$ at 350°C and hydrogen-to-helium selectivities of (40–360):1. A novel microreactor design strategy was also presented to employ resultant mini-channel membrane networks together with precision-machined distributors capable of addressing each individual channel as part of a thermally integrated portable reformer for high-purity hydrogen production, as shown in Figure 8.7. This new design enables series and/or parallel integration of multiple unique processing steps within a single structure for both heat and mass integration, and provides an inexpensive and rapid means of constructing MMRs.

8.3 APPLICATIONS OF MEMBRANE MICROREACTORS

As mentioned above, MMRs benefit from fast and thermodynamically limited reactions, for which mass transfer limits the reaction processes to a large extent. Owing to their compact features and high degree of control, MMRs are especially suitable for portable chemical processing. Table 8.1 summarizes typical applications of MMRs for catalytic reactions.

8.3.1 Pd-MMRs for Hydrogenation/Dehydrogenation Reactions

Pd-based membranes possess high hydrogen permeation fluxes and infinite hydrogen selectivity, and have found considerable application in a variety of dehydrogenation reactions [40]. Applying microchannels in the membrane

Table 8.1 Applications of membrane microreactors

Reactor configuration	Membrane	Reaction	Results	Ref.
Annular tube	Pd/Al_2O_3	Dehydrogenation of cyclohexane to benzene	Yield increased by 200%	[29]
Plate with coated catalyst	Pd–Ag	Partial oxidation of methanol to H_2	$Y_{H2} = 34.2\%$ with improved conversion and selectivity	[38]
Hollow fiber with impregnated catalyst	Pd/Al_2O_3	WGS	17% above equilibrium conversion	[6, 34]
Hollow fiber with impregnated catalyst	Pd–Ag/YSZ	ESR	53% H_2 yield higher than in ESR	[35]
Oxidized porous silicon plate with no catalyst	Pd	Hydroxylation of benzene to phenol	20% phenol yield and 54% benzene conversion	[4]
Stainless steel plate with coated zeolite catalyst	ZSM-5 and/or NaA zeolite	Knoevenagel condensation	Supra-equilibrium conversion	[3, 28]
Stainless steel plate with coated TS-1 catalyst	ZSM-5 zeolite	Oxidation of aniline by H_2O_2 to azoxybenzene	Improved yield and selectivity	[39]
Hollow fiber with coated Pd catalyst	Porous Al_2O_3	Reduction of nitrite (NO_2^-) by H_2 in water	Drastically enhanced performance	[5, 21]
Hollow fiber with impregnated Pd catalyst	PSS	Reduction of nitrite (NO_2^-) by H_2 in water	Improved performance	[30]

reactors can accelerate the thermodynamically limited dehydrogenation of cyclohexane to benzene, leading to significantly improved yields [29]:

$$C_6H_{12} \Leftrightarrow C_6H_6 + 3H_2$$

The significant interest in the development of portable systems powered by polymeric proton-exchange fuel cells (PEMFCs) drives the development of Pd-MMRs for hydrogen production, since they can integrate fuel reforming and hydrogen purification within a single small device, and produce high-purity hydrogen with a high hydrogen recovery [7, 8]. Wilhite *et al.* [38] demonstrated a composite catalytic micromembrane comprising a Pd/Ag micromembrane washcoated with reforming catalyst ($LaNi_{0.95}Co_{0.05}O_3$) to combine partial oxidation of methanol with hydrogen purification. The highest extracted hydrogen yield of 32.4% was achieved at an oxygen-to-fuel feed ratio of 0.43.

Recently, García-García *et al.* and Rahman *et al.* carried out ESR and WGS reactions in catalytic hollow fiber Pd-MMRs [34–36]. For the WGS reaction, asymmetric Al_2O_3 hollow fibers produced by a phase-inversion and sintering technique were employed as a single substrate for coating the Pd membrane on the outer surface and impregnation of the 30% CuO/CeO_2 catalyst on the porous fiber wall. The results indicate that the catalytic hollow fiber MMR performed much better than the fixed bed reactor and the catalytic hollow fiber microreactor under the same conditions. A conversion of 17% higher than the corresponding thermodynamic equilibrium conversion was achieved in the hollow fiber MMR and a high purity of hydrogen was obtained. Figure 8.8 shows the catalytic hollow fiber MMR for high-purity hydrogen production with the ESR reaction, where YSZ hollow fibers were used as substrate for the deposition of Pd/Ag membrane on the outer surface of the hollow fiber

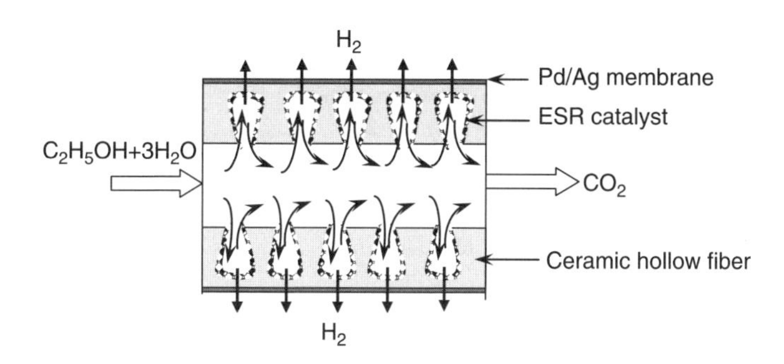

Figure 8.8 Schematic representation of the catalytic hollow fiber MMR for hydrogen production using ethanol steam reforming reaction.

and impregnation of the 10 wt% $NiO/MgO–CeO_2$ ESR catalyst inside the hollow fiber. The reactants enter the conical microchannels in which the ESR takes place. H_2 will be separated using the Pd/Ag membrane, while CO_2 will remain in the lumen. The results indicate that the flow rate of hydrogen produced in the hollow fiber MMR was three times higher than that in the fixed bed reactor and twice as high as that in the hollow fiber microreactor despite less catalyst being used. During operation, high-purity hydrogen was obtained and the yield of high-purity hydrogen is more than 53% of the total hydrogen produced in the ESR reaction.

The MEMS-based Pd-MMR was demonstrated by Ye *et al.* [4] for one-step conversion of benzene to phenol. The structure of the MMR is shown schematically in Figure 8.9. It comprised three layers: top glass cap, Si substrate, and bottom glass cap. H_2 and reaction gas are supplied

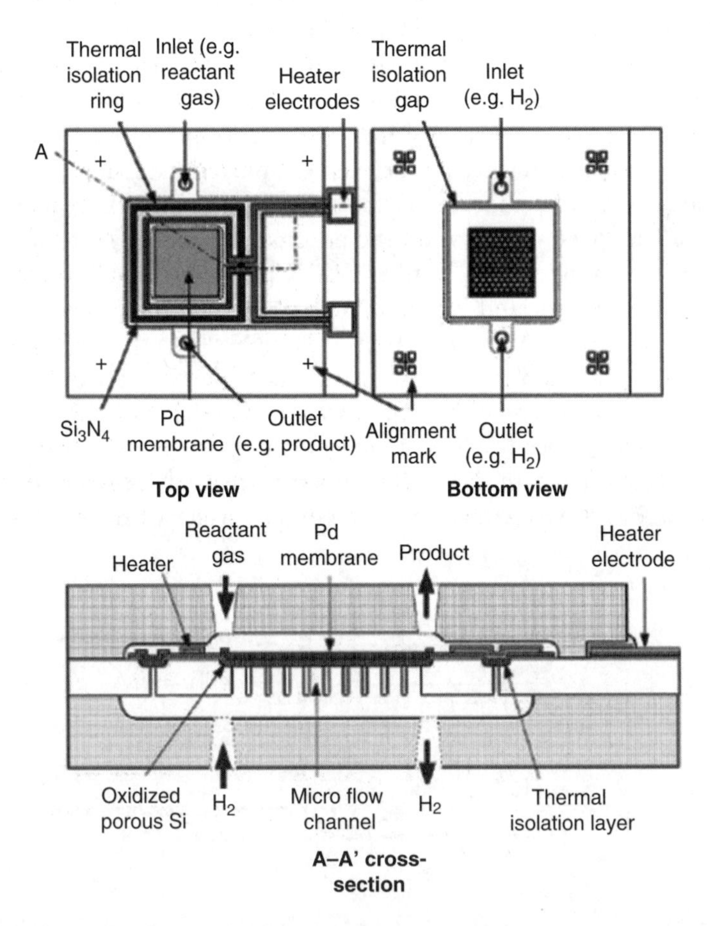

Figure 8.9 Structure of the MEMS-based Pd-MMR for direct hydroxylation of benzene. Reproduced from [4]. With permission from Elsevier.

to the bottom and top side of the Pd membrane, respectively. Reaction products are obtained from the top side. Oxidized porous silicon (PS) serves as a structural support for the Pd membrane. Under the oxidized PS support, microflow channels with a diameter of 100 μm are opened up. A Pt/Ti microheater is formed on an Si_3N_4 insulation layer for heating the Pd membrane to improve H_2 permeability and prevent H_2 embrittlement. The oxidized PS ring is also formed around the Pd membrane for thermal isolation. The thermal isolation gap under the oxidized PS ring is 100 μm in width.

In operation, the active hydrogen species, which appear on the surface via the Pd membrane, can easily react with molecules of oxygen and form active species like HO* or HOO* radicals. Then, the active species attack benzene and hydroxylation occurs readily to form phenol. No additional catalyst is needed, since the Pd acts as both catalyst and separator. Since Pd-MMRs have a smaller dimension and larger surface-to-volume ratio compared with Pd-MRs, the concentration of reactant gases is almost uniform on the surface of the membrane and, as a result, active hydrogen species can be used more effectively for hydroxylation of benzene on the Pd surface of the Pd-MMRs. Therefore, the Pd-MMRs can finish the transformation of a dominating reaction from oxidation to hydroxylation of the benzene and achieve the largest yield of phenol at a lower ratio of H_2/O_2. A phenol yield of 20% and benzene conversion of 54% were obtained at 200°C. In addition, the product distribution in Pd-MMRs is very different from that in Pd-MRs. Phenol and dihydric phenols dominate the product distribution in Pd-MMRs, and the hydrogenation products of cyclohexane (C_6H_{12}) and cyclohexanone ($C_6H_{10}O$) from benzene and phenol were not detected, whereas these are the main by-products in Pd-MRs.

8.3.2 Zeolite-MMRs for Knoevenagel Condensation and Selective Oxidation Reactions

Zeolite membranes have been applied for gas permeation and separation, and liquid pervaporation. A clear advantage of microscale zeolite membranes is the higher probability of obtaining a defect-free interface, since this probability increases for smaller membrane areas [41]. In zeolite MMRs, the zeolites are incorporated as a catalyst for reaction and a membrane for separation, as well as structural material of the reactors. Reactions conducted in MMRs include mainly Knoevenagel condensation [3, 42, 43] and selective oxidation reactions [39]. Supra-equilibrium conversion may be obtained in the former, while the latter displays improved performance against catalyst deactivation.

The Knoevenagel condensation reaction is an important C–C bond-forming reaction commonly used for the production of fine chemical intermediates and pharmaceuticals. It involves the condensation of methylene compounds (i.e., Z–CH_2–Z′ or Z–CHR–Z′) with ketones or aldehydes [3], which is constrained by unfavorable thermodynamics. A typical example is the Knoevenagel condensation reaction between benzaldehyde and ethyl cyanoacetate to produce ethyl 2-cyano-3-phenylacrylate:

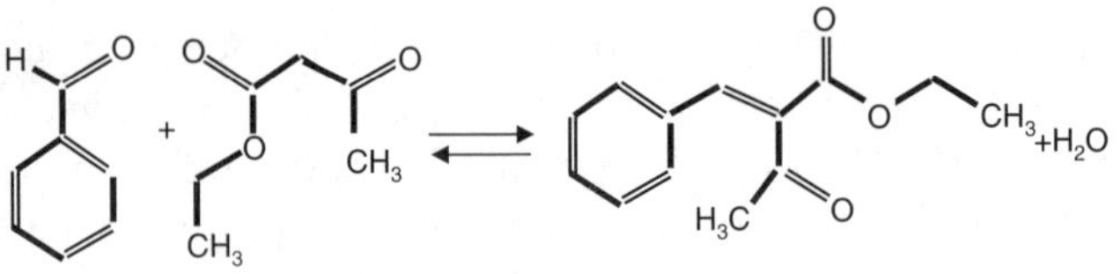

In addition to the condensation product (2-acetyl-3-phenylacrylic acid ethyl ester), numerous by-products may also be produced from various competing side-reactions. Selective water removal by membrane pervaporation may increase the reaction conversion significantly and suppress side-reactions, leading to an overall increase in product yield and purity. Furthermore, the enormous surface area-to-volume ratio and the short diffusion distance facilitate the selective removal of water by-product and less catalyst deactivation. Yeung *et al.* [28] investigated the above reaction in MEMS-based stainless steel MMRs. The ZSM-5 and NaA miniature membranes grown on the back of the multi-channel plate were used for selective membrane pervaporation of water by-product from the condensation reactions. A hybrid NaA–CsNaX bilayer membrane catalyst film was coated on the micro-channels to further enhance the reaction conversions and selectivity. The undesired reaction was suppressed using catalyst film instead of powder, because of the smaller external catalyst surfaces in contact with the solution, preventing further product reaction. Supra-equilibrium conversion and high product purity were obtained by selective removal of water during the reaction. Besides, the catalyst life was also prolonged by removing the water by-product, which is a poisoning substance for the zeolite catalyst. The MMR performed consistently better than the microreactor and the packed bed membrane reactor at a comparable catalyst loading and residence time. The best reaction yield was obtained by locating the CsNaX catalyst immediately adjacent to the membrane, because the composite NaA–Faujasite micromembrane combined successfully the good catalytic properties of Cs–Faujasite for Knoevenagel condensation reaction and the excellent separation properties of NaA membrane for water, resulting in an enhanced performance. An increase in microchannel height-to-width aspect ratio led to a larger volume of the reaction solution being in contact with the catalyst, resulting in a proportional increase in the reaction conversion.

Wan *et al.* [39] investigated the selective oxidation of aniline by hydrogen peroxide to azoxybenzene in a multi-channel MMR, with or without water removal, employing TS-1 nanozeolite as catalyst. The reaction was conducted at different residence times and temperatures. A hydrophilic ZSM-5 membrane was used to remove water selectively from the reaction mixture by membrane pervaporation. The results indicate that catalyst deactivation was reduced during the reaction. An improvement in the product yield and selectivity toward azoxybenzene was also observed. Increasing temperature was beneficial for both yield and selectivity, but beyond 340 K, microreactor operation was ineffective due to bubble formation and hydrogen peroxide decomposition.

8.3.3 Catalytic MMRs for G–L–S Reactions

MMRs can be applied for heterogeneously catalyzed G–L–S reactions. Catalysts are incorporated with the membrane by coating on the membrane surface or depositing inside the membrane wall. In the operation, the gas and liquid phases are introduced into the reaction zone from the opposite sides of the membrane and meet precisely on the catalyst layer. Such MMR systems exhibit many advantages, such as: (i) the gas, liquid, and solid phases can be controlled independently during the process and the gas/liquid interface is well-defined and stabilized by the membrane; (ii) the gaseous reactant is distributed homogeneously along the full reactor length, which makes the reactors advantageous for processes with low gaseous reactant solubility or high gas consumption, preventing a possible depletion of this reactant in the reactor; (iii) mass transfer limitations in the liquid phase (reactant diffusion from bulk to the catalytic wall) can be overcome by miniaturizing the characteristic dimension of the reactor channels.

Aran *et al.* [5, 21, 30] used porous Al_2O_3 and stainless steel hollow fiber MMRs for the removal of nitrite (NO_2^-) from water by the catalytic hydrogenation reaction over Pd catalysts, where the aqueous solution of NO_2^- was pumped into the lumen while the gaseous hydrogen reactant was delivered to the shell side of the reactor. The nitrite comes in direct contact with hydrogen in the catalytic regions to form nitrogen:

$$2NO_2^- + 3H_2 \xrightarrow{Pd} N_2 + 2OH^- + 2H_2O$$

$$2NO_2^- + 6H_2 \xrightarrow{Pd} 2NH_4^+ + 4OH^-$$

In the porous α-Al_2O_3 hollow fiber MMR a Pd-γ-Al_2O_3 catalyst layer was coated on the inner surface of the fibers [5, 21], as shown in Figure 8.3(a). In order to realize liquid flow inside the intrinsically hydrophilic porous reactor channel and to obtain a stabilized G–L–S interface, the α-Al_2O_3 hollow fiber support was hydrophobized through the treatment with a FOTS, whereas the Pd–γ-Al_2O_3 catalyst layer remained hydorphilic. Experimental results showed that the performance of the reactor could be enhanced drastically by tuning the surface properties, and remained high even at low partial pressures of hydrogen due to the high transfer efficiency. As the thickness of the catalyst support layer was increased, the NO^{-2} reaction rate per Pd catalyst decreased, indicating internal mass transfer limitations. Reducing the inner diameter of the reactor and also integrating slug flow enhanced the performance by improving its external mass transfer. In the stainless steel HFMMR [30], CNFs were grown inside the hollow fibers as support for the Pd catalyst, as shown in Figure 8.3(b). The outer surface of the porous hollow fiber was covered with a dense, gas-permeable polymeric coating in order to stabilize the G–L–S interface. The results showed the high nitrite reduction performance of the reactors, even without the presence of Pd or additional hydrogen supply. In addition to the high surface area and mechanical strength, the intrinsic reducing properties and catalytic activity of the reactors make them very suitable for hydrogenation reactions.

8.4 FLUID FLOW IN MEMBRANE MICROREACTORS

Microreactors have a small characteristic length (or hydraulic diameter), which makes the Reynold's number (Re) very small, hence the flow in the microchannels of MMRs is predominantly laminar. Secondary "molecular" effects can be significant as the characteristic length decreases to a point where properties like density, temperature, and velocity are not continuous functions of position, and the fluid–surface interaction becomes more important than the fluid–fluid interactions. At this moment, the continuum assumptions for the normal reactors are no longer valid and a velocity slip at boundaries has to be implemented to describe the fluid flows. Furthermore, the selective permeation of a fluid component through the membrane also influences significantly the overall velocity field in the MMRs. Therefore, the flow in MMRs is far more complicated than that in macroscale reactors.

The departure of the flow in microchannels from the continuum assumption can be measured by the Knudsen number, that is, the ratio of the mean free path λ to the hydraulic diameter h_D of the microchannel [44]:

$$K_n = \frac{\lambda}{h_D} = \frac{3\nu\sqrt{\pi M_W/(2RT)}}{2h_D}$$

where M_W is the average molecular weight of the fluid, R is the universal gas constant, T is the absolute temperature, and ν is the kinematic viscosity.

It is generally agreed that the continuum assumption holds only with $K_n \leq 10^{-3}$ and the flow can be described by the application of the Navier–Stokes equations with their customarily used no-slip boundary conditions (that is, the velocity on the wall is zero), whereas the behavior of the fluid for $K_n \geq 10$ has to be described with molecular-level modeling. For most microchannel systems, K_n is less than 0.1 and the fluid flow is in the slip regime ($10^{-3} \leq K_n \leq 0.1$) [45]. In this case, the fluid flow can still be described using the continuum conservative equations, but slip boundary conditions must be incorporated since velocity slip and temperature jump may occur at the microchannel surfaces [46].

Computational modeling and simulation are important tools for studying the flow and reaction behavior of MMRs. Alfadhel and Kothanre [44] set up a mathematical model to describe isothermal microfluidic steady flow in a silicon MMR employing the Navier–Stokes equations with first-order tangential velocity slip at the walls to account for high Knudsen number and fluid permeation drift velocity boundary conditions for fluid permeation through the membrane. The effects of fluid permeation through the membrane and the Knudsen number on the velocity profile and pressure drop were evaluated. An analytical solution of the velocity profiles in MMRs, incorporating both slip and fluid permeation at the boundary, was derived. The general pressure gradient formula derived in this work is in agreement with results from published literature for the limiting cases of no slip ($K_n \rightarrow 0$) and no permeation. Jani *et al.* [47] provided a two-dimensional CFD model using the continuum conservative equations with a normal parabolic velocity profile (no-slip boundary condition) to study the mass transport and heterogeneously catalyzed reactions in a porous hollow fiber MMR. Boundary conditions were derived which represent the reactant concentration at the microreactor inner wall as a function of catalytic layer properties to obtain an optimized reactor geometry. An optimum in conversion was found by varying the catalytic membrane layer thickness. Simulation results also indicate that reactor miniaturization significantly improves the mass transfer and reaction performances in MMRs [48].

8.5 PROSPECTS AND CHALLENGES

MMR represents a promising technology for chemical processing when external mass and heat transfer limitations cannot be ignored. It exhibits great potential for process intensification by integrating multiple process streams within one compact unit for mass and heat transfer applications, and thus is especially suitable for portable applications [49, 50]. Although considerable progress has been made in recent years in this field, there are still many challenges to be faced:

1. The design and fabrication of commercially viable MMRs require low-cost and standardized fabrication techniques. The fabrication of microchemical systems remains extremely difficult, non-standard, and based on customized designs that use microfabrication techniques borrowed from the microelectronics industries. Alternative fabrication methods amenable for mass production, such as casting or stamping, must be developed to reduce their cost and increase their productivity. In addition, strategies for precise and selective incorporation of active elements such as catalysts, membranes, adsorbents, contactors, and sensors must be established. Using the phase-inversion/sintering technique enables us to simplify the fabrication process and reduce the costs of hollow fiber-type MMRs significantly, but it still has to be improved so that the morphology and microstructure of the hollow fiber membranes can be controlled as required.
2. Highly resistant materials are required to fabricate microchannel structures that are stable under reaction conditions and to reduce costs. Substitution of stainless steel for silicon is one step toward cost reduction, but this is currently offset by the cost of micromachining and fabrication. Ceramics such as cordierite, mullite ($3Al_2O_3$–$2SiO_2$), alumina (Al_2O_3), and so on are ideal to reduce heat losses due to

their low thermal conductivity, but their brittleness (especially under harsh reaction environments) limits their practical applications. Utilization of porous ceramic monoliths to realize large networks of microchannel membranes will be a promising way to obtain superior mechanical stability.

3. Better materials need to be developed to address problems of membrane stability, and to reach the ultimate goal of high permeation rate combined with high perm-selectivity. Although inorganic membranes can be used in high-temperature and chemically harsh environments, their long-term stability under technical conditions has to be improved significantly. Using a Pd membrane made by CVD or a Pd/Ag alloyed membrane made by sputtering can improve the stability of Pd-MMRs for hydrogen production.
4. The theory and fundamental physical understanding of microchemical systems remains an area of ongoing research. Computational modeling and simulation are important tools for the design and operation of microreactors. However, adequate models involve many physical parameters that can only be found experimentally and thus are classically complicated because of the coupling of heat, mass, and momentum transfer in confined geometries. Moreover, the match of membrane permeability with catalytic activity, which determines the performance of membrane reactors, needs a deep theoretical investigation.

REFERENCES

[1] Mills, P.L., Quiram, D.J. and Ryley, J.F. (2007) Microreactor technology and process miniaturization for catalytic reactions–a perspective on recent developments and emerging technologies. *Chemical Engineering Science*, 62, 6992–7010.

[2] Leung, Y.L.A. and Yeung, K.L. (2004) Microfabricated ZSM-5 zeolite micromembranes. *Chemical Engineering Science*, 59, 4809–4817.

[3] Lau, W.N., Yeung, K.L. and Martin-Aranda, R. (2008) Knoevenagel condensation reaction between benzaldehyde and ethyl acetoacetate in microreactor and membrane microreactor. *Microporous and Mesoporous Materials*, 115, 156–163.

[4] Ye, S.-Y., Hamakawa, S., Tanaka, S., Sato, K., Esashi, M. and Mizukami, F. (2009) A one-step conversion of benzene to phenol using MEMS-based Pd membrane microreactors. *Chemical Engineering Journal*, 155, 829–837.

[5] Aran, H.C., Chinthaginjala, J.K., Groote, R. *et al.* (2011) Porous ceramic mesoreactors: A new approach for gas–liquid contacting in multiphase microreaction technology. *Chemical Engineering Journal*, 169, 239–246.

[6] García-García, F.R., Kingsbury, B.F.K., Rahman, M.A. and Li, K. (2012) Asymmetric ceramic hollow fibres applied in heterogeneous catalytic gas phase reactions. *Catalysis Today*, 193, 20–30.

[7] Shah, K., Ouyang, X. and Besser, R.S. (2005) Microreaction for microfuel processing: Challenges and prospects. *Chemical Engineering Technology*, 28, 303–313.

[8] Moharana, M.K., Peel, N.R., Khandekar, S. and Kunzru, D. (2011) Distributed hydrogen production from ethanol in a microfuel processor: Issues and challenges. *Renewable and Sustainable Energy Reviews*, 15, 524–533.

[9] Wan, Y.S.S., Chau, J.L.H., Gavriilidis, A. and Yeung, K.L. (2001) Design and fabrication of zeolite-based microreactors and membrane microseparators. *Microporous and Mesoporous Materials*, 42, 157–175.

[10] Wise, K.D. and Najafi, K. (1991) Microfabrication techniques for integrated sensors and microsystems. *Science*, 254, 1335–2342.

[11] Ehrfeld, W., Golbig, K., Hessel, V., Löwe, H. and Richter, T. (1999) Characterization of mixing in micromixers by a test reaction: Single mixing units and mixer arrays. *Industrial and Engineering Chemistry Research*, 38, 1075–1082.

[12] Piner, R.D., Zhu, J., Xu, F., Hong, S. and Mirkin, C.A. (1999) "Dip-pen" nanolithography. *Science*, 283, 661–663.

[13] Thomas, J.M. and Thomas, W.J. (1997) *Principles and Practice of Heterogeneous Catalysis*, VCH Publishers, Weinheim.

[14] Conant, T., Karim, A. and Datye, A. (2007) Coating of steam reforming catalysts in non-porous multi-channeled microreactors. *Catalysis Today*, 125, 11–15.

[15] Conant, T., Karim, A., Rogers, S., Samms, S., Randolph, G. and Datye, A. (2006) Wall coating behavior of catalyst slurries in non-porous ceramic microstructures. *Chemical Engineering Science*, 61, 5678–5685.

[16] Men, Y., Gnaser, H., Zapf, R., Hessel, V. and Ziegler, C. (2004) Parallel screening of $Cu/CeO_2/\gamma\text{-}Al_2O_3$ catalysts for steam reforming of methanol in a 10-channel microstructured reactor. *Catalysis Communications*, 5, 671–675.

[17] Stefanescu, A., van Veen, A.C., Mirodatos, C., Beziat, J.C. and Duval-Brunel, E. (2007) Wall coating optimization for microchannel reactors. *Catalysis Today*, 125, 16–23.

[18] Cui, T.H., Fang, J., Zheng, A.P., Jones, F. and Reppond, A. (2000) Fabrication of microreactors for dehydrogenation of cyclohexane to benzene. *Sensors and Actuators B: Chemical*, 71, 228–231.

[19] Caro, J. (2008) Catalysis in micro-structured membrane reactors with nano-designed membranes. *Chinese Journal of Catalysis*, 29, 1169–1177.

[20] Kwan, S.M., Leung, A.Y.L. and Yeung, K.L. (2010) Gas permeation and separation in ZSM-5 micromembranes. *Separation and Purification Technology*, 73, 44–50.

[21] Aran, H.C., Klooster, H., Jani, J.M., Wessling, M., Lefferts, L. and Lammertink, R.G.H. (2012) Influence of geometrical and operational parameters on the performance of porous catalytic membrane reactors. *Chemical Engineering Journal*, 207/208, 814–821.

[22] de la Iglesia, O., Sebastián, V., Mallada, R. *et al.* (2007) Preparation of Pt/ZSM-5 films on stainless steel microreactors. *Catalysis Today*, 125, 2–10.

[23] Sebastián, V., de la Iglesia, O., Mallada, R. *et al.* (2008) Preparation of zeolite films as catalytic coatings on microreactor channels. *Microporous and Mesoporous Materials*, 115, 147–155.

[24] Zamaro, J.M. and Miró, E.E. (2009) Confined growth of thin mordenite films into microreactor channels. *Catalysis Communications*, 10, 1574–1576.

[25] Mies, M.J.M., Rebrov, E.V., Jansen, J.C., de Croon, M.H.J.M. and Schouten, J.C. (2007) Method for the in situ preparation of a single layer of zeolite Beta crystals on

a molybdenum substrate for microreactor applications. *Journal of Catalysis*, 247, 328–338.

[26] Pina, M.P., Mallada, R., Arruebo, M. *et al.* (2011) Zeolite films and membranes: Emerging applications. *Microporous and Mesoporous Materials*, 144, 19–27.

[27] Chau, J.L.H., Wan, Y.S.S., Gavriilidis, A. and Yeung, K.L. (2002) Incorporating zeolites in microchemical systems. *Chemical Engineering Journal*, 88, 187–200.

[28] Yeung, K.L., Zhang, X., Lau, W.N. and Martin-Aranda, R. (2005) Experiments and modeling of membrane microreactors. *Catalysis Today*, 110, 26–37.

[29] Yamamoto, S., Hanaoka, T., Hamakawa, S., Sato, K. and Mizukami, F. (2006) Application of a microchannel to catalytic dehydrogenation of cyclohexane on Pd membrane. *Catalysis Today*, 118, 2–6.

[30] Aran, H.C., Pacheco Benito, S., Luiten-Olieman, M.W.J. *et al.* (2011) Carbon nanofibers in catalytic membrane microreactors. *Journal of Membrane Science*, 381, 244–250.

[31] Tan, X., Liu, S. and Li, K. (2001) Preparation and characterization of inorganic hollow fiber membranes. *Journal of Membrane Science*, 188, 87–95.

[32] Kingsbury, B.F.K. and Li, K. (2009) A morphological study of ceramic hollow fibre membranes. *Journal of Membrane Science*, 328, 134–140.

[33] Tan, X., Liu, N., Meng, B. and Liu, S. (2011) Morphology control of the perovskite hollow fibre membranes for oxygen separation using different bore fluids. *Journal of Membrane Science*, 378, 308–318.

[34] García-García, F.R., Rahman, M.A., Kingsbury, B.F.K. and Li, K. (2010) A novel catalytic membrane microreactor for CO_x free H_2 production. *Catalysis Communications*, 12, 161–164.

[35] Rahman, M.A., García-García, F.R., Hatim, M.D.I., Kingsbury, B.F.K. and Li, K. (2011) Development of a catalytic hollow fibre membrane micro-reactor for high purity H_2 production. *Journal of Membrane Science*, 368, 116–123.

[36] Rahman, M.A., García-García, F.R. and Li, K. (2012) Development of a catalytic hollow fibre membrane microreactor as a microreformer unit for automotive application. *Journal of Membrane Science*, 390/391, 68–75.

[37] Kim, D., Kellogg, A., Livaich, E. and Wilhite, B.A. (2009) Towards an integrated ceramic micro-membrane network: Electroless-plated palladium membranes in cordierite supports. *Journal of Membrane Science*, 340, 109–116.

[38] Wilhite, B.A., Weiss, S.E., Ying, J.Y., Schmidt, M.A. and Jensen, K.F. (2006) High-purity hydrogen generation in a microfabricated 23 wt% Ag–Pd membrane device integrated with 8:1 $LaNi_{0.95}Co_{0.05}O_3/Al_2O_3$ catalyst. *Advanced Materials*, 18, 1701–1704.

[39] Wan, Y.S.S., Yeung, K.L. and Gavriilidis, A. (2005) TS-1 oxidation of aniline to azoxybenzene in a microstructured reactor. *Applied Catalysis A: General*, 281, 285–293.

[40] Tosti, S. (2010) Overview of Pd-based membranes for producing pure hydrogen and state of art at ENEA laboratories. *International Journal of Hydrogen Energy*, 35, 12650–12659.

[41] Coronas, J. and Santamaria, J. (2004) The use of zeolite films in small-scale and micro-scale applications. *Chemical Engineering Science*, 59, 4879–4885.

[42] Lai, S.M., Ng, C.P., Martin-Aranda, R. and Yeung, K.L. (2003) Knoevenagel condensation reaction in zeolite membrane microreactor. *Microporous and Mesoporous Materials*, 66, 239–252.

[43] Lau W.N., Yeung K.L., Zhang X.F., Martin-Aranda R. (2007) Zeolite membrane microreactors and their performance. From Zeolites to Porous MOF Materials: The 40th Anniversary of International Zeolite Conference, Proceedings of the 15th International Zeolite Conference, August 12–17, Beijing, Vol. 170, pp. 1460–1465.

[44] Alfadhel, K.A. and Kothare, M.V. (2005) Microfluidic modeling and simulation of flow in membrane microreactors. *Chemical Engineering Science*, 60, 2911–2926.

[45] Alfadhel, K. and Kothare, M.V. (2005) Modeling of multicomponent concentration profiles in membrane microreactors. *Industrial and Engineering Chemistry Research*, 44, 9794–9804.

[46] Jie, D., Diao, X., Cheong, K.B. and Yong, L.K. (2000) Navier–Stokes simulations of gas flow in microdevices. *Journal of Micromechanics Microengineering*, 10, 372–379.

[47] Jani, J.M., Can Aran, H., Wessling, M. and Lammertink, R.G.H. (2012) Modeling of gas–liquid reactions in porous membrane microreactors. *Journal of Membrane Science*, 419/420, 57–64.

[48] Chasanis, P., Kenig, E.Y., Hessel, V. and Schmitt, S. (2008) Modelling and simulation of a membrane microreactor using computational fluid dynamics. *Computer Aided Chemical Engineering*, 25, 751–756.

[49] Gallucci, F., Fernandez, E., Corengia, P. and van Sint Annaland, M. (2013) Recent advances on membranes and membrane reactors for hydrogen production. *Chemical Engineering Science*, 92, 40–66.

[50] Kolb, G. (2013) Review: Microstructured reactors for distributed and renewable production of fuels and electrical energy. *Chemical Engineering Process*, 65, 1–44.

9

Design of Membrane Reactors

9.1 INTRODUCTION

Membrane reactor design is based on an understanding of the relations between various parameters such as reactor configuration and dimensions, membrane properties (permeability and selectivity), operating conditions (temperature, feed flow rate, concentration, etc.), as well as the reaction kinetics to the reactor performance. This can be achieved by experimental study combined with modeling work, which is based on a description of the phenomena of mass, heat, and momentum transfer and the reactions occurring in the reactor. Through validated mathematical modeling, systematic process understanding, optimal design, and scale-up of membrane reactors can be realized. This chapter involves mainly the general modeling process for several typical membrane reactors such as PBMRs, FTCMRs, and FBMRs, followed by their applications to a few reaction systems.

9.2 DESIGN EQUATIONS FOR MEMBRANE REACTORS

The design equations of a membrane reactor consist of the mass, heat, and momentum balances in combination with the membrane permeation and the reaction kinetics conducted on the feed side and permeate side, respectively. Table 9.1 summarizes the design equations to describe the phenomena occurring in membrane reactors. Among these equations,

Inorganic Membrane Reactors: Fundamentals and Applications, First Edition. Xiaoyao Tan and Kang Li.

Table 9.1 Design equations of membrane reactors

Phenomenon	Related to	Number of equations	Used for
Mass balance	MR configuration	2 × number of species in the reaction system	Conversion, selectivity, and yield
Heat balance	MR configuration	2, for both the feed and the permeate side	Temperature profiles
Momentum balance	MR configuration	2, for both the feed and the permeate side	Pressure profiles
Membrane transport kinetics	Membrane properties	Number of permeated species through the membrane	In combination with the mass, heat, and momentum balances
Reaction kinetics	Catalyst	Number of reactions	

the permeation one is determined by the membrane properties and the transport mechanism in the membrane, while the reaction kinetics can be given by the reaction system related to the catalyst. The mass balance, heat balance, and momentum balance equations are highly dependent on the membrane reactor configuration, and are used to calculate the conversion/selectivity/yield of reactions, the temperature and pressure profiles in the membrane reactor, respectively. If the temperature or the pressure profiles are not considered, which corresponds to the isothermal and the isobaric operation modes, respectively, the heat balance or the momentum balance equations can be ignored in the modeling process.

9.2.1 Packed Bed Membrane Reactors

It is well known that most membrane reactors are based on the fixed bed and tubular configuration because this is relatively easy to construct and scale up in applications. For such membrane reactors, two types of models have generally been adopted: one-dimensional and two-dimensional, according to whether or not the radial concentration and temperature gradients are taken into consideration. Also noteworthy is that all the following modeling work is based on the ideal gases (applicable at high temperature and low pressure) and steady-state operation.

9.2.1.1 One-Dimensional Model

Figure 9.1 shows schematically the flow pattern in a general PBMR. The tubular membrane is housed co-axially in a shell tube. Different catalysts are packed in the membrane tube and the annular region (shell side),

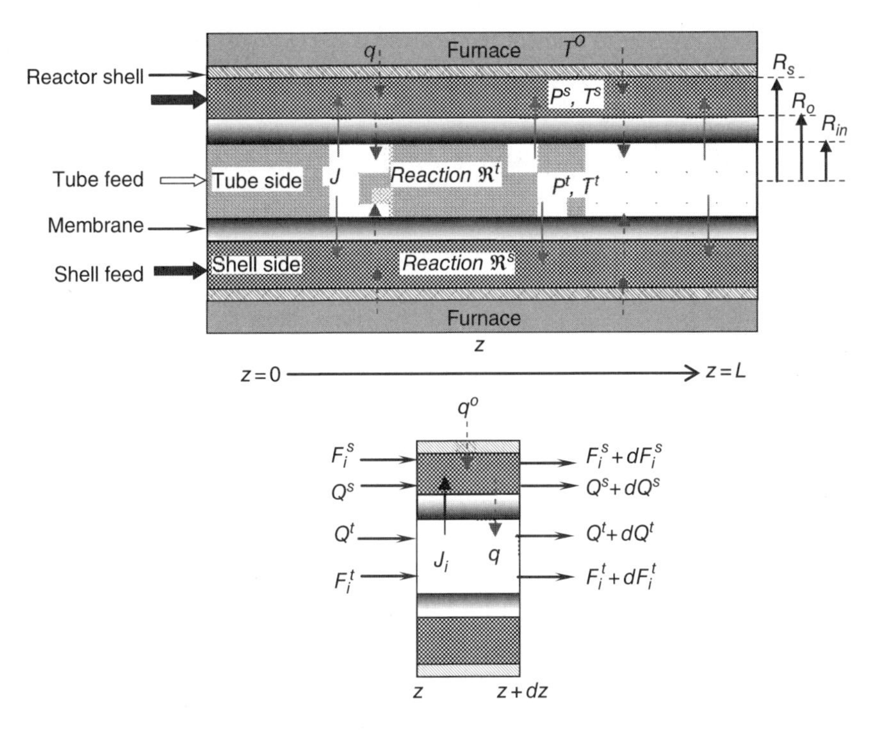

Figure 9.1 Schematic diagram showing the flow pattern in PBMRs (co-current flow mode) (a) and the differential element for mass and heat balance (b).

respectively, for different reactions. Two reactant feeds are introduced concurrently into the tube and the shell side of the reactor. The reactor is placed in a furnace with a constant temperature T^o. It is noted that the mass and heat transfer directions inside the membrane reactor can also be inversed to those shown in the figure.

The following assumptions are made for the formation of design equations:

1. The radial concentration, temperature, and pressure gradients are negligible (for a one-dimensional model).
2. Both the tube and the shell streams are in plug flow. This assumption is valid for large length/diameter and tube/particle size ratios (>10).
3. Axial mass and heat dispersions are negligible (axial mass and heat Peclet number $\gg$ 1.0[1]).
4. A pseudo-homogeneous model for the catalyst bed (negligible intra-particle and inter-phase mass- and heat-transfer resistance).

The mass balance with respect to species i is performed in the differential element, giving the mass conservation equation as

$$\frac{dF_i^t}{dz} = -2\pi R_m \cdot J_i + \pi R_{in}^2 \cdot \left(1-\varepsilon_b^t\right)\rho_c^t \sum_j \nu_{ij} \Re_j^t \left(T^t, c_1^t, c_1^t, \ldots\right) \quad (9.1)$$

with the boundary condition

$$z = 0, \quad F_i^t = F_{i,0}^t \quad (9.1a)$$

where superscript t refers to the tube phase; F_i is the molar flow rate of species i (mol s^{-1}); R_m is the logarithmic mean radius of the membrane, $R_m = \frac{R_o - R_{in}}{\ln\left(R_o / R_{in}\right)}$, in which R_o and R_{in} are the outer and inner radius of the membrane tube (m), respectively; J_i is the permeation flux of gas species i through the membrane; ε_b is the porosity of the catalyst bed; ρ_c is the catalyst density (kg m^{-3}); ν_{ji} is the stoichiometric coefficient of species i in reaction j (positive for product and negative for reactant); and $\Re_j^t$ is the apparent reaction rate (mol kg^{-1} s^{-1}) as a function of temperature and component concentrations c_i^t (mol m^{-3}). It is related to the gas flow rate by

$$c_i^t = \frac{p^t}{R_G T^t} \cdot \frac{F_i^t}{\sum_i F_i^t} \quad (9.2)$$

where p and T are the local pressure and temperature, respectively, in the catalyst bed; and R_G is the ideal gas constant.

The energy balance equation on the tube side is given by

$$\sum_i F_i^t c_{pi} \frac{dT^t}{dz} = 2\pi R_m \cdot \sum_{i=1}^{n} J_i \int_{T^{ref}}^{T^t} c_{pi} dT + 2\pi R_m \cdot U_{mem} \left(T^s - T^t\right) + \pi R_{in}^2 \left(1-\varepsilon_b^t\right)\rho_c^t \cdot \sum_j \left(-\Delta H_{r,j}\right) \cdot \Re_j^t \quad (9.3)$$

with the boundary condition

$$z = 0, \quad T^t = T_0^t \quad (9.3a)$$

where c_{pi} is the specific heat capacity of species i (J mol^{-1} K^{-1}); T^{ref} equals T^t when the transmembrane flux of component i is from the tube side to the shell side, whereas T^{ref} equals T^s when the flux is from the shell side to the tube side [2]. U_{mem} is the overall heat transfer coefficient of the membrane (J m^{-2} K^{-1} s^{-1}) and ΔH_{rj} is the heat of reaction j (J mol^{-1}).

Ergun's equation is applied generally to describe the pressure drop in the catalyst bed:

$$\frac{dp^t}{dz} = -\frac{u_g^t}{d_p^t}\frac{1-\varepsilon_b^t}{\left(\varepsilon_b^t\right)^3}\left(\frac{150\mu^t\left(1-\varepsilon_b^t\right)}{d_p^t} + 1.75\rho_g^t u_g^t\right) \tag{9.4}$$

where d_p is the equivalent diameter of the catalyst particles; μ (Pa s) and ρ_g (kg m^{-3}) are the viscosity and density of the gas mixture, respectively; and u_g (m s^{-1}) is the superficial average velocity of the gas stream, given by

$$u_g^t = \frac{1}{\pi R_{in}^2} \cdot \frac{R_g T^t \sum F_i^t}{p^t} \tag{9.5}$$

Similarly, the mass, heat, and momentum conservation equations on the shell side are given by, respectively:

$$\frac{dF_i^s}{dz} = 2\pi R_m \cdot J_i + \pi R_{in}^2 \cdot \left(1-\varepsilon_b^s\right)\rho_c^s \sum_j v_{ij}\Re_j^s\left(T^s, c_1^s, c_1^s, \ldots\right) \tag{9.6}$$

$$\sum_i F_i^s c_{pi} \frac{dT^s}{dz} = 2\pi R_m \cdot \sum_{i=1}^{n} J_i \int_{T^{ref}}^{T^s} c_{pi} dT - 2\pi R_m \cdot U_{mem}\left(T^s - T^t\right) + 2\pi R_s \cdot U_s\left(T^o - T^s\right)$$
$$+ \pi\left(R_s^2 - R_o^2\right)\left(1-\varepsilon_b^s\right)\rho_c^s \cdot \sum_j\left(-\Delta H_{r,j}\right)\cdot \Re_j^s \tag{9.7}$$

$$\frac{dp^s}{dz} = -\frac{u_g^s}{d_p^s}\frac{1-\varepsilon_b^s}{\left(\varepsilon_b^s\right)^3}\left(\frac{150\mu^s\left(1-\varepsilon_b^s\right)}{d_p^s} + 1.75\rho_g^s u_g^s\right) \tag{9.8}$$

with the boundary conditions

$$z = 0, \quad F_i^s = F_{i,0}^s, \quad T^s = T_0^s, \quad p^s = p_0^s \tag{9.9}$$

where R_s is the inner radius of the reactor shell; U_s is the overall heat transfer coefficient of the membrane reactor shell (W m^{-2} K^{-1}); and T^o is the heating temperature of the furnace.

As can be seen, the design equations for the PBMRs in a one-dimensional model are a group of ordinary differential equations. The normal Runge–Kutta method can be applied to solve these equations numerically and obtain the concentration, temperature, and pressure profiles inside the membrane reactor.

9.2.1.2 *Two-Dimensional Model*

In a two-dimensional model, the radial concentration and temperature gradients are taken into consideration, whereas the radial convective flow as well as the radial pressure drop can still be neglected. Figure 9.2 shows the differential element for mass and heat balances. Assumptions 1–4 for the one-dimensional model are still applicable for the two-dimensional model.

The mass conservation equation for species i on the shell side (catalyst bed) is given by

$$\frac{1}{A_c^s}\frac{\partial F_i^s}{\partial z} = D_{ei}^s\left(\frac{1}{r}\frac{\partial c_i^s}{\partial r} + \frac{\partial^2 c_i^s}{\partial r^2}\right) + \rho_c^s\left(1-\varepsilon_b^s\right)\sum_j \nu_{ij}\Re_j^s\left(T^s, c_1^s, c_2^s, \ldots\right) \quad (9.10)$$

with boundary conditions

$$z = 0, \quad c_i^s = c_{i,0}^s, \quad F_i^s = F_{i,0}^s \quad (9.10a)$$

$$r = R_o, \quad -D_{ei}^s\frac{\partial c_i^s}{\partial r} = \frac{R_m}{R_o}\cdot J_i \quad (9.10b)$$

$$r = R_s, \quad \frac{\partial c_i^s}{\partial r} = 0 \quad (9.10c)$$

where A_c is the cross-sectional area of the fluid, $A_c^s = \pi\left(R_s^2 - R_o^2\right)$, and D_{ei} is the effective diffusion coefficient of species i in the catalyst bed ($m^2\ s^{-1}$).

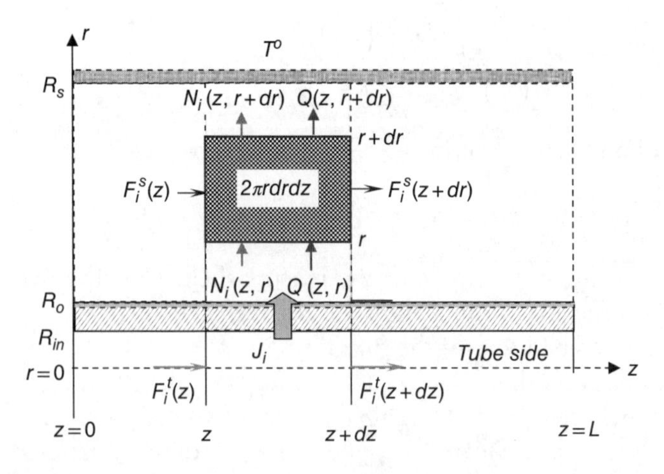

Figure 9.2 Differential element for mass and heat balance in the two-dimensional model of PBMRs.

The energy conservation equation in a catalyst bed is given by [3]

$$\frac{1}{A_c^s}\sum_{i=1}^{n} F_i^s c_{pi} \frac{\partial T^s}{\partial z} = \lambda_{er}^s \left(\frac{1}{r}\frac{\partial T^s}{\partial r} + \frac{\partial^2 T^s}{\partial r^2}\right) + \rho_c^s \left(1-\varepsilon_b^s\right)\sum_j \left(-\Delta H_{r,j}\right)\cdot \Re_j \quad (9.11)$$

with boundary conditions

$$z = 0, \quad T^s = T_0^s \quad (9.11a)$$

$$r = R_o, \quad -\lambda_{er}^s \frac{\partial T^s}{\partial r} = \frac{R_m}{R_o}\left[U_{mem}\left(T^t - T_{r=R_o}^s\right) + \sum_{i=1}^{n} J_i H_i\right] \quad (9.11b)$$

$$r = R_s, \quad -\lambda_{er}^s \frac{\partial T^s}{\partial r} = U_s\left(T_{r=R_s}^s - T^o\right) \quad (9.11c)$$

where λ_{er} is the effective radial thermal conductivity of the catalyst bed (W m^{-1}K^{-1}) and H_i is the enthalpy of component i (J mol^{-1}).

The pressure drop is formed due to catalyst packing and can be expressed by Ergun's equation (Eq. (9.8)), in which the average gas velocity is calculated as

$$u_g^s = \frac{1}{A_c^2}\cdot \sum_i \int_{R_o}^{R_s} F_i^s \cdot \frac{R_g T^s}{p^s}\cdot 2\pi r dr \quad (9.12)$$

The mass and energy conservation equations on the tube side are given in the form of a one-dimensional model, respectively, as

$$\frac{dF_i^t}{dz} = -2\pi R_m \cdot J_i \quad (9.13)$$

$$\sum_{i=1}^{n} F_i^t c_{pi} \frac{dT^t}{dz} = 2\pi R_m \left[U_{mem}\left(T_{r=R_o}^s - T^t\right) - \sum_{i=1}^{n} H_i J_i\right] \quad (9.14)$$

The model equations (9.10)–(9.14) are a group of non-linear partial differential equations. The finite difference method is applied to the partial derivative terms with respect to the radius in the equations, for example

$$\frac{\partial c}{\partial r} = \frac{c_{k+1} - c_{k-1}}{2\Delta r} \quad (0 < k < m) \quad (9.15a)$$

$$\frac{\partial^2 c}{\partial r^2} = \frac{c_{k+1} - 2c_k + c_{k-1}}{\Delta r^2} \quad (0 < k < m) \quad (9.15b)$$

where k is the differential mesh order, m is the total number of difference meshes, and Δr is the radial differential distance: $\Delta r = R_{in}/m$. As a result, the model equations may be changed into a set of ordinary differential equations and can be solved numerically using the Runge–Kutta method with a variable step.

Since the concentration and flow rate of the species are a function of radius, the overall conversion of reactants should be given by integration over the fluid cross-sectional area, that is

$$X_i = \left(1 - \frac{1}{F_{i,0} A_c} \int_{R_o}^{R_s} F_i^s \cdot 2\pi r dr\right) \times 100\% \qquad (9.16)$$

Simpson's rule can be applied to the above integration.

In addition to the reaction kinetics and the membrane permeation parameters, values of the mass and heat transfer coefficients are necessary for the design of membrane reactors. These can be calculated using the local physical properties of the reaction system (Table 9.2).

Table 9.2 Mass and heat transfer coefficients in PBMRs

Parameter	Equation	Ref.
Effective diffusion coefficient of gas species i	$D_{ei} = \varepsilon_b \cdot \dfrac{1 - y_i}{\sum_{j \neq i} y_i / D_{i,j}}$	[4]
Binary molecular diffusion coefficient	$D_{i,j} = \dfrac{4.36 \times 10^{-5} T^{3/2} \left(1/M_i + 1/M_j\right)^{0.5}}{p\left(\upsilon_i^{1/3} + \upsilon_j^{1/3}\right)^2}$	
Effective thermal conductivity of catalyst bed	$\lambda_{er} = \varepsilon_b \sum_i y_i \lambda_{gi} + (1 - \varepsilon_b)\lambda_c$	[5]
Overall heat transfer coefficient of the reactor shell	$U_s = \left(\dfrac{R_s}{\lambda_s} \cdot \ln(R_e/R_s) + \dfrac{1}{\alpha_c}\right)^{-1}$	
Overall heat transfer coefficient of the membrane tube	$U_{mem} = \left(\dfrac{R_m}{R_{in}} \cdot \dfrac{1}{\alpha_t} + \dfrac{R_m \ln(R_o/R_{in})}{\lambda_m} + \dfrac{R_m}{R_o} \cdot \dfrac{1}{\alpha_c}\right)^{-1}$	
Gas/wall convective heat transport coefficient on the catalyst side	$\dfrac{\alpha_c d_p}{\lambda_g} = 2 + 1.1\left(\dfrac{d_p u_g \rho_g}{\mu}\right)^{0.6} \left(\dfrac{c_p \mu}{\lambda_g}\right)^{1/3}$	[6]
Gas/membrane convective heat transfer coefficient on the catalyst side	$\alpha_t = \dfrac{\lambda_g}{d_e} \cdot 0.023\left(\dfrac{d_e u_g \rho_g}{\mu}\right)^{0.8} \left(\dfrac{c_p \mu}{\lambda_g}\right)^{1/3}$	[7]

9.3 FLOW-THROUGH CATALYTIC MEMBRANE REACTORS

Figure 9.3 depicts the mass and heat transport process in an FTCMR. Feed 1 containing reactants (A + B) is introduced into the membrane tube while the sweep gas flows on the shell side. The reaction takes place as the reactants flow through the catalytic membrane wall. Since no catalysts are packed in the tube and shell side, the pressure drop in the axial direction can be neglected and the temperature and concentration changes only occur in the axial direction.

In the elemental length dz depicted in Figure 9.3, the differential equations describing the mass and heat transfer in the axial direction are:

$$\text{tube side material balance}: \frac{dF_i^t}{dz} = -2\pi R_{in} \cdot N_i^t \quad (9.17)$$

$$\text{shell side material balance}: \frac{dF_i^s}{dz} = 2\pi R_o \cdot N_i^s \quad (9.18)$$

$$\text{tube side energy balance}: \sum_i F_i^t c_{pi} \frac{dT^t}{dz} = -2\pi R_{in} \cdot \alpha_t \left(T^t - T^m_{r=R_{in}}\right) \quad (9.19)$$

$$\text{shell side energy balance}: \sum_i F_i^s c_{pi} \frac{dT^s}{dz} = 2\pi R_s \cdot U_s \left(T^o - T^s\right) + 2\pi R_o \cdot \alpha_s \left(T^m_{r=R_o} - T^s\right) \quad (9.20)$$

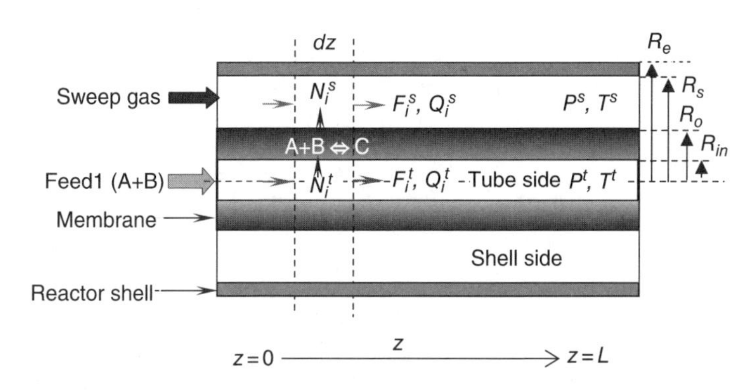

Figure 9.3 Schematic transport diagram in FTCMRs.

In the catalytically active membrane wall, the differential material and energy balances are expressed as

$$\frac{1}{r}\frac{\partial}{\partial r}(rN_i) = \rho_{mem}\sum_{j=1} \nu_{i,j}\Re_j \quad (9.21)$$

$$\frac{1}{r}\frac{\partial}{\partial r}\left(r\lambda_{mem}\frac{\partial T}{\partial r}\right) + \rho_{mem}\sum_{j=1}\left(-\Delta H_{r,j}\right)\Re_j = 0 \quad (9.22)$$

where ρ_{mem} and λ_{mem} are the apparent density and thermal conductivity of the membrane, respectively.

Transport of components within the porous membrane can be expressed using the dusty gas model, which is based on the Maxwell–Stefan equations for multi-component molecular diffusion [8]:

$$\frac{p}{R_G T}\frac{\partial y_i}{\partial r} + \frac{y_i}{R_G T}\left(1 + \frac{B_o p}{\mu D_{Ki}^e}\right)\frac{\partial p}{\partial r} = \sum_{k=1,k\neq i}^{n} \frac{y_i N_k - y_k N_i}{D_{i,k}^e} - \frac{N_i}{D_{Ki}^e} \quad (9.23)$$

Here, B_o is the Poiseuille constant (m^2), D_{Ki} and $D_{i,k}$ are the Knudsen diffusion coefficient and molecular diffusion coefficient ($m^2 s^{-1}$), respectively. Both the Knudsen and molecular diffusion coefficients are then multiplied by the square of the porosity ε_{mem}^2 to allow for the combined effect of porosity and tortuosity [9].

The mass transfer flux from the bulk gas to the membrane interface on the tube side and shell side of the membrane can be given by

$$N_i = \sum_{j=1}^{n-1} k_{i,j}^b \left(c_i^b - c_i^I\right) + y_i^b N_{tot} \quad (9.24)$$

where $k_{i,j}$ ($m\ s^{-1}$) is the finite flux mass transfer coefficient between the gas bulk and the membrane interface, which can be determined by the correlation

$$\frac{k_{i,j} d_e}{D_{i,j}} = \kappa\left(\frac{d_p u_g \rho_g}{\mu}\right)^{0.8}\left(\frac{\mu\rho_g}{D_{i,j}}\right)^{0.33} \quad (9.25)$$

where κ is dependent on the tube length-to-diameter ratio. Its appropriate values are 0.032 for the tube side and 0.05 for the shell side.

In order to solve the axial mass and energy balances, the differential equations (9.21) and (9.22) are first arranged by replacing first- and second-order derivatives with finite difference approximations. As a result, the model equations will finally be changed into a system of linearized differential and algebraic equations which can be solved numerically [10].

9.3.1 Fluidized Bed Membrane Reactors

There are two main models commonly applied to describe the prevailing flow phenomena in a fluidized bed reactor: the two-phase model [11–13] and the bubble assemblage model [14, 15]. The two-phase model assumes that the emulsion phase is well mixed and the bubble phase flows in a plug flow through the emulsion phase. Mass exchange between the bubble phase and emulsion phase is accounted for by assuming a constant average bubble diameter. In the bubble assemblage model, the fluidized bed is divided in the axial direction into a number of continuously ideally stirred tank reactors (CISTRs) for the bubble phase as well as for the emulsion phase, where the size of the CISTRs is related to the local bubble size.

FBMRs can usually be divided into two sectors, that is, a lower bubbling fluidized bed and an upper freeboard region. Figure 9.4 shows a schematic diagram of the FBMR in which tubular Pd membranes are

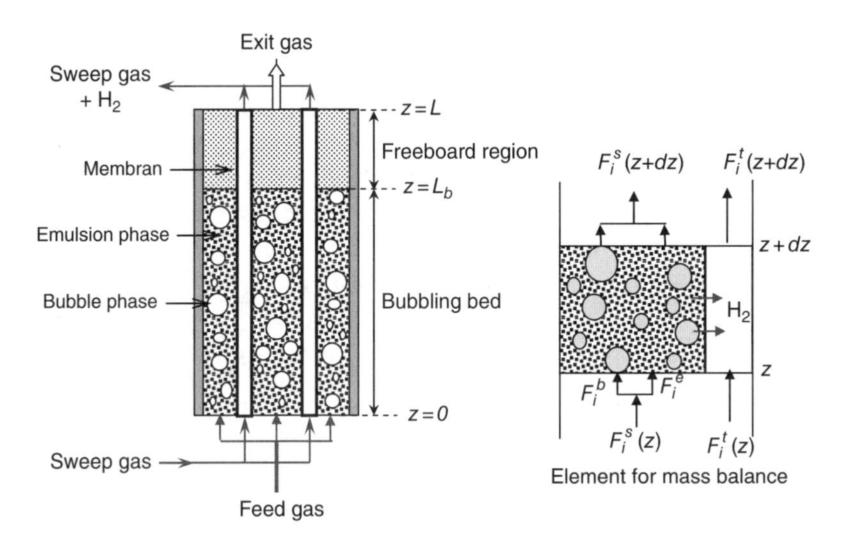

Figure 9.4 Schematic diagram of the FBMR.

used for H_2 separation. The two-phase model for FBMRs is based on the following assumptions:

1. The lower bubbling catalyst bed is composed of only an emulsion phase and a bubble phase.
2. There is plug flow with negligible axial dispersion of gas for the emulsion phase, the bubble phase, the tube phase, and in the free-board region.
3. Reaction occurs mostly in the dense phase, but the bubble phase also contains some catalyst particles and contributes to the overall reactions.
4. Bubbles are spherical with constant size.
5. The inter-phase and intra-particle diffusion resistances inside the catalyst particles can be neglected.
6. There is isothermal operation due to rapid mixing of the solid.

Under these assumptions, the modeling equations at steady state can be given for the lower bubbling bed and the freeboard region respectively as follows [16, 17].

- **Lower bubbling fluidized bed**

  Bubble phase. A differential molar balance on the *i*th component in the bubble phase gives

$$\frac{dF_i^b}{dz} = k_{be,i} a^b \varepsilon_b^b A_b \left(c_i^e - c_i^b\right) + \Phi^b \rho_c A_b \sum_j \nu_{ji} \Re_j^b - \varepsilon_b^b J_i^b \tag{9.26}$$

  where superscripts *b* and *e* refer to bubble and emulsion phases, respectively; k_{bei} is the mass transfer coefficient between the bubble phase and the emulsion phase (m s^{-1}); *a* is the bubble specific area (m^{-1}); A_b is the cross-sectional area occupied by the catalyst bed (m^2); and Φ is the volume fraction of the catalyst bed occupied by solid particles.

  Emulsion phase

$$\frac{dF_i^e}{dz} = k_{be,i} a^b \varepsilon_b^b A_b \left(c_i^b - c_i^e\right) + \Phi^e \rho_c A_b \sum_j \nu_{ji} \Re_j^e - \varepsilon_b^e J_i^e \tag{9.27}$$

  Tube phase

$$\frac{dF_{H_2}^s}{dz} = \varepsilon_b^b J_{H_2}^b + \varepsilon_b^e J_{H_2}^e \tag{9.28}$$

The boundary conditions for Eqs (9.26)–(9.28) are

$$z = 0; F_i^b = F_{i0} \cdot \frac{u - u_{mf}}{u}, F_i^e = F_{i0} \cdot \frac{u_{mf}}{u}, F_i^s = 0 \tag{9.29}$$

where it is assumed that the gas flow through the dense phase is that needed for minimum fluidization; u_{mf} is the superficial gas velocity at minimum fluidizing conditions (m s^{-1}).

- **Freeboard region (upper lean phase)**
  A material balance of species on the reaction side over the elemental thickness of the freeboard gives

$$\frac{dF_i^f}{dz} = \theta_z A \sum_j \nu_{ji} \Re_j^f - J_i^f \tag{9.36}$$

  where superscript f refers to the freeboard. The solids mass concentration (θ_z) falls off exponentially from the value at the lower bed surface and is given by

$$\theta_z = \frac{E_\infty + (E_0 - E_\infty)\exp(-a_c z)}{u_0} \left(\text{kgm}^{-3}\right) \tag{9.37}$$

  where E_0 is the entrainment flux of solids at the mean surface of the lower bed, E_∞ is the entrainment flux of solids above the transport disengagement height (TDH), and a_c is the overall decay constant (m^{-1}).

A differential molar balance equation for hydrogen on the separation side of the freeboard can be written as

$$\frac{dF_{H_2}^s}{dz} = J_{H_2}^f \tag{9.38}$$

The boundary conditions are given by

$$z = L_b; F_i^f = F_i^b + F_i^e, F_{H_2}^s = \left[F_{H_2}^s\right]_{z=L_b} \tag{9.39}$$

As can be seen, the modeling equations are a group of ordinary differential equations, and can be solved by the Runge–Kutta method. The hydrodynamic parameters for the fluidized bed reactors are summarized in Table 9.3.

Table 9.3 Hydrodynamic parameters for the fluidized bed reactors [18]

Parameter	Equation
Superficial velocity at minimum fluidization	$\frac{1.75}{\varepsilon_{mf}^3}\left(\frac{d_p u_{mf}\rho_g}{\mu}\right)^2 + \frac{150(1-\varepsilon_{mf})}{\varepsilon_{mf}^3}\left(\frac{d_p u_{mf}\rho_g}{\mu}\right) = \frac{d_p^3\rho_g(\rho_s-\rho_g)g}{\mu^2}$
Bubble diameter just above the distributor	$d_{b0} = 0.376(u-u_{mf})^2$
Maximum bubble diameter	$d_{bm} = 1.64(A_b(u-u_{mf}))^{0.4}$
Bubble diameter	$d_b = d_{bm} - (d_{bm} - d_{b0})\exp(-0.3z/D)$
Mass transfer coefficient (bubble emulsion phase)	$k_{be,i} = \frac{u_{mf}}{3} + \left[\frac{4D_{im}\varepsilon_{mf}u_b}{\pi d_b}\right]^{1/2}$
Bubble rising velocity	$u_b = u - u_{mf} + 0.711\sqrt{gd_b}$
Volume fraction of bubble phase to overall bed	$\varepsilon_b^b = \frac{u-u_{mf}}{u_b}$
Specific surface area for bubble	$a^b = 6\varepsilon_b^b / d_b$
Density for emulsion phase	$\Phi^e = (1-\varepsilon^b)(1-\varepsilon_{mf})$

9.4 MODELING APPLICATIONS

9.4.1 Oxidative Dehydrogenation of n-Butane in a Porous Membrane Reactor

The oxidative dehydrogenation of *n*-butane to butane and butadiene is accompanied by side-reactions of deep oxidation of products and reactant to CO and CO_2. The reaction network is shown in Figure 9.5.

The membrane reactor for this reaction consists of an α-Al_2O_3 porous membrane tube housed in a stainless steel shell. The V/MgO catalyst (24 wt% V_2O_5) is packed on the annulus side [3]. Figure 9.6 shows a schematic diagram of the membrane reactor and the flow pattern. A two-dimensional model can be established by performing mass balance and energy balance under the assumptions of negligible pressure drop, constant reactor wall temperature, and catalytically inactive membrane. The modeling equations are summarized in Table 9.4.

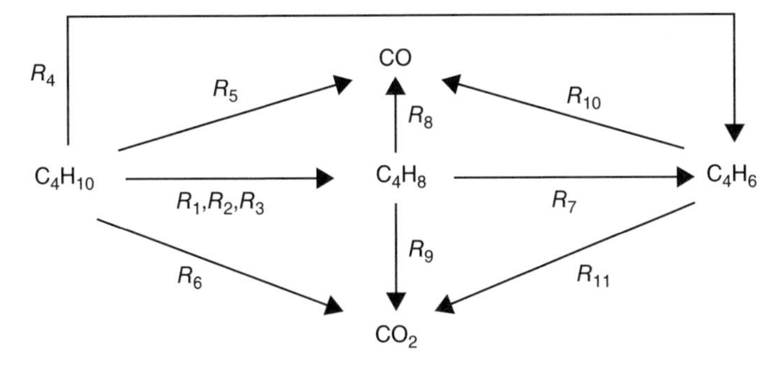

Figure 9.5 Schematic network of the oxidative dehydrogenation of *n*-butane. Reproduced from [19]. With permission from Elsevier.

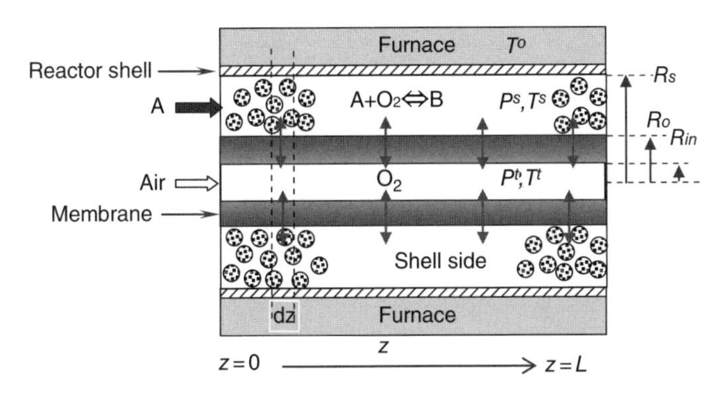

Figure 9.6 Schematic diagram of the PBMR for oxidative dehydrogenation of butane.

9.4.2 Coupled Dehydrogenation and Hydrogenation Reactions in a Pd/Ag Membrane Reactor

The integrated Pd/Ag membrane reactor for the coupling of hydrogenation/dehydrogenation reactions is shown schematically in Figure 9.7. It consists of a shell compartment surrounding a Pd/Ag membrane tube with different catalyst on the shell and tube sides. Catalytic dehydrogenation of ethylbenzene to styrene takes place on the shell side, while hydrogenation of nitrobenzene to aniline occurs inside the tube. Heat is transferred continuously through the membrane from the exothermic reaction inside the tubes (hydrogenation of nitrobenzene to aniline) to the endothermic reaction on the shell side (dehydrogenation of ethylbenzene to styrene), while the reactor shell is well insulated. At the same time,

Table 9.4 Design equations for the membrane reactor for oxidative dehydrogenation of butane

Design equation	Mathematical expression	Ref.
Mass balance on shell side	$\frac{1}{A_c}\frac{\partial F_i^s}{\partial z} = D_{ei}^s\left(\frac{1}{r}\frac{\partial c_i^s}{\partial r} + \frac{\partial^2 c_i^s}{\partial r^2}\right) + \rho_c(1-\varepsilon_b)\sum_j \nu_{ij}\Re_j$	[3]
	B.C. $z = 0, \quad c_i^s = c_{i,0}^s, \quad F_i^s = F_{i,0}^s$	
	$z>0, \quad r = R_o, \quad -D_{ei}^s \frac{\partial c_i^s}{\partial r} = \frac{R_m}{R_o}\cdot J_i; \quad r = R_s, \quad \frac{\partial c_i^s}{\partial r} = 0$	
Mass balance on tube side	$\frac{dF_i^t}{dz} = -2\pi R_m \cdot J_i$	
	B.C. $z = 0, \quad F_i^t = F_{i,0}^t$	
Energy balance on shell side	$\frac{1}{A_c}\sum_{i=1}^{n} F_i^s c_{pi}\frac{\partial T^s}{\partial z} = \lambda_{er}\left(\frac{1}{r}\frac{\partial T^s}{\partial r} + \frac{\partial^2 T^s}{\partial r^2}\right) + \rho_c(1-\varepsilon_b)\sum_j(-\Delta H_{r,j})\cdot\Re_j$	
	B.C. $z = 0, \quad T^s = T_0^s$	
	$z>0, r = R_o, \quad -\lambda_{er}\frac{\delta T^s}{\delta r} = \frac{R_m}{R_o}\left[U_m\left(T^t - T^s\right) + \sum_{i=1}^{n} J_i H_i\right]$	
	$r = R_s, \quad -\lambda_{er}\frac{\partial T^s}{\partial r} = U_s(T^s - T^o)$	
Energy balance on tube side	$\sum_{i=1}^{n} F_i^t c_{pi}\frac{dT^t}{dz} = 2\pi R_m\left[U_m(T^s\big\vert_{R_o} - T^t) - \sum_{i=1}^{n} H_i J_i\right]$	
	B.C.: $z = 0, \quad T_i^t = T_0^t$	

Membrane permeation	$J_i = \frac{\beta_K (p^t y_i^t - p^s y_i^s)}{\sqrt{M_i T}} + \frac{\beta_V y_i^t ((p^t)^2 - (p^s)^2)}{2\mu_g T}$		[20]
Reaction kinetics	$C_4H_{10} + X_O \rightarrow 1\text{-}C_4H_8 + H_2O + X$	$\Re_1 = k_1 p_{C_4H_{10}} \theta_O$	[19]
	$C_4H_{10} + X_O \rightarrow \text{Trans-2-}C_4H_8 + H_2O + X$	$\Re_3 = k_3 p_{C_4H_{10}} \theta_O$	
	$C_4H_{10} + X_O \rightarrow \text{Cis-2-}C_4H_8 + H_2O + X$	$\Re_2 = k_2 p_{C_4H_{10}} \theta_O$	
	$C_4H_{10} + X_O \rightarrow C_4H_6 + 2H_2O + 2X$	$\Re_4 = k_4 p_{C_4H_{10}} \theta_O$	
	$C_4H_{10} + 9Z_O \rightarrow 4CO + 5H_2O + 9Z$	$\Re_5 = k_5 p_{C_4H_{10}} \theta_O$	
	$C_4H_{10} + 13Z_O \rightarrow 4CO_2 + 5H_2O + 13Z$	$\Re_6 = k_6 p_{C_4H_{10}} \theta_O$	
	$C_4H_8 + X_O \rightarrow C_4H_6 + H_2O + X$	$\Re_7 = k_7 p_{C_4H_8} \theta_O$	
	$C_4H_8 + 8Z_O \rightarrow 4CO + 4H_2O + 8Z$	$\Re_8 = k_8 p_{C_4H_8} \theta_O$	
	$C_4H_8 + 12Z_O \rightarrow 4CO_2 + 4H_2O + 12Z$	$\Re_9 = k_9 p_{C_4H_8} \lambda_O$	
	$C_4H_6 + 7Z_O \rightarrow 4CO + 3H_2O + 7Z$	$\Re_{10} = k_{10} p_{C_4H_6} \lambda_O$	
	$C_4H_6 + 11Z_O \rightarrow 4CO_2 + 3H_2O + 11Z$	$\Re_{11} = k_{11} p_{C_4H_6} \lambda_O$	
	$O_2 + 2X \rightarrow 2X_O$	$\Re_{12} = k_{12} p_{O_2} (1 - \theta_O)$	
	$O_2 + 2Z \rightarrow 2Z_O$	$\Re_{13} = k_{13} p_{O_2} (1 - \lambda_O)$	
	where $\theta_O = 2k_{12} p_{O_2} / \{2k_{12} p_{O_2} + (k_1 + k_2 + k_3 + 2k_4) p_{C_4H_{10}} + k_7 p_{C_4H_8}\}$		
	$\lambda_O = 2k_{13} p_{O_2} / \{2k_{13} p_{O_2} + (9k_5 + 13k_6) p_{C_4H_{10}} + (8k_8 + 12k_9) p_{C_4H_8} + (7k_{10} + 11k_{11}) p_{C_4H_6}\}$		

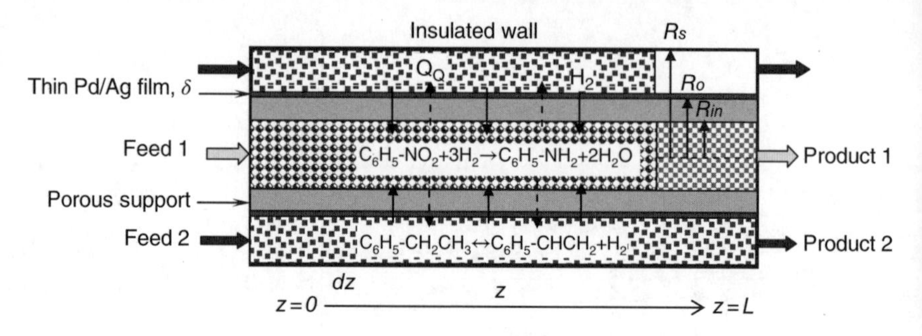

Figure 9.7 Schematic diagram of the integrated Pd/Ag membrane reactor for the coupling reaction of ethylbenzene dehydrogenation and nitrobenzene hydrogenation.

hydrogen is transferred continuously through the hydrogen-selective membrane from the dehydrogenation reaction on the shell side to the hydrogenation reaction on the tube side [1]. A one-dimensional model can be established by performing the mass balance, energy balance, and pressure drop on both sides of the integrated membrane reactor, with the resulting governing equations summarized in Table 9.5.

9.4.3 POM in a Dense Ceramic Oxygen-Permeable Membrane Reactor

Figure 9.8 shows schematically the dense ceramic membrane reactor with an oxygen-permeable membrane housed co-axially in a shell tube for partial oxidation of methane. The mixed conducting ceramic is coated on a porous support alumina tube to form a dense film with a thickness of δ while a typical POM catalyst such as Ni/γ-Al_2O_3 is packed in the tube side. Methane is fed into the tube side (catalyst bed) while air is introduced co-currently into the shell side of the membrane reactor. A two-dimensional model can be established by performing mass balance, energy balance, and pressure drop on both sides of the membrane reactor, with the resulting governing equations summarized in Table 9.6.

The membrane reactor performance in terms of methane conversion and CO selectivity is analyzed in the following using the given model based on $La_{0.6}Sr_{0.4}Co_{0.2}Fe_{0.8}O_{3-\delta}$ [23].

9.4.3.1 *Effect of Temperature*

Figure 9.9 shows the methane conversion (X_{CH4}), CO selectivity (S_{CO}), and oxygen permeation rate (Pr_{O2}) as a function of temperature.

Table 9.5 Design equations for the integrated Pd/Ag membrane reactor for the coupling reaction of ethylbenzene dehydrogenation and nitrobenzene hydrogenation

Design equation	Mathematical expression	Ref.
Mass balance on tube side	$\frac{dF_i^t}{dz} = 2\pi R_m \cdot J_i + \pi R_{in}^2 \cdot (1-\varepsilon_b^t)\rho_c^t \sum_j \nu_{ij}\Re_j^t$	[1]
Energy balance on tube side	$\frac{dT^t}{dz} = \frac{1}{\sum_i F_i^t c_{pi}} \left\{ 2\pi R_m \cdot \sum_{i=1}^{n} J_i \int_{T^{ref}}^{T^t} c_{pi} dT + 2\pi R_m \cdot U_{mem}(T^s - T^t) + \pi R_{in}^2 (1-\varepsilon_b^t)\rho_c^t \cdot \sum_j (-\Delta H_{r,j}) \cdot \Re_j^t \right\}$	
Pressure drop on tube side	$\frac{dp^t}{dz} = -\frac{u_g^t}{d_p^t} \frac{1-\varepsilon_b^t}{(\varepsilon_b^t)^3} \left(\frac{150\mu^t(1-\varepsilon_b^t)}{d_p^t} + 1.75\rho_g^t u_g^t \right)$	
Mass balance on shell side	$\frac{dF_i^s}{dz} = 2\pi R_m \cdot J_i + \pi R_{in}^2 \cdot (1-\varepsilon_b^s)\rho_c^s \sum_j \nu_{ij}\Re_j^s(T^s, c_1^s, c_1^s, \ldots)$	
Energy balance on shell side	$\frac{dT^s}{dz} = \frac{1}{\sum_i F_i^s c_{pi}} \left\{ 2\pi R_m \cdot \sum_{i=1}^{n} J_i \int_{T^{ref}}^{T^s} c_{pi} dT - 2\pi R_m \cdot U_{mem}(T^s - T^t) + 2\pi R_s \cdot U_s (T^o - T^s) + \pi (R_s^2 - R_o^2)(1-\varepsilon_b^s)\rho_c^s \cdot \sum_j (-\Delta H_{r,j}) \cdot \Re_j^s \right\}$	
Pressure drop on shell side	$\frac{dp^s}{dz} = -\frac{u_g^s}{d_p^s} \frac{1-\varepsilon_b^s}{(\varepsilon_b^s)^3} \left(\frac{150\mu^s(1-\varepsilon_b^s)}{d_p^s} + 1.75\rho_g^s u_g^s \right)$	
Permeation equation	$J_{H_2} = \frac{P_{0,H_2} \exp(-E_a / RT)}{\delta} \cdot \left(\sqrt{p_{H_2}^s} - \sqrt{p_{H_2}^t} \right)$	

(*Continued*)

Table 9.5 (*Cont'd*)

Design equation	Mathematical expression	Ref.
Reaction kinetics	Reaction on dehydrogenation side (partial pressures given in bar)	[21]
	$C_6H_5CH_2CH_3 \Leftrightarrow C_6H_5CHCH_2 + H_2$ $\quad \Re_1^s = k_1(p_{EB} - p_{ST}p_{H_2/K_A})$	
	$C_6H_5CH_2CH_3 \rightarrow C_6H_6 + C_2H_4$ $\quad \Re_2^s = k_2 p_{EB}$	
	$C_6H_5CH_2CH_3 + H_2 \rightarrow C_6H_5CH_3 + CH_4$ $\quad \Re_3^s = k_3 p_{EB} p_{H_2}$	
	$2H_2O + C_2H_4 \rightarrow 2CO + 4H_2$ $\quad \Re_4^s = k_4 p_{H_2O} p_{C_2H_4}^{1/2}$	
	$H_2O + CH_4 \rightarrow CO + 3H_2$ $\quad \Re_5^s = k_5 p_{H_2O} p_{CH_4}$	
	$H_2O + CO \rightarrow CO_2 + H_2$ $\quad \Re_6^s = k_6(p/T^3) p_{H_2O} p_{CO}$	
	Reaction on hydrogenation side (partial pressures given in kPa)	[22]
	$C_6H_5NO_2 + 3H_2 \rightarrow C_6H_5NH_2 + H_2O$ $\Delta H_{298} = -443.0$ kJ mol^{-1} $\quad \Re^t = \dfrac{kK_{NB}K_{H_2}p_{NB}p_{H_2}^{1/2}}{(1 + K_{NB}p_{NB} + K_{H_2}p_{H_2}^{1/2})^2}$	
Boundary conditions	$z = 0,\ F_i^t = F_{i,0}^t,\ T^t = T_0^t,\ p^t = p_0^t,\ F_i^s = F_{i,0}^s,\ T^s = T_0^s,\ p^s = p_0^s$	

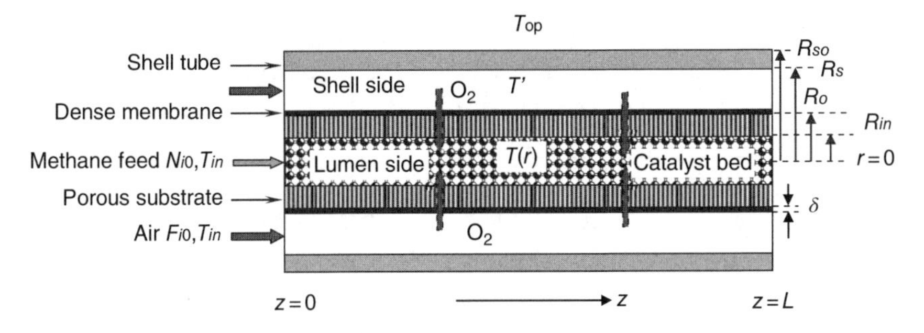

Figure 9.8 Schematic of the tubular mixed conducting ceramic membrane reactor for POM.

Both the methane conversion and the oxygen permeation rate increase as the operating temperature is increased. At lower temperatures, the methane conversion is much lower than 100% and the CO selectivity is close to 100%. As the temperature is increased to higher than 870°C, the methane conversion approaches 100% and the CO selectivity is decreased noticeably. This indicates that a small amount of methane present in the product stream is essential to high CO selectivity. The optimum methane conversion for the maximum CO yield is around 98%. The CO selectivity can be very high as long as the methane concentration is kept to a certain value, which can be maintained by increasing the methane feed flow rate for a given membrane area.

9.4.3.2 *Effect of the Amount of Catalyst*

Figure 9.10 illustrates the membrane reactor performance as a function of the void fraction of the catalyst bed. It can be seen that the methane conversion, CO selectivity, and oxygen permeation rate remain almost constant even if the porosity reaches 0.98. This implies that the POM reactions on the catalysts are so fast that the effect of the amount of catalyst can be negligible in the studied range. In other words, the POM process is controlled mainly by the oxygen permeation rate but not by the reaction on the catalyst. Therefore, the amount of catalyst required in practice in the membrane reactor is very small and can be integrated into the porous support, minimizing hydrodynamic problems and thus giving better reactor performance.

Table 9.6 Design equations for the tubular mixed conducting ceramic membrane reactor for POM

Design equation	Mathematical expression	Ref.
Mass balance on tube side	$\frac{\partial N_i^t}{\partial z} = D_{ei}\left(\frac{1}{r}\frac{\partial c_i^t}{\partial r} + \frac{\partial^2 c_i^t}{\partial r^2}\right) + \rho_c(1-\varepsilon_b)\sum_j \nu_{ji}\Re_j^t$ B.C. $z=0,\quad N_i^t = N_{i0}^t$ $z>0, r=0,\quad \frac{\partial c_i^t}{\partial r}=0;\quad r=R_{in},\quad D_{ei}\frac{\partial c_i^t}{\partial r} = \frac{R_m}{R_{in}}\cdot J_i$	[23]
Energy balance on tube side	$\sum_i N_i c_{p,i}\frac{\partial T^t}{\partial z} = \lambda_{er}\left(\frac{1}{r}\frac{\partial T^t}{\partial r} + \frac{\partial^2 T^t}{\partial r^2}\right) + \rho_c(1-\varepsilon_b)\sum_j \Re_j^t(-\Delta H_j)$ B.C. $z=0,\quad T^t = T_0^t$ $r=0,\quad \frac{\partial T^t}{\partial r}=0,\quad r=R_{in},\quad \lambda_{ec}\frac{\partial T^t}{\partial r} = U_{mem}(T_{R_{in}}^t - T^s)$	
Pressure drop on tube side	$\frac{dp^t}{dz} = -\frac{u_g^t}{d_p^t}\frac{1-\varepsilon_b^t}{(\varepsilon_b^t)^3}\left(\frac{150\mu^t(1-\varepsilon_b^t)}{d_p^t} + 1.75\rho_g^t u_g^t\right)$ B.C. $z=L,\quad p^t = p_a$	
Mass balance on shell side	$\frac{dF_i^s}{dz} = -2\pi R_m\cdot J_i\quad (i = N_2, O_2)$ B.C. $z=0,\quad F_i^s = F_{i0}^s$	
Energy balance on shell side	$\sum_i F_i^s c_{pi}\frac{dT^s}{dz} = U_s\cdot 2\pi R_s(T^o - T^s) - U_{mem}\cdot 2\pi R_m(T^s - T_{R_{in}}^t)$ B.C. $z=0,\quad T^s = T_0^s$	
Permeation equation	$J_{O_2} = \dfrac{k_r\left[(p_{O_2}^s)^{0.5} - (p_{O_2(r=R_{in})}^t)^{0.5}\right]}{\frac{R_m}{R_o+\delta}\cdot(p_{O_2(r=R_{in})}^t)^{0.5} + \frac{2k_f\delta}{D_V}\cdot(p_{O_2(r=R_{in})}^t p_{O_2}^s)^{0.5} + \frac{R_m}{R_o}\cdot(p_{O_2}^s)^{0.5}}$	[24]
Reaction kinetics	$CH_4 + 2O_2 \rightarrow CO_2 + 2H_2O\quad \Re_1 = k_1 p_{CH_4} p_{O_2}$ $CH_4 + H_2O \Leftrightarrow CO + 3H_2\quad \Re_2 = k_2 p_{CH_4} p_{H_2O}\left(1 - \frac{p_{CO}p_{H_2}^3}{K_2 p_{CH_4} p_{H_2O}}\right)$ $CH_4 + CO_2 \Leftrightarrow 2CO + 2H_2\quad \Re_3 = k_3 p_{CH_4} p_{CO_2}\left(1 - \frac{p_{CO}^2 p_{H_2}^2}{K_3 p_{CH_4} p_{CO_2}}\right)$	[25]

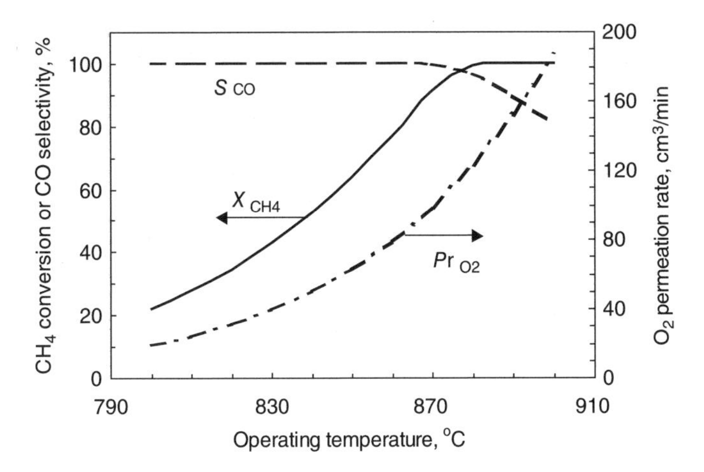

Figure 9.9 Effect of temperature on the MIEC membrane reactor performance. Reproduced with permission from [23]. John Wiley & Sons.

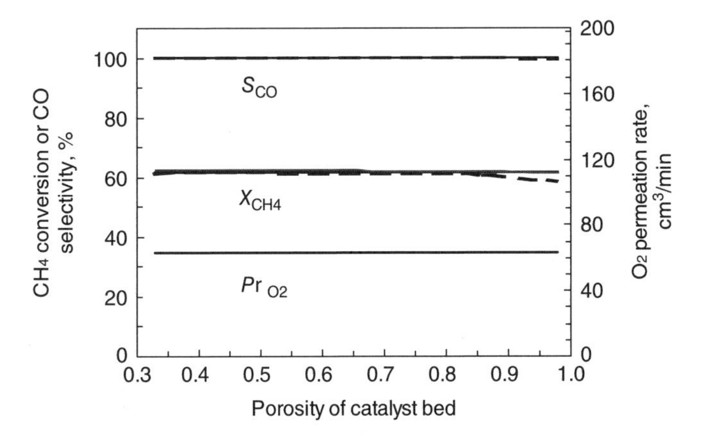

Figure 9.10 Effect of void fraction of catalyst bed on the membrane reactor performance. Reproduced with permission from [23]. John Wiley & Sons.

9.4.3.3 Effect of Membrane Dimensions

Figure 9.11 shows the membrane reactor performance with varying inner radius of the membrane tube, where the membrane wall thickness is kept at 0.2 cm while the carbon space velocity is kept constant. It can be seen that the methane conversion will decrease noticeably as the inner radius is increased to 0.5 cm. This suggests that a membrane tube larger than 0.5 cm ID cannot provide a high enough oxygen permeation rate for the POM reactions, although the oxygen permeation rate increases

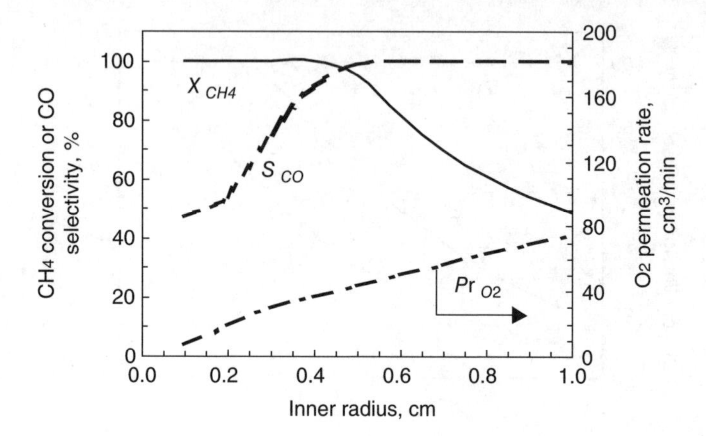

Figure 9.11 Effect of tube diameter on the membrane reactor performance. Reproduced with permission from [23]. John Wiley & Sons.

with the tube size all the time. The hollow fiber membrane exhibits an obvious advantage since it can provide a much higher area/volume ratio (500–9000 $m^2\ m^{-3}$) compared with other configurations.

9.5 CONCLUDING REMARKS

Modeling is an effective tool for designers to improve their understanding of the process and to optimize the operating conditions of membrane reactors without the need for long and expensive experimental work. Various modeling strategies, classified in decreasing order of complexity from complex 2D heterogeneous models to simple 1D pseudo-homogeneous models, can be used. The selection of mathematical model category depends on the quality of information needed by designers to properly dimension the reactor. To develop a membrane reactor design tool, designers need:

1. a reliable reaction-kinetics scheme;
2. an expression of the membrane properties in terms of permeation flux;
3. an expression of all components appearing in the overall heat-transfer coefficients;
4. the physical properties of the components, such as heat capacity, viscosity, density, and so on.

NOTATION

Symbol	Description
A	specific area (m^{-1})
A_b	cross-sectional area occupied by the catalyst bed (m^2)
A_c	cross-sectional area of the fluid stream, $A_c = \pi(R_s^2 - R_o^2)$, on the shell side
a_c	overall decay constant (m^{-1})
B_0	Poiseuille constant in dusty gas model (m^2)
c_i	concentration of species i (mol m^{-3})
c_{pi}	specific heat capacity of species i (J $mol^{-1} K^{-1}$)
D	diameter of fluidized bed reactor (m)
D_{ei}	effective diffusion coefficient of species i (m^2 s^{-1})
$D_{i,j}$	molecular diffusion coefficient of the pair i, j (m^2 s^{-1})
D_{Ki}	Knudsen diffusion coefficient (m^2 s^{-1})
d_b	bubble diameter (m)
d_{b0}	bubble diameter just above the distributor (m)
d_{bm}	maximum bubble diameter (m)
d_e	equivalent diameter of the fluid cross-sectional area (m)
d_p	diameter of the catalyst particle (m)
E_0, E_∞	entrainment flux of solids at the mean surface of the bubbling bed and above the transport disengagement height, respectively (kg $m^{-2} s^{-1}$)
F_i	molar flow rate of species i (mol s^{-1})
ΔH_r	heat of reaction (J mol^{-1})
H_i	enthalpy of component i (J mol^{-1})
J_i	permeation molar flux of species i (mol $m^{-2} s^{-1}$)
$k_{i,j}$	finite mass transfer coefficient (m s^{-1})
$k_{be,i}$	mass transfer coefficient of component i between the bubble phase and the emulsion phase (m s^{-1})
k_f	rate constant for the forward surface exchange reaction (cm $Pa^{-0.5} s^{-1}$)
k_r	rate constant for the reverse surface exchange reaction (mol $cm^{-2} s^{-1}$)
L	reactor length (m)
L_b	height of the bubbling catalyst bed (m)
M_i	molecular weight of gas species i (kg mol^{-1})
m	total number of differential meshes
N_i	molar flux of species i in gas stream (mol $m^{-2} s^{-1}$)
p	pressure (Pa)

R_{in}, R_o inner and outer radius of membrane tube (m)
R_G ideal gas constant, 8.314 J mol^{-1} K^{-1}
R_m logarithmic mean radius, $R_m = (R_o - R_{in})/\ln(R_o/R_{in})$
R_s, R_e inner and outer radius of the reactor shell (m)
r radius variable of membrane tube (m)
T temperature (K)
U overall heat transfer coefficient (W m^{-2} K^{-1})
u superficial gas velocity (m s^{-1})
u_b bubble rising velocity (m s^{-1})
u_{mf} superficial gas velocity at minimum fluidizing conditions (m s^{-1})
X conversion
y_i mole fraction of species i
z length variable (m)

Greeks

α convective heat transfer coefficient (W m^{-2} K^{-1})
β_K Knudsen parameter (mol K$^{1/2}$ m^{-1} s^{-1} kPa^{-1})
β_V viscous parameter (mol K m^{-1} kPa^{-1})
ε_b porosity of catalyst bed
ε_{mf} bed voidage at minimum fluidizing condition
ρ_c catalyst density (kg m^{-3})
ρ_{mem} density of membrane (kg m^{-3})
$\Re_j$ apparent reaction rate of reaction j (mol kg^{-1} s^{-1})
υ_{ji} stoichiometric coefficient of species i in reaction j
δ membrane thickness (m)
μ dynamic viscosity (Pa s)
λ heat conductivity (W m^{-1} K^{-1})
λ_{er} effective radial thermal conductivity of catalyst bed (W m^{-1} K^{-1})
υ molecular volume of gas species i (m^3 mol^{-1})
κ constant in Eq. (9.25)
Φ volume fraction of the catalyst bed occupied by solid particles
θ_z solids mass concentration in Eq. (9.37) (kg m^{-3})
δ membrane thickness (m)

Superscripts

b bulk or bubble phase
f freeboard in fluidized bed reactor
i interface
e emulsion phase in fluidized bed reactor
o oven
s shell phase
t tube phase

Subscripts

b	bed
c	catalyst
g	gas phase
i, j	component i, j
mem	membrane
s	reactor shell
0	initial value

REFERENCES

[1] Abo-Ghander, N.S., Logist, F., Grace, J.R., Van Impe, J.F.M., Elnashaie, S.S.E.H. and Lim, C.J. (2010) Optimal design of an autothermal membrane reactor coupling the dehydrogenation of ethylbenzene to styrene with the hydrogenation of nitrobenzene to aniline. *Chemical Engineering Science*, 65, 3113–3127.

[2] Yu, W., Ohmori, T., Yamamoto, T. *et al.* (2005) Simulation of a porous ceramic membrane reactor for hydrogen production. *International Journal of Hydrogen Energy*, 30, 1071–1079.

[3] Assabumrungrat, S., Rienchalanusarn, T., Praserthdam, P. and Goto, S. (2002) Theoretical study of the application of porous membrane reactor to oxidative dehydrogenation of *n*-butane. *Chemical Engineering Journal*, 85, 69–79.

[4] Patel, K.S. and Sunol, A.K. (2007) Modeling and simulation of methane steam reforming in a thermally coupled membrane reactor. *International Journal of Hydrogen Energy*, 32, 2344–2358.

[5] De Falco, M., Di Paola, L. and Marrelli, L. (2007) Heat transfer and hydrogen permeability in modelling industrial membrane reactors for methane steam reforming. *International Journal of Hydrogen Energy*, 32, 2902–2913.

[6] Avci, A.K., Trimm, D.L. and İlsen Önsan, Z. (2001) Heterogeneous reactor modeling for simulation of catalytic oxidation and steam reforming of methane. *Chemical Engineering Science*, 56, 641–649.

[7] De Falco, M., Di Paola, L., Marrelli, L. and Nardella, P. (2007) Simulation of large-scale membrane reformers by a two-dimensional model. *Chemical Engineering Journal*, 128, 115–125.

[8] Taylor, R. and Krishna, R. (1993) *Multicomponent Mass Transfer*, John Wiley & Sons, Inc., New York.

[9] Froment, G.F. and Bischoff, K.B. (1990) *Chemical Reactor Analysis and Design*, John Wiley & Sons, Inc., New York.

[10] Brinkmann, T., Perera, S.P. and Thomas, W.J. (2001) An experimental and theoretical investigation of a catalytic membrane reactor for the oxidative dehydrogenation of methanol. *Chemical Engineering Science*, 56, 2047–2061.

[11] Rahimpour, M.R. and Elekaei, H. (2009) A comparative study of combination of Fischer–Tropsch synthesis reactors with hydrogen-permselective membrane in GTL technology. *Fuel Processing Technology*, 90, 747–761.

[12] Abashar, M. (2004) Coupling of steam and dry reforming of methane in catalytic fluidized bed membrane reactors. *International Journal of Hydrogen Energy*, 29, 799–808.

[13] Rahimpour, M.R., Rahmani, F. and Bayat, M. (2010) Contribution to emission reduction of CO_2 by a fluidized-bed membrane dual-type reactor in methanol synthesis process. *Chemical Engineering and Processing: Process Intensification*, 49, 589–598.

[14] Dehkordi, A.M., Savari, C. and Ghasemi, M. (2011) Steam reforming of methane in a tapered membrane–assisted fluidizedbed reactor: Modeling and simulation. *International Journal of Hydrogen Energy*, 36, 490–504.

[15] Dehkordi, A. and Memari, M. (2009) Compartment model for steam reforming of methane in a membrane-assisted bubbling fluidized-bed reactor. *International Journal of Hydrogen Energy*, 34, 1275–1291.

[16] Adris, A.M., Lim, C.J. and Grace, J.R. (1997) The fluidized-bed membrane reactor for steam methane reforming: Model verification and parametric study. *Chemical Engineering Science*, 52, 1609–1622.

[17] Abashar, M.E.E., Alhumaizi, K.I. and Adris, A.M. (2003) Investigation of methane-steam reforming in fluidized bed membrane reactors. *Chemical Engineering Research and Design*, 81, 251–158.

[18] Rahimpour, M.R., Bayat, M. and Rahmani, F. (2010) Enhancement of methanol production in a novel cascading fluidized-bed hydrogen permselective membrane methanol reactor. *Chemical Engineering Journal*, 157, 520–529.

[19] Téllez, C., Menéndez, M. and Santamaría, J. (1999) Kinetic study of the oxidative dehydrogenation of butane on V/MgO catalysts. *Journal of Catalysis*, 183, 210–221.

[20] Assabumrungrat, S. and White, D.A. (1998) Permeation of acetone and isopropanol vapours through a porous alumina membrane. *Chemical Engineering Science*, 53, 1367–1374.

[21] Abo-Ghander, N.S., Grace, J.R., Elnashaie, S.S.E.H. and Lim, C.J. (2008) Modeling of a novel membrane reactor to integrate dehydrogenation of ethylbenzene to styrene with hydrogenation of nitrobenzene to aniline. *Chemical Engineering Science*, 63, 1817–1826.

[22] Amon, B., Redlingshöer, H., Klemm, E., Dieterich, E. and Emig, G. (1999) Kinetic investigations of the deactivation by coking of a noble metal catalyst in the catalytic hydrogenation of nitrobenzene using a catalytic wall reactor. *Chemical Engineering and Processing*, 38, 395–404.

[23] Tan, X. and Li, K. (2009) Design of mixed conducting ceramic membranes/reactors for the partial oxidation of methane (POM) to syngas. *AIChE Journal*, 55, 2675–2685.

[24] Tan, X. and Li, K. (2002) Modeling of air separation in a $La_{0.6}Sr_{0.4}Co_{0.2}Fe_{0.8}O_{3-\delta}$ hollow fiber membrane module. *AIChE Journal*, 48, 1469–1477.

[25] Jin, W., Gu, X., Li, S., Huang, P., Xu, N. and Shi, J. (2000) Experimental and simulation study on a catalyst packed tubular dense membrane reactor for partial oxidation of methane to syngas. *Chemical Engineering Science*, 55, 2617–2625.

Index

Note: Locators in *italics* denote tables and figures (when outside page ranges).

Inorganic Membrane Reactors: Fundamentals and Applications, First Edition. Xiaoyao Tan and Kang Li.